Deep Learning, Machine Learning and IoT in Biomedical and Health Informatics

Biomedical Engineering: Techniques and Applications
Series Editors: Donald R. Peterson, Northern Illinois University, DeKalb, IL, USA

For more information about this series, please visit:
https://www.routledge.com/Biomedical-Engineering/book-series/CRCBIOMEDENG

Deep Learning, Machine Learning and IoT in Biomedical and Health Informatics

Techniques and Applications

Edited by
Sujata Dash
Subhendu Kumar Pani
Joel J. P. C. Rodrigues
Babita Majhi

CRC Press
Taylor & Francis Group
Boca Raton London New York

CRC Press is an imprint of the
Taylor & Francis Group, an **informa** business

First edition published 2022
by CRC Press
6000 Broken Sound Parkway NW, Suite 300, Boca Raton, FL 33487-2742

and by CRC Press
2 Park Square, Milton Park, Abingdon, Oxon, OX14 4RN

CRC Press is an imprint of Taylor & Francis Group, LLC

Library of Congress Cataloging-in-Publication Data
Names: Dash, Sujata, 1964- editor. | Pani, Subhendu Kumar, 1980- editor. |
Rodrigues, Joel, 1972- editor. | Majhi, Babita, editor.
Title: Deep learning, machine learning and IoT in biomedical and health informatics : techniques and applications / edited by Sujata Dash, Subhendu Kumar Pani, Joel Rodrigues, Babita Majhi.
Description: First edition. | Boca Raton : CRC Press, 2022. | Includes bibliographical references and index. | Summary: "Biomedical and Health Informatics is an important field that brings tremendous opportunities and helps address challenges due to an abundance of available biomedical data. This book examines and demonstrates state-of-the-art approaches for IOT and Machine Learning based biomedical and health related applications"-- Provided by publisher.
Identifiers: LCCN 2021043589 (print) | LCCN 2021043590 (ebook) | ISBN 9780367544256 (hardback) | ISBN 9780367548469 (paperback) | ISBN 9780367548445 (ebook)
Subjects: LCSH: Artificial intelligence--Medical applications. | Machine learning. | Internet of things. | Bioinformatics. | Medical informatics.
Classification: LCC R859.7.A78 D43 2022 (print) | LCC R859.7.A78 (ebook) | DDC 610.285--dc23/eng/20211104
LC record available at https://lccn.loc.gov/2021043589
LC ebook record available at https://lccn.loc.gov/2021043590

ISBN: 978-0-367-54425-6 (hbk)
ISBN: 978-0-367-54846-9 (pbk)
ISBN: 978-0-367-54844-5 (ebk)

DOI: 10.1201/9780367548445

Typeset in Times
by SPi Technologies India Pvt Ltd (Straive)

Contents

PART I   Machine Learning Techniques in Biomedical and Health Informatics

PART II Deep Learning Techniques in Biomedical and Health Informatics

PART III   Internet of Things (IoT) in Biomedical and Health Informatics

Preface

OVERVIEW

The novel applications of Internet of Things (IoT) in health care data analytics can be regarded as an emerging field in computer science, medicine, biology application, natural environmental engineering, and pattern recognition. The use of various techniques of IOT systems for health care data analytics are nowadays successfully implemented in many domestic, commercial, and industrial applications due to the low-cost and high performance of various tools. Biomedical and Health Informatics is a new era that brings tremendous opportunities and challenges because of the easily available abundance of biomedical data. Deep Learning (DL) has been showing tremendous improvement in accuracy, robustness, and cross-language generalizability over conventional approaches. The aim of healthcare informatics is to ensure high quality, efficient healthcare, with better treatment and quality of life, by efficiently analysing this abundant biomedical and healthcare data. Earlier, a common requirement was to have a domain expert develop a model for biomedical or healthcare; but now the patterns for prediction are learned automatically. Due to the rapid advances in wearable sensors and actuators, the IoT and intelligent algorithms have growing significance in healthcare data analytics. The IoT focuses on the common idea of things that are recognizable, readable, locatable, controllable, and addressable via the Internet. It covers devices, sensors, people, data, and machines.

Machine Learning (ML) can provide computational methods for accumulating, updating, and changing knowledge in the intelligent systems, especially learning mechanisms that help us induce knowledge from the data. ML is helpful in the cases where direct algorithmic solutions are unavailable, lack formal models, or require knowledge about the application domain that is skimpy. IoT is anticipated to have the capability to change the way we work and live. These computing methods also play a significant role in the design and optimization of diverse engineering disciplines. With the influence and development of the IoT concept, the need for Artificial Intelligence (AI) techniques has become more significant than ever. The aim of these techniques is to accept imprecision, uncertainties, and approximations to arrive at a rapid solution. However, recent advancements in the representation of Intelligent IoT systems suggest a more intelligent and robust system that could provide a human-interpretable, low-cost, rough solution. Intelligent IoT systems have demonstrated great performance in a variety of areas, including big data analytics, time series, biomedical and health informatics, etc.

This book will play a vital role in improving human life. Researchers and practitioners will be benefited by those who are working in the fields of biomedical, health informatics, IoT, and Machine Learning. This book contains a collection of state-of-the-art approaches for Machine Learning, Deep Learning, and IoT-based biomedical and health related applications. Here, researchers and practitioners working in the field can quickly find the most functional models. They can compare different approaches and carry forward their own research in this area to directly impact human life and health.

OBJECTIVE

Deep Learning, Machine Learning and IoT in Biomedical and Health Informatics Techniques and Applications reports the latest advances of the state-of-the-art methods of Deep Learning, Machine Learning and IoT in the field of biomedical and health informatics. Its nineteen chapters are divided into three sections:

- **Machine Learning Techniques in Biomedical and Health Informatics**
- **Deep Learning Techniques in Biomedical and Health Informatics**
- **Internet of Things (IoT) in Biomedical and Health Informatics**

ORGANIZATION

- *Part 1: Machine Learning Techniques in Biomedical and Health Informatics*
  This section focuses on applying Machine Learning techniques to early diagnosis of diseases such as COVID 19, heart disease, diabetes, epilepsy, the prediction of diseases, and removing noise from medical images. Chapter 1 explains the impact of socioeconomic and environmental factors on the growth of 2019-nCoV. Chapter 2 shows an exhaustive application of Machine Learning in healthcare. Chapters 3 and 4 discuss advanced Machine Learning techniques for assessing heart disease and diabetes. Chapter 5 explains some prediction techniques for COVID 19. Chapter 6 presents the variational mode decomposition-based, automated diagnosis method for epilepsy using EEG signals. Chapters 7, 8, and 9 describe Clinical Decision Support Systems (CDSM) and an assessment of a set of adaptive median filters for eliminating noise from medical images.
- *Part 2: Deep Learning Techniques in Biomedical and Health Informatics*
  This section focuses on five chapters, wherein state-of-the-art deep learning techniques are discussed for diagnosis and prognosis of different healthcare problems. Chapter 10 explains the performance of Convolutional Neural Networks (CNNs) on SPECT scan classification, both for binary and multi-class images. Chapter 11 presents an exoteric review of the introduction to Deep Learning and discusses various Deep Learning applications in bioinformatics. Chapter 12 studies the problem of retrieval and classification of appropriate proteins associated with schizophrenia from available online molecular data by using amino acid descriptors and deep neural network-based classifiers. In Chapter 13, implementation of Deep Learning architectures for various major application domains that may be supervised, semi-supervised, or unsupervised or Reinforcement Learning are discussed. In this chapter, a detailed survey and discussion of DL architectures, libraries, and frameworks are shown, which will be useful for applications in healthcare informatics. Finally, Chapter 14 aims to design, develop a deployable melanoma detection model using a CNN. The model is easily accessible and tunable to assist patients, doctors, and the medical community.
- *Part 3: Internet of Things (IoT) in Biomedical and Health Informatics*
  This section includes five chapters that deal with IoT based healthcare systems. Chapter 15 shows how artificial network neurons are highly effective in cardiology. Hospitals have well-equipped ICUs with this technology for checking all

normal parameters and predicting heart health deterioration of patients under intensive care. The healthcare sector has a vital role in assessing the patient's physiological parameters, and IoT application has a huge role in assessing these parameters for better diagnosis. Chapter 16 is focused on recent advances and innovation in the development and applications of computational intelligence in human healthcare, environmental healthcare, and agricultural healthcare sectors. Chapter 17 focuses on the diverse Machine Learning algorithms and IoT involved in the high-performance computing for IoT applications in healthcare. Chapter 18 presents an improved Particle Swarm Optimization (PSO) clustering algorithm by perceiving the sores present in the retina and then studying the affected retina as compared to different types of hypertensive retinopathy. The current methodology recognizes the related manifestations of retinopathy for hypertension. This chapter gives an appraisal of hypertensive retinopathy strategies that apply a range of image handling methodology, utilized to highlight extraction and characterization. Lastly, Chapter 19 presents the structure of the healthcare system and a structure of the elderly patient care system which is proposed by the author. The health care system has been divided into two groups; the healthcare system that allows user mobility and the healthcare system without user mobility support. Many architectures of the healthcare system also are compared in this volume.

TARGET AUDIENCES

The current volume is a reference text aimed to support several potential audiences, including the following:

- Researchers in this field who wish to have up-to-date knowledge of the current practice, mechanisms, and research developments.
- Students and academicians of the biomedical and computer science field who have an interest in further enhancing the knowledge of the current developments.
- Industry and staff of technical institutes, R & D organizations, and those working in the field of machine learning, IoT, cloud computing, biomedical engineering, health informatics, and related fields.

Sujata Dash
Maharaja Srirama Chandra Bhanja Deo University, Baripada, India

Subhendu Kumar Pani
*Professor, Department of Computer Science Engineering and
Research Coordinator at Krupajal Engineering College (KEC),
Bhubaneswar, Odisha, India*

Joel J. P. C. Rodrigues
*Senac Faculty of Ceará, Fortaleza-CE, Brazil
Instituto de Telecomunicações, Covilhã, Portugal*

Babita Majhi
Guru Ghasidas Vishwavidyalaya, Central University, Bilaspur, India

Acknowledgement

The editors would like to acknowledge the help of all the people involved in this project and, more specifically, to the reviewers who took part in the review process. Without their support, this book would not have become a reality.

First, the editors would like to thank each one of the authors for their time, contribution, and understanding during the preparation of the book.

Second, the editors wish to acknowledge the valuable contributions of the reviewers regarding the improvement of quality, coherence, and content presentation of chapters.

Last but not the least, the editors wish to acknowledge the love, understanding, and support of their family members during the preparation of the book.

Sujata Dash
Maharaja Srirama Chandra Bhanja Deo University, Baripada, India

Subhendu Kumar Pani
Professor ,Department of Computer Science Engineering and Research Coordinator at Krupajal Engineering College (KEC), Bhubaneswar,Odisha,India

Joel J. P. C. Rodrigues
Senac Faculty of Ceará, Fortaleza-CE, Brazil Instituto de Telecomunicações, Covilhã, Portugal

Babita Majhi
Guru Ghasidas Vishwavidyalaya, Central University Bilaspur, India

Editors

Sujata Dash received her Ph.D. degree in Computational Modeling from Berhampur University, Orissa, India in 1995. She is an associate professor in the P.G. Department of Computer Science & Application, North Orissa University, at Baripada, India. She has published more than 150 technical papers in international journals, conferences, and books of reputed publications. She is a recipient of prestigious Association of Commonwealth Fellowship. She has guided many scholars for their Ph.D. degrees in computer science. Also, she is associated with many professional bodies (i.e., IEEE, CSI, ISTE, OITS, OMS, IACSIT, IMS and IAENG), is on the editorial board of several international journals, and a reviewer in many international journals. Her current research interests include Machine Learning, Distributed Data Mining, Bioinformatics, Intelligent Agent, Web Data Mining, Recommender System, and Image Processing.

Subhendu Kumar Pani received his Ph.D. from Utkal University, Odisha, India in 2013. He is a professor in the Department of Computer Science Engineering and research coordinator at Krupajal Engineering College (KEC), Bhubaneswar. With more than fifteen years of teaching and research experience, his research interests include Data Mining, Big Data Analysis, Web Data Analytics, Fuzzy Decision Making, and Computational Intelligence. He is the recipient of five researcher awards. In addition to research, he has guided two PhD students and 31 MTech students. He has published fifty-one International Journal papers (twenty-five Scopus index). His professional activities include roles as associate editor, editorial board member and/or reviewer of various international journals. Also, he is an associate with several conference societies, has more than 200 international publications, five authored books, 20 edited books, and 25 books to his credit, and is a fellow in SSARSC and life member in IE, ISTE, ISCA, OBA. OMS, SMIACSIT, SMUACEE, and CSI.

Joel J. P. C. Rodrigues [Fellow, IEEE & AAIA] is with Senac Faculty of Ceará, Brazil, head of research, development, and innovation; and senior researcher at the Instituto de Telecomunicações, Portugal. Prof. Rodrigues is an Highly Cited Researcher, the leader of the Next Generation Networks and Applications (NetGNA) research group (CNPq), an IEEE Distinguished Lecturer, Member Representative of the IEEE Communications Society on the IEEE Biometrics Council, and the President of the scientific council at ParkUrbis – Covilhã Science and Technology Park. He was Director for Conference Development - IEEE ComSoc Board of Governors, Technical Activities Committee Chair of the IEEE ComSoc Latin America Region Board, a Past-Chair of the IEEE ComSoc Technical Committee (TC) on eHealth and the TC on Communications Software, a Steering Committee member of the IEEE Life Sciences Technical Community and Publications co-Chair. He is the editor-in-chief of the International Journal of E-Health and Medical Communications and editorial board member of several high-reputed journals (mainly, from IEEE). He has been general chair and TPC Chair of many international conferences, including

IEEE ICC, IEEE GLOBECOM, IEEE HEALTHCOM, and IEEE LatinCom. He has authored or co-authored about 1000 papers in refereed international journals and conferences, three books, two patents, and one ITU-T Recommendation. He had been awarded several Outstanding Leadership and Outstanding Service Awards by IEEE Communications Society and several best papers awards. Prof. Rodrigues is a member of the Internet Society, a senior member ACM, and Fellow of AAIA and IEEE.

Babita Majhi earned her Ph.D. in 2009 from the National Institute of Technology Rourkela, Odisha, India and did post-doctoral research work at the University of Sheffield, UK (December 2011–December 2012) under the prestigious BOYSCAST Fellowship of DST, Government of India. She is presently working as an assistant professor in the Department of Computer Science and Information Technology, Guru Ghasidas Vishwavidyalaya, Central University, Bilaspur, India. She has guided six Ph.D. and eight MTech theses in the fields of adaptive signal processing, bioinformatics, data mining, computational finance, and Machine Learning and seventy MCA and MSc IT theses. She has published 128 research papers in various referred international journals and conferences. Her total number of citations are 1439, with an h-index of eighteen and an i10-index of thirty-five. She is a senior member of IEEE. Her research interests are Adaptive Signal Processing, Machine Learning, Computational Finance, Distributed Signal Processing, and Data Mining.

Contributors

Mayowa J. Adeniyi
Department of Physiology
Edo State University Uzairue
Iyamho, Nigeria

Charles O. Adetunji
Applied Microbiology, Biotechnology
    and Nanotechnology Laboratory
Department of Microbiology
Edo State University Uzairue
Iyamho, Nigeria

Santosini Bhutia
Siksha O Anusandhan (Deemed to be)
    University
Bhubaneswar, India

Ahmed Chowdhary
Jagannath International Management
    School
Kalkaji, New Delhi

Sujata Dash
Maharaja Srirama Chandra Bhanjs Deo
    University
Baripada, India

Durga Bhavani Dasari
K L Deemed to be University
Vaddeswaram, India

Prajna Parimita Dash
Birla Institute of Technology – Mesra
Ranchi, India

Tusar Kanti Dash
Department of Electronics &
    Telecommunication
CV Raman Global University
Bhubaneswar, India

Ruby Dhiman
SRM University
Delhi-NCR
Sonepat, India

Gajala Deethamvali Ghousepeer
SRM University
Delhi-NCR, Sonepat, India

D. Haritha
K L Deemed to be University
Vaddeswaram, India

Daniel Ingo Hefft
Edgbaston Campus, University of
    Birmingham
School of Chemical Engineering
Birmingham, United Kingdom

M. A. Jabbar
Vardhaman College of Engineering
Hyderabad, India

Lambodar Jena
Siksha O Anusandhan (Deemed to be)
    University
Bhubaneswar, India

Biswajit Karan
Department of Electronics and
    Communication Engineering
Birla Institute of Technology – Mesra
Ranchi, India

Aarti Kashyap
Guru Ghasidas Vishwavidyalaya
Central University
Bilaspur, India

Harman Kaur
Jagannath International Management
    School
New Delhi, India

Vasantham Vijay Kumar
K L Deemed to be University
Vaddeswaram, India

Jyoti Kukreja
Jagannath International Management
    School
New Delhi, India

Venkata Rao Maddumala
Vignan's Nirula Institute of Technology
    & Science for Women, Guntur
Andhra Pradesh, India

Babita Majhi
Department of CSIT
Guru Ghasidas Vishwavidyalaya
Central University
Bilaspur, India

Snehasis Mallick
Department of Bioinformatics
Odisha University of Agriculture and
    Technology
Bhubaneswar, India

Sushma Rani Martha
Bioinformatics Department
Odisha University of Agriculture and
    Technology
Bhubaneswar, India

Soumya Mishra
Department of Electronics &
    Telecommunication
CV Raman Global University
Bhubaneswar, India

Sudhansu Kumar Mishra
Department of EEE
Birla Institute of Technology – Mesra
Ranchi, India

Riya Mukherjee
SRM University
Delhi-NCR
Sonepat, India

Ram Bilas Pachori
Indian Institute of Technology Indore
Indore, India

Trilok Nath Pandey
Siksha O Anusandhan (Deemed to be)
    University
Bhubaneswar, India

Ganapati Panda
Department of Electronics &
    Telecommunication
CV Raman Global University
Bhubaneswar, India

Trilochan Panigrahi
Department of Electronics and
    Communication Engineering
National Institute of Technology Goa
Goa, India

Bichitrananda Patra
Siksha O Anusandhan (Deemed to be)
    University
Bhubaneswar, India

Manorama Patri
Ravenshaw University
Cuttack, India

Anjali Priyadarshini
SRM University
Delhi-NCR
Sonepat, India

P. Pujari
Guru Ghasidas Vishwavidyalya
Central University
Bilaspur, India

Ashutosh Rath
ICFAI Business School
Mumbai, India

Vishal Rathod
Department of Electronics and
    Communication Engineering
National Institute of Technology Goa
Goa, India

Samrat L. Sabat
Centre for Advanced Studies
    in Electronics Science and
    Technology
School of Physics
University of Hyderabad
Hyderabad, India

Sitanshu Sekhar Sahu
Department of Electronics and
    Communication Engineering
Birla Institute of Technology – Mesra
Ranchi, India

Arindam Sarkar
Ramakrishna Mission Vidyamandira
Belur Math-711202
Howrah, India

Animesh Sharma
Department of Electronics and
    Communication Engineering
Birla Institute of Technology – Mesra
Ranchi, India

Nongmeikapam Brajabidhu Singh
North Eastern Regional Institute of
    Science & Technology
Nirjuli, India

Moirangthem Marjit Singh
North Eastern Regional Institute of
    Science & Technology
Nirjuli, India

K. Rupabanta Singh
Maharaja Srirama Chandra Bhanjs Deo
    University
Baripada, India

Ananta Charan Ojha
School of CS & IT, Jain (Deemed to be
    University)
Bangalore, India

Olugbemi T. Olaniyan
Laboratory for Reproductive Biology
    and Developmental Programming
Department of Physiology
Edo State University Uzairue
Iyamho, Edo State

Akshith Ullal
Vanderbilt University

Utkarsh Umarye
Department of Electronics and
    Communication Engineering
National Institute of Technology Goa
Goa, India

Bhimavarapu Usharani
Koneru Lakshmaiah Education
    Foundation
Vaddeswaram, India

C. Vinitha
School of CS & IT
Jain (Deemed to be University)
Bangalore, India

Part I

Machine Learning Techniques in Biomedical and Health Informatics

1 Effect of Socio-economic and Environmental Factors on the Growth Rate of COVID-19 with an Overview of Speech Data for Its Early Diagnosis

Soumya Mishra, Tusar Kanti Dash, and Ganapati Panda
CV Raman Global University, Bhubaneswar, India

CONTENTS

DOI: 10.1201/9780367548445-2

1.1 INTRODUCTION

A major concern for human health is the development of the 2019-nCoV virus in China. It has been reported as a pandemic by the World Health Organization (WHO) [1]. By 19th September 2020, there have been 30,295,744 humans affected by the 2019-nCoV [2]. The growth rate of the virus is rapid compared to two of its ancestors, SARS-CoV and Middle East respiratory syndrome coronavirus (MERS-CoV) [3]. Numerous researchers are working to identify the clinical predictors and features of mortality based on the analysis of data patients from Wuhan, China [4, 5]. The virus has become a challenge for critical care in the whole world, due to the small number of available resources [6]. In this section, a brief literature review of the COVID-19 growth rate analysis is presented.

The growth rate of this virus is varying from country to country. Research is ongoing to identify the factors that can influence the growth rate. In [7], the authors have given an analysis of the growth rate of COVID-19 on the Diamond Princess cruise ship and found out that the median (with 95% confidence) interval of the growth rate of COVID-19 was 2.28 during the early stage and that probable outbreak size is largely dependent on the change of growth rate. The growth rate of COVID-19, as compared with the SARS coronavirus, shows that the reproduction number of COVID-19 is considerably higher than the SARS coronavirus [8]. In another interesting paper [9], the authors have estimated the basic growth rate, and the infection, recovery, and mortality rates by using the Susceptible-Infected-Recovered-Dead model. They have found out that there is a gap between the actual and the reported cases. The numbers of infected cases are twenty times more than the reported ones. Similarly, the recovered cases are forty times higher as compared to the reported cases. For an estimation of the growth pattern of the COVID 2019, a novel fractional time delays dynamic system with fractional derivative is used in [10]. The simulation results proved that the proposed model provides a satisfactory estimation of the available actual data. In [11], Susceptible-Exposed-Infected-Removed model is used to generate the dynamics of the widespread diffusion of COVID-19 by incorporating the daily intercity migration data of China. The factors used in the analysis are the rate of infection and recovery, with the final percentage of the infested residents for more than 350 cities in China, and predicted that the growth rate will crown from the middle of February to the initial week of March 2020 for China.

The three mathematical models (i.e., Logistic model, Bertalanffy model, and Gompertz model) are used in [12] for analysis of the number of people expected to be affected by COVID-19 in Wuhan and the non-Hubei areas of China. They have predicted that the COVID-19 infection will be over by late April 2020 in Wuhan and by the end of March 2020 in Non-Hubei areas. In [13], the multivariate logistic regression and sensitivity analyses were carried out to identify the risk factors for developing severe Novel Coronavirus Pneumonia and found out that the early admission and surveillance by CT should be used for the improvement of clinical outcomes. By using the susceptible-exposed-infected-removed compartment model, the growth rate of COVID-19 was calculated in [14] and found out to be between 2.8 and 3.3 for China and between 3.2 and 3.9 for the international cases. The simulation results proved that the proposed model provides a satisfactory estimation of the available actual data.

1.1.1 MOTIVATION AND RESEARCH OBJECTIVE

From the brief literature review, it has been observed that in the recent past, research has been mainly carried out on the influencing factors of the medical field, but very little data is available on the analysis of the socio-economic and environmental factors. To provide an accurate analysis of the effect of these factors on the growth rate of COVID-19, the research has been carried out in this chapter. The main research objectives are:

- Calculation of the growth rate of COVID-19 for different countries and classification of them into the low, medium, and high categories.
- Preparation of the Socioeconomic and Environmental factors based on the data set of different countries.
- Application of feature selection algorithms to identify the importance of each of these factors on the growth rate of COVID-19
- Analysis of different techniques and databases used for Non-invasive COVID-19 detection using speech signals

The remaining sections of the chapter have been organized as follows: Section 1.2 deals with the preparation of the database and extraction of the Socioeconomic and Environmental features. The Growth rate calculation and feature selection methods are dealt with in Section 1.3. In Section 1.4, speech-based COVID-19 detection is analyzed in detail. Section 1.5 concludes the chapter.

1.2 DATABASES AND SOCIOECONOMIC, ENVIRONMENTAL FEATURES

In this Section, the details of the collection and preparation of databases along with the Socioeconomic, Environmental Features are discussed. Several online resources are used for this purpose [15–24]. The details are listed in Table 1.1 Thus, the factors possibly influencing lung health, and immunity have been collected.

1.2.1 TEMPERATURE (F1)

Ambient temperature is one of the important influential factors of diseases related to the human lungs [15–18]. Analysis of the effect of temperature on influenza A(H7N9) outbreaks in China in the year 2013–2014 [15], shows that a high risk of infection was found for the minimum and maximum temperature in the range of 5 to 9 °C, and 13 to 18 °C respectively. In [16], the relationship between exposure to cold temperature and acute respiratory tract infections is studied, showing that the cold temperature was associated with an increased occurrence of respiratory tract infections. It is reported in [17], that cold or low temperature has always been correlated to flu, respiratory congestion, and inflammation, requiring urgent hospitalization, and sometimes death. To evaluate the precise relationship between influenza cycles and periodic temperature variation, a non-parametric classification methodology is used in [18].

TABLE 1.1
Details of Features and Data Collection Sources

Sl. No	Feature Name	Data Collection Source
f1	Temperature (Temp) [15–18]	Weather-Atlas [19]
f2	Happiness Index (HI) [20–22]	Global Economy [23]
f3	Cleanliness Index (CI) [24–27]	Envirocenter [28]
f4	Gross Domestic Product (GDP) [29–31]	Global Economy [32]
f5	Pollution Index (PI) [33–36]	Numbeo [37]
f6	Number of Care-Givers/ Nurses (CG) [38–40]	World-Bank-Indicators [41]
f7	Number of Physicians (Phy) [42, 43]	World-Bank-Indicators [41]
f8	Diabetes Prevalence (Diab) [44–47]	World-Bank-Indicators [41]
f9	Population of age over 65 (PoP 65) [48–50]	World-Bank-Indicators [41]
f10	Smokers above age 15 (SMO) [51–54]	World-Bank-Indicators [41]

1.2.2 HAPPINESS INDEX (F2)

There have been several findings to substantiate that long-term or acute psychological stress has somehow been responsible for causing pneumonia and relevant lung diseases [20–22]. During experiments under chronic and perennial stress, the subject's metabolism, normal behavior, and immune system have been seriously affected [20]. In [21], it is found that chronic stress is seen as a prospective co-factor in communicable disease pathogenicity. An exploratory study on the Stress Response (SR) [22] continues to show that many of the relevant biological modifications are triggered by centrally mediated biochemical fluctuations.

1.2.3 CLEANLINESS INDEX (F3)

The human lung is one of the vital organs. Proper sanitation allows greater security of the lungs from virus attacks and infections [24–27]. Different study results have shown that community-based prevention methodologies for periodic and pandemic influenza are critical in order to eliminate their potential threat to public health. The study [24] shows that the blend of hand hygiene and face masks together has a significant impact on influenza. It is observed from a survey [25] that promoting the practice of frequent hand hygiene at favorable times could massively improve overall health. In another study, it has been observed that if at least fifty percent of the total population started using medicated surgical masks with N-95 specifications, a rapidly spreading influenza could be reduced immediately by thirty percent [26, 27].

1.2.4 GROSS DOMESTIC PRODUCT (F4)

A nation experiencing a rising GDP has a stronger backbone in fighting an epidemic than a nation with a stagnant or the falling GDP due to the expenses of hospitalization, immunization, or commercial production of pharmaceuticals [29–31]. A study on the surge of heart disease in countries with low economies has shown that over ninety percent of fatal autoimmune disease-affected citizens are living in resource poverty, with minimal availability of necessary treatment [29]. A gravity model was proposed to predict the spread of the virus [30]. The model shows that the day-to-day established cases of the viral flu calculated cumulatively in each municipality of the United States were positively associated with GDP. Respiratory infections are mostly preventable, and protection costs just a portion of the cost of treatment. The need to handle and eradicate respiratory ailments largely depends on one important parameter, such as GDP [31].

1.2.5 POLLUTION INDEX (F5)

Air pollutants cause deadly respiratory and pulmonary defects, and exposure to higher levels of air pollution increases the prospect of getting infected with the deadly flu virus [33–36]. It is seen in the third wave of the Chinese Longitudinal Healthy Longevity Survey [33], that an increase in urbanization led to an increase in the air pollution index in areas with a high income.

Environmental and occupational contaminants in developed countries have identified a statistical link between exposure to asbestos and leukemia and lung cancer [35]. There is ample evidence on the impact of arsenic toxicity on all vital organs. A large pollution survey [36] found that air pollution may increase the vulnerability of viral illness and increase the likelihood of serious complications.

1.2.6 NUMBER OF CAREGIVERS/NURSES PER 1000 PEOPLE (F6)

Owing to the threat of being infected due to direct contact with the patients, the availability of skilled hospital-based caregivers and nurses during a pandemic is a huge factor in determining the quality of treatment and prevention of spreading the viral disease [38–40]. Crisis readiness would be strengthened if authorities have a clearer idea of why medical caregivers might be unavailable, and what could inspire the latter to serve during the flu pandemic [39]. Examining probable forecasting of the willingness of nursing assistants to work during the swine-flu pandemic [40], more than one thousand nurses were randomly chosen for the questionnaire-based survey. Most replied less likely to attend to patients if lacking sufficient protective gears and suits, as they worried that members of their family may become infected by the flu virus. The potential degree of danger faced by a flu outbreak is inversely proportional to the consent of the nurses to work.

1.2.7 NUMBER OF PHYSICIANS PER 1000 PEOPLE (F7)

Preparedness of a hospital or primary healthcare facility is essential to cope up with any emergency. The availability of an adequate number of health personals plays a crucial role in dealing with the pandemic situation [42, 43]. Proper planning, deployment, and positive consent of the health professionals are truly essential to gather a bigger health care workforce to mitigate the pandemic war. The bigger workforce will be able to effectively control the tragedy breaking up the nation's backbone.

1.2.8 Diabetes Prevalence (f8)

Improper management of insulin levels has been correlated with multiple vital organ dysfunctions. Diabetes prevalence has been shown to affect negatively the endocrine and immune systems of the human body [44–47]. There have been indications of an increased frequency of coronavirus disease in insulin patients [44]. It was estimated that ten percent of lung and flu-related fatalities were due to diabetes exclusively [45]. The effect of diabetes on pneumonia- and influenza-related deaths was found to be significant. The existence of insulin deficiencies or diabetes in patients hospitalized for pneumonia, and its effect on the duration of stay and in-hospital mortality, is studied [46]. In yet another study [47] on community transmitted pneumonia in the European countries, it was found that almost fifty percent of the admitted patients had undiagnosed diabetes or pre-diabetes.

1.2.9 Population Aged over Sixty-five (f9)

Age determines the anabolic and catabolic activity of a human being. Surveys conducted to evaluate the potential factor of an age above sixty-five had evidence to substantiate that age is a feature to be considered during research on lung infections and respiratory tract infections [48–50]. Research on the effect of age on abnormalities and consequences showed cases of diagnosed elderly patients (over sixty-five years of age) with pneumonia to investigate the effects of age on infectious disease [48]. The study showed that the elderly had a longer hospital stay for the treatment of pneumonia. In yet another study [49], it is observed that the mortality rate due to pneumonia increased dramatically with aging and was especially high for people over seventy-five years of age. A thorough study was aimed at assessing the chronicity of pneumonia infection [50]. It was observed that in older persons with an average age of seventy-four years, the higher chronicity predictor of the pneumonia infection was easily noticeable relative to the younger population.

1.2.10 Smokers above Age Fifteen (f10)

Smoking is highly injurious to the cardiovascular, pulmonary, and immune systems of the human body. It drastically affects the proper functioning of the lungs over prolonged use and degrades the pulmonic wellbeing over time, making it vulnerable to infections [51–54]. In a case study, it was found that consumption of any form of tobacco had a positive slope of elevated risk of catching pneumonia, with a rise in addiction length, daily smoking frequency, and an aggregate volume of cigarettes [51]. More than one hundred heavy smokers with distinct lung disorders were identified as prospective participants. For this objective, a multinomial logistic regression has been used. The result showed that nearly 80% of smokers have been shown to suffer from emphysema [52]. Dosage-based data analysis from several studies showed that smokers consuming more cigarettes are more likely to be infected by pneumococcal pneumonia [54]. Adults older than sixty-five years of age who are passive smokers face an even greater risk of lung disease. A strong dose-response association is apparent for active smokers.

The listed features were collected from internet sources and prepared in the form of a dataset containing the numeric values of features corresponding to each of the included countries. The total database is prepared for eighty-four countries, with the data updated until March 22, 2020.

1.3   GROWTH RATE CALCULATION AND FEATURE SELECTION

In this section, the methods used for the COVID-19 growth rate calculation and the effect of the ten features are discussed.

1.3.1   GROWTH RATE CALCULATION

The growth curve of COVID-19 for different countries is determined using the polynomial regression [55] through the least square algorithm. First, the cumulative sum of the number of infected patients is prepared, and then the slope is estimated. The general polynomial regression model [56] has been developed using the method of least squares. The least squares algorithm aims to minimize the variance between the values estimated from the polynomial and the expected values from the dataset. Considering the polynomial regression model with $\mathbf{X}$, $\mathbf{P}$, $\mathbf{C}$, and $\mathbf{Y}$ are vectors of input data, polynomial coefficients, intercepts, and output respectively. The polynomial coefficients can be calculated using equations (1.1), (1.2), and (1.3).

$$\begin{bmatrix} y_1 \\ y_2 \\ \cdot \\ \cdot \\ \cdot \\ \cdot \\ \cdot \\ y_n \end{bmatrix} = \begin{bmatrix} 1 \; x_1 \ldots \ldots x_1^m \\ 1 \; x_2 \ldots \ldots \ldots x_2^m \\ 1 \; x_3 \ldots \ldots \ldots x_3^m \\ \cdot \\ \ddots \\ \cdot \\ 1 \; x_n \ldots \ldots \ldots x_n^m \end{bmatrix} \begin{bmatrix} p_0 \\ p_1 \\ p_2 \\ \cdot \\ \cdot \\ \cdot \\ \cdot \\ p_m \end{bmatrix} + \begin{bmatrix} c_1 \\ c_2 \\ c_3 \\ \cdot \\ \cdot \\ \cdot \\ \cdot \\ c_n \end{bmatrix} \tag{1.1}$$

$$Y = X.P + C \tag{1.2}$$

$$P = \left(X^T X \right)^{-1} X^T Y \tag{1.3}$$

The growth rate has been calculated from the polynomial coefficients for all the COVID-19 affected countries. The span of the growth rate across all countries was huge, starting from 1.15 (Srilanka) to 55,635 (China). The threshold levels have been set appropriately to categorize the nations based on their growth rate as low, medium, and high. A country is considered to be under the low-growth rate of COVID-19 if it has a slope less than 60. The slope from 60 to 500 is considered to be medium and a high-growth rate is considered for any country having a slope greater than 500. An illustration of the growth curve in all three categories has been shown in Figures 1.1 through 1.3 for low, medium, and high growth-rate countries respectively. The resulting growth curve

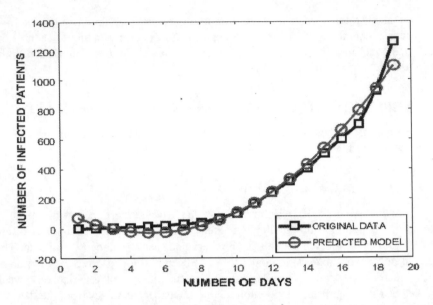

FIGURE 1.1 Comparative analysis between the original and predicted model for a low slope growth rate valued country: Greece having Slope-56.8421 (low).

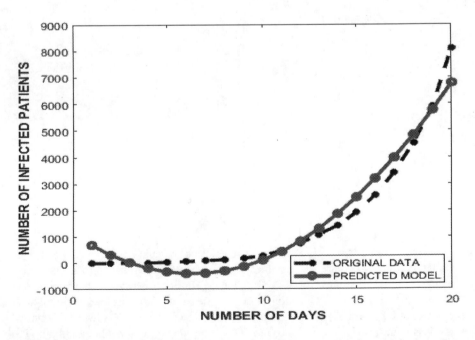

FIGURE 1.2 Comparative analysis between the original and predicted model for a medium slope growth rate valued country: Switzerland having Slope-319.7173 (medium).

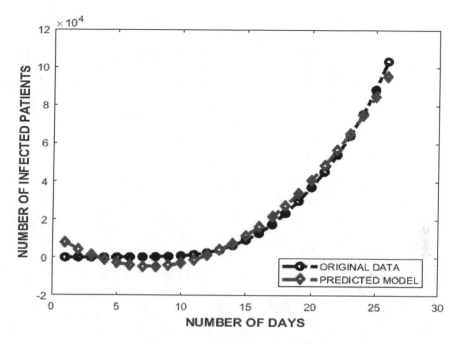

FIGURE 1.3 Comparative analysis between original and predicted data for a high slope growth rate valued country: Iran having Slope-3496.2 (high).

estimate from the polynomial regression model of the 2nd order for Greece (low growth rate) in Figure 1.1 achieves an R-squared value (coefficient of determination) of 0.9768, which is appreciable. The estimate for Switzerland (middle growth rate) in Figure 1.2 achieves an R-squared value of 0.9527, and for Iran (high growth rate) in Figure 1.3, it achieves an R-squared value of 0.9841.

1.3.2 FEATURE SELECTION

With the advent of a massive amount of high-dimensional data for different research purposes, including data acquisition, automatic classification techniques, combines science and bioengineering, feature importance plays a crucial role. Features extraction is a strategy for choosing a specific sample from a provided range of attributes by deleting obsolete and potential substitutes. Proper functionality selection not only decreases the dimension of the functionality and thus the volume of information used in training, but also reduces the severity of the dimensionality scourge to increase the generalizability of findings of algorithms. The feature selection algorithms are used to calculate the importance of the features in the form of feature weights, and the Neighborhood Component Feature Selection (NCFS) method is one of the popular and effective tools for this [57, 58]. A simple and effective non-linear decision-making rule for classification is to exploit the Nearest Neighborhood concept. Here, to extract the features and assess the best ones, Neighborhood Component Feature Selection (NCFS) is used. The goal of NCFS is to find the weighting factor 'w' that will help to select the feature subset by optimizing the nearest neighbor classification method [57]. Assuming:

T = {(x1, y1), ..., (xi, yi)..., (xN, yN)} is a set of training samples, where xi is a d-dimensional feature vector, yi ∈ {1, 2, 3, ..., C} is its corresponding class label and N is the number of samples. The weighted distance between two samples x_i and x_j is calculated by:

$$D_w\left(x_i,x_j\right) = \sum_{l=1}^{d} w_l^2 \left|x_{il} - x_{jl}\right| \tag{1.4}$$

where, w_l is a weight associated with lth feature. The probability that the query point x_i and x_j is correctly classified, is given by:

$$p_i = \sum_j y_{ij} p_{ij} \tag{1.5}$$

where, the probability of x_i selects x_j as its reference point is defined as:

$$p_{ij} = \begin{cases} \dfrac{k\left(D_w\left(x_i,x_j\right)\right)}{\sum_{k \neq i} k\left(D_w\left(x_i,x_j\right)\right)}, & \text{if } i \neq j \\ 0, & \text{if } i = j \end{cases} \tag{1.6}$$

where k(z) is the kernel function. The approximate leave-one-out classification accuracy is given by:

$$\xi\left(w\right) = \sum_i \sum_j y_{ij}\, p_{ij} - \lambda \sum_{l=1}^{d} w_l^2 \tag{1.7}$$

where $\lambda > 0$ is a regularization parameter. The objective function is differentiable and its derivative with respect to wl is written as:

$$\Delta_l = 2\left(\frac{1}{\sigma}\sum_i\left(p_i \sum_{j \neq l} p_{ij}\left|x_{il} - x_{il}\right| - \sum_j y_{ij}\, p_{ij}\left|x_{il} - x_{il}\right|\right) - \lambda\right) w_l^{(t)} \tag{1.8}$$

The gradient ascent weight update equation is given by:

$$w^{(t)} = w^{t-1} + \alpha\Delta$$

$$\epsilon^{(t)} = \xi\left(w^{(t-1)}\right) \tag{1.9}$$

The processing of weights update continues until:

$$\left|\epsilon^{(t)} - \epsilon^{(t-1)}\right| < \eta \tag{1.10}$$

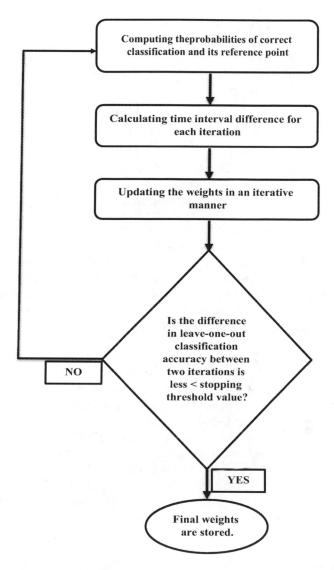

FIGURE 1.4   Flow chart depiction of Neighborhood Component Feature Selection algorithm.

where η represents the stopping threshold value. The flow chart of the Neighborhood Component Feature Selection algorithm is shown in Figure 1.4.

The simulations have been carried out in the MATLAB platform. At first, the features have been extracted for all COVID-19 affected countries. The growth rate of all the listed countries is estimated using polynomial regression and divided into the low, medium, and high categories. This feature vector is analyzed using the NCFS algorithm, and the importance of each of the features is calculated and plotted in Figure 1.5. It is demonstrated that the name of features in increasing order of importance score are cleanliness, GDP, pollution index, number of caregivers per 1000

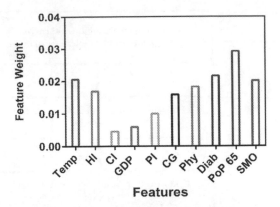

FIGURE 1.5   Plot of Feature Importance Score.

people, happiness index, number of physicians per 1000 people, smoker population above age 15, temperature, diabetes prevalence, the population of people over the age 65 (CI<GDP<PI<CG<HI<Phy<SMO<Temp<Diab<PoP65). The top three influential factors in the high growth rate of COVID-19 are the population of people over age 65 (PoP65), diabetes (Diab), and temperature (Temp).

1.4   COVID-19 SPEECH ANALYSIS

After analyzing the socio-economic and environmental factors on the growth rate of COVID-19, the early detection of the disease is discussed in this section. There have been several invasive and non-invasive methods of detection of COVID-19, out of which speech-based detection is one of the simplest and cost-effective tools [59]. The relationship between vocal fold oscillations and COVID-19 is analyzed in [60]. It is observed from the Phase space trajectories of vocal folds of COVID-19 positive and negative individuals for different vowel pronunciations that vocal fold oscillation patterns can be a distinguishing feature of COVID-19. Some of the recent works for the detection of COVID-19 from speech signals are presented here.

- In [61], it is found out that cough and breathing difficulties appear to be the most common speech symptoms for COVID-19 detection. These respiratory sounds, along with Machine Learning techniques, can be used to develop a diagnostic tool. In this research, cough, breath, and voice sounds are used to generate a speech database named "COSWARA". The sound samples are collected via worldwide crowdsourcing, using a website application. This includes speech samples in nine categories, including breathing, coughing, vowel phonation, and digit counting. There are around 6,000 audio files in the data set, which correspond to respiratory sound samples from over 900 people. Spectral contrast, Mel-Frequency Cepstral Coefficients, spectral roll-off, spectral centroid, mean square energy, polynomial fit to the spectrum, zero-crossing rate, spectral bandwidth, and spectral flatness were among the temporal and spectral acoustic features extracted from speech signals. Using these extracted features, and a random forest classifier, the accuracy of test data is calculated to be 66.74%.

- In another paper [62], data analysis of a large-scale crowdsourced dataset of respiratory sounds have been collected for detection of COVID-19. This data set contains additional samples in the asthma category to identify COVID-19 from asthma. Cough and breathing sounds are used to differentiate among COVID-19, asthma, or healthy categories. Various handcrafted audio features like onset, tempo, period, RMS energy, spectral centroid, Roll-off Frequency, Zero crossing, and Mel-Frequency Cepstral Coefficients are used with Logistic Regression, Gradient Boosting Trees, and Support Vector Machines. Across all tasks, these models obtain an AUC of more than 80%.

- Analysis of Short-Duration speech samples recorded from a Smartphone is used to detect COVID-19 in [63]. Simulations were performed using glottal, prosodic, and spectral features, along with an n-best feature selection scheme with a decision tree classifier. With n-best feature selection and speech task fusion, 82% classification accuracy has been achieved. It is also noticed that the vowel task outperformed the pataka and nasal phrase task speech analysis.

- A bio-inspired based cepstral feature-based COVID-19 detection from speech signal is proposed in [64]. The Mel Frequency Cepstral Coefficient analysis is modified using the conversion between a normal frequency scale to perceptual frequency scale and the frequency range of the filters. To further improve the detection performance, speech enhancement has been carried out. The proposed features are used in two datasets, along with an SVM classifier. The highest accuracy of 85% is achieved.

- In [65], analysis of the differential dynamics of the glottal flow waveform during voice production, the significant features for the detection of COVID-19 from voice are performed. A CNN-based 2-step attention model is used to identify the anomalies in the time-feature space of the glottal flow waveforms. It is observed that the residual and the phase difference are two of the relevant features. The COVID-19 detection yields an ROC-AUC of 0.9 for the extended vowels /u/ + /i/.

- Intelligent Speech analysis for COVID-19 patients is carried out in [66] by analyzing the speech recordings from these patients. The model classifies the health state of patients in four aspects, including the severity of illness, sleep quality, fatigue, and anxiety. This model uses two established acoustic feature sets and support vector machines, with an average accuracy of .69 in estimating the severity of illness. Experimental results demonstrate that this analysis has the potential of using speech-based analysis.

- In [67], Artificial Intelligence based diagnosis using cough recordings is proposed. Mel Frequency Cepstral Coefficient features and the CNN-based model have shown an AUC of 0.97. A total 4,256 subjects are used for training and 1,064 subjects are used for testing. This low-cost model has also shown potential in group-based outbreak detection, which could pre-screen whole populations daily.

- To detect COVID-19 in asymptomatic and symptomatic stages, a speech modeling framework is proposed in [68]. This model uses the distinct nature of COVID-19 in lower and upper respiratory tract inflammation. This model proposes vocal biomarkers of COVID-19 based on speech production, including respiration, phonation, and articulation.

1.5 CONCLUSION

Analysis of the probable factors that could accelerate the spread of a COVID-19-like pandemic is very important to control the growth rate. In this chapter, several socio-economic and environmental factors are analyzed that can have a prominent effect on the growth rate of COVID-19. A total of ten features are analyzed for eighty-four countries, using polynomial regression and feature selection techniques. From the analysis, it was found that the top three significant features contributing to COVID-19 are the vulnerability of the population of people over age sixty-five, diabetes, and temperature. This analysis can be useful in being prepared to deal with the pandemic from one country to another. In the future, analysis can be extended using updated databases with more available features. In the second part of the chapter, a brief overview is provided of the recently developed non-invasive methods of speech-based detection of COVID-19. The speech-based analysis can be one of the effective and potential tools in the coming future for automatic detection of COVID-19.

REFERENCES

1. Lai, C.-C., Shih, T.-P., Ko, W.-C., Tang, H.-J., and Hsueh, P.-R., "Severe acute respiratory syndrome coronavirus 2 (SARS-CoV-2) and corona virus disease-2019 (COVID-19): The epidemic and the challenges," *International Journal of Antimicrobial Agents* 55, 2020: 105924.
2. "world health organization." [Online]. Available: https://www.who.int/emergencies/diseases/novel-coronavirus-2019
3. Singhal, T., "A review of coronavirus disease-2019 (COVID-19)," *The Indian Journal of Pediatrics* 1–6, 2020.
4. Ruan, Q., Yang, K., Wang, W., Jiang, L., and Song, J., "Clinical predictors of mortality due to COVID-19 based on an analysis of data of 150 patients from Wuhan, China," *Intensive Care Medicine* 1–3, 2020.
5. Huang, C., Wang, Y., Li, X., Ren, L., Zhao, J., Hu, Y., Zhang, L., Fan, G., Xu, J., and Gu, X., "Clinical features of patients infected with 2019 novel coronavirus in Wuhan, China," *The Lancet* 395(10223) (2020): 497–506.
6. Arabi, Y. M., Murthy, S., and Webb, S., "COVID-19: a novel coronavirus and a novel challenge for critical care," *Intensive Care Medicine* 1–4, 2020.
7. Zhang, S., Diao, M., Yu, W., Pei, L., Lin, Z., and Chen, D., "Estimation of the reproductive number of Novel Coronavirus (COVID-19) and the probable outbreak size on the Diamond Princess cruise ship: A data-driven analysis," *International Journal of Infectious Diseases* 2020, doi: 10.1016/j.ijid.2020.02.033.
8. Liu, Y., Gayle, A. A., A. Wilder-Smith, and J. Rocklöv, "The reproductive number of COVID-19 is higher compared to SARS coronavirus," *Journal of Travel Medicine* 27, 2020: taaa021.

9. Anastassopoulou, C., Russo, L., Tsakris, A., and Siettos, C., "Data-based analysis, modelling and forecasting of the novel coronavirus (COVID-19) outbreak," *PloS one* 15(3) (2020): e0230405.

10. Chen, Y., Cheng, J., Jiang, X., and Xu, X. "The reconstruction and prediction algorithm of the fractional TDD for the local outbreak of COVID-19," arXiv preprint arXiv:2002.10302, (2020).

11. Zhan, C., Tse, C., Fu, Y., Lai, Z., and Zhang, H., "modelling and prediction of the 2019 coronavirus disease spreading in China incorporating human migration data," Available at SSRN 3546051, (2020).

12. Jia, L., Li, K., Jiang, Y., and Guo, X., "Prediction and analysis of Coronavirus Disease 2019," arXiv preprint arXiv:2003.05447, (2020).

13. Feng, Z., Yu, Q., Yao, S., Luo, L., Duan, J., Yan, Z., Yang, M., Tan, H., Ma, M., and Li, T., "Early prediction of disease progression in 2019 novel coronavirus pneumonia patients outside Wuhan with CT and clinical characteristics," medRxiv, (2020).

14. Zhou, T., Liu, Q., Z. Yang, J. Liao, K. Yang, W. Bai, X. Lu, and W. Zhang, "Preliminary prediction of the basic reproduction number of the Wuhan novel coronavirus COVID-19," *Journal of Evidence-Based Medicine* 13(1) (2020): 3–7.

15. Zhang, Y., et al. "The impact of temperature and humidity measures on influenza A (H7N9) outbreaks—evidence from China." *International Journal of Infectious Diseases* 30 (2015): 122–124.

16. Mäkinen, T.M., et al. "Cold temperature and low humidity are associated with increased occurrence of respiratory tract infections." *Respiratory Medicine* 103(3) (2009): 456–462.

17. Lin, Y.K., et al. "Relationships between cold-temperature indices and all causes and cardiopulmonary morbidity and mortality in a subtropical island." *Science of the Total Environment* 461 (2013): 627–635.

18. Tamerius, J.D., et al. "Environmental predictors of seasonal influenza epidemics across temperate and tropical climates." *PLoS Pathogens* 9(3) (2013): e1003194.

19. https://www.weather-atlas.com/

20. Kiank, C., Daeschlein, G., and Schuett, C. "Pneumonia as a long-term consequence of chronic psychological stress in BALB/c mice." *Brain, Behavior, and Immunity* 22(8) (2008): 1173–1177.

21. Biondi, M., and Zannino, L.G. "Psychological stress, neuroimmunomodulation, and susceptibility to infectious diseases in animals and man: A review." *Psychotherapy and Psychosomatics* 66(1) (1997): 3–26.

22. Dusek, J.A., and Benson, H. "Mind-body medicine: A model of the comparative clinical impact of the acute stress and relaxation responses." *Minnesota Medicine* 92(5) (2009): 47.

23. https://www.theglobaleconomy.com/rankings/happiness/

24. Wong, V.W.Y., Cowling, B.J., and Aiello, A.E. "Hand hygiene and risk of influenza virus infections in the community: A systematic review and meta-analysis." *Epidemiology & Infection* 142(5) (2014): 922–932.

25. Bloomfield, Sally F., et al. "The effectiveness of hand hygiene procedures in reducing the risks of infections in home and community settings including handwashing and alcohol-based hand sanitizers." *American Journal of Infection Control* 35(10) (2007): S27–S64.

26. Warren-Gash, C., Fragaszy, E., and Hayward, A. "Hand hygiene to reduce community transmission of influenza and acute respiratory tract infection: A systematic review." *Influenza and Other Respiratory Viruses* 7(5) (2013): 738–749.

27. Tracht, S.M., Del Valle, S.Y., and Hyman, J.M. "Mathematical modeling of the effectiveness of facemasks in reducing the spread of novel influenza A (H1N1)." *PloS one* 5(2) (2010): e9018

28. https://epi.envirocenter.yale.edu,https://photius.com/rankings/2019/population/sanitation_facility_access_improved_total_2019_1.Html

29. Gaziano, T.A., et al. "Growing epidemic of coronary heart disease in low-and middle-income countries." *Current Problems in Cardiology* 35(2) (2010): 72–115.

30. Li, X. et al. "Validation of the gravity model in predicting the global spread of influenza." *International Journal of Environmental Research and Public Health* 8(8) (2011): 3134–43. doi:10.3390/ijerph8083134

31. Ferkol, T., and Schraufnagel, D. "The global burden of respiratory disease." *Annals of the American Thoracic Society* 11(3) (2014): 404–406.

32. https://www.theglobaleconomy.com/rankings/GDP_constant_dollars/

33. Sun, R., and Danan, G. "Air pollution, economic development of communities, and health status among the elderly in urban China." *American Journal of Epidemiology* 168(11) (2008): 1311–1318.

34. Hu, G., and Ran, P. "Indoor air pollution as a lung health hazard: focus on populous countries." *Current Opinion in Pulmonary Medicine* 15(2) (2009): 158–164.

35. Hashim, D., and Boffetta, P. "Occupational and environmental exposures and cancers in developing countries." *Annals of Global Health* 80(5) (2014): 393–411.

36. Clay, K., Lewis, J., and Severnini, E. "Pollution, infectious disease, and mortality: evidence from the 1918 Spanish influenza pandemic." *The Journal of Economic History* 78(4) (2018): 1179–1209.

37. www.numbeo.com

38. Balicer, R.D., Barnett, D.J., Thompson, C.B. et al. Characterizing hospital workers' willingness to report to duty in an influenza pandemic through threat- and efficacy-based assessment. *BMC Public Health* 10 (2010): 436.

39. Ives, J., et al. "Healthcare workers' attitudes to working during pandemic influenza: A qualitative study." *BMC Public Health* 9(1) (2009): 1–13.

40. Martin, S. D., Brown, L.M., and Michael Reid, W. "Predictors of nurses' intentions to work during the 2009 influenza A (H1N1) pandemic." *AJN The American Journal of Nursing* 113(12) (2013): 24–31.

41. https://data.humdata.org/dataset/world-bank-indicators-of-interest-to-the-covid-19-outbreak

42. Sprung, C.L., Zimmerman, J.L., Christian, M.D. et al. Recommendations for intensive care unit and hospital preparations for an influenza epidemic or mass disaster: summary report of the European Society of Intensive Care Medicine's Task Force for intensive care unit triage during an influenza epidemic or mass disaster. *Intensive Care Medicine* 36 (2010): 428–443 doi:10.1007/s00134-010-1759-y

43. Irvin, C.B., et al. "Survey of hospital healthcare personnel response during a potential avian influenza pandemic: will they come to work." *Prehosp Disaster Medicine* 23(4) (2008): 328–335.

44. Singh, A. K., et al. "Diabetes in COVID-19: Prevalence, pathophysiology, prognosis and practical considerations." *Diabetes & Metabolic Syndrome: Clinical Research & Reviews* 14(4) (2020): 303–310.

45. Hussain, A., Bhowmik, B., and do Vale Moreira, N.C. "COVID-19 and diabetes: Knowledge in progress." *Diabetes Research and Clinical Practice* 162 (2020): 108142.

46. Valdez, R., et al. "Impact of diabetes mellitus on mortality associated with pneumonia and influenza among non-Hispanic black and white US adults." *American Journal of Public Health* 89(11) (1999): 1715–1721.

47. Martins, M., et al. "Diabetes hinders community-acquired pneumonia outcomes in hospitalized patients." *BMJ Open Diabetes Research and Care* 4(1) (2016): 1–7.
48. Luna, C.M., et al. "The impact of age and comorbidities on the mortality of patients of different age groups admitted with community-acquired pneumonia." *Annals of the American Thoracic Society* 13(9) (2016): 1519–1526.
49. Gutiérrez, F., et al. "The influence of age and gender on the population-based incidence of community-acquired pneumonia caused by different microbial pathogens." *Journal of Infection* 53(3) (2006): 166–174.
50. Kelly, E., et al. "Community-acquired pneumonia in older patients: Does age influence systemic cytokine levels in community-acquired pneumonia" *Respirology* 14(2) (2009): 210–216.
51. Almirall, J., et al. "Proportion of community-acquired pneumonia cases attributable to tobacco smoking." *Chest* 116.2 (1999): 375–379.
52. Marten, K., et al. "Non-specific interstitial pneumonia in cigarette smokers: A CT study." *European Radiology* 19(7) (2009): 1679–1685.
53. Díaz, A., et al. "Etiology of community-acquired pneumonia in hospitalized patients in Chile: the increasing prevalence of respiratory viruses among classic pathogens." *Chest* 131.3 (2007): 779–787.
54. Baskaran, V., et al. "Effect of tobacco smoking on the risk of developing community acquired pneumonia: A systematic review and meta-analysis." *PloS one* 14(7) (2019): e0220204.
55. Murty, V.N., and Studden, W.J. "Optimal designs for estimating the slope of a polynomial regression." *Journal of the American Statistical Association* 67.340 (1972): 869–873.
56. Melas, V.B., Shpilev, P.V. Constructing c-optimal designs for polynomial regression without an intercept. *Vestnik St. Petersburg University, Mathematics* 53 (2020): 223–231. doi:10.1134/S1063454120020120
57. Yang, W., Wang, K. and Zuo, W. Neighborhood component feature selection for high-dimensional data. *JCP*, 7(1) (2012): 161–168.
58. Dash, T.K. and Solanki, S.S., 2019. Investigation on the effect of the input features in the noise level classification of noisy speech. *Journal of Scientific & Industrial Research* 78 (2019): 868–72.
59. Ramesh, G., Covid or just a Cough? AI for detecting COVID-19 from Cough Sounds.
60. Al Ismail, M., Deshmukh, S. and Singh, R. "Detection of COVID-19 through the analysis of vocal fold oscillations" In *ICASSP 2021-2021 IEEE International Conference on Acoustics, Speech and Signal Processing (ICASSP)* (2021), (pp. 1035–1039). IEEE.
61. Sharma, N., Krishnan, P., Kumar, R., Ramoji, S., Chetupalli, S.R., Ghosh, P.K., Ganapathy, S. Coswara. A database of breathing, cough, and voice sounds for COVID-19 diagnosis. *Proc. Interspeech 2020* (2020): 4811–4815, doi: 10.21437/Interspeech.2020-2768.
62. C. Brown, J. Chauhan, A. Grammenos, J. Han, A. Hasthanasombat, D. Spathis, T. Xia, P. Cicuta, and C. Mascolo, "Exploring Automatic Diagnosis of COVID-19 from Crowdsourced Respiratory Sound Data," *Proceedings of the ACM SIGKDD International Conference on Knowledge Discovery and Data Mining*, August 2020. [Online]. Available: 10.1145/3394486.3412865
63. Stasak, B., Huang, Z., Razavi, S., Joachim, D. and Epps, J. Automatic detection of COVID-19 based on short-duration acoustic smartphone speech analysis. *Journal of Healthcare Informatics Research* 5.2 (2021): 201–217.
64. Dash, T.K., Mishra, S., Panda, G. and Satapathy, S.C. Detection of COVID-19 from speech signal using bio-inspired based cepstral features. *Pattern Recognition* 117 (2021): 107999.

65. Deshmukh, S., Al Ismail, M. and Singh, R. "Interpreting glottal flow dynamics for detecting COVID-19 from voice". In *ICASSP 2021–2021 IEEE International Conference on Acoustics, Speech and Signal Processing (ICASSP)* (2021), (pp. 1055–1059). IEEE.

66. Han, J., Qian, K., Song, M., Yang, Z., Ren, Z., Liu, S., Liu, J., Zheng, H., Ji, W., Koike, T. and Li, X., An Early study on intelligent analysis of speech under COVID-19: Severity, sleep quality, fatigue, and anxiety. age [years], 64 (9.5) 2020: 62–67.

67. Laguarta, J., Hueto, F. and Subirana, B. COVID-19 artificial intelligence diagnosis using only cough recordings. *IEEE Open Journal of Engineering in Medicine and Biology* 1 (2020): 275–281.

68. Quatieri, T.F., Talkar, T. and Palmer, J.S. A framework for biomarkers of COVID-19 based on coordination of speech-production subsystems. *IEEE Open Journal of Engineering in Medicine and Biology* 1 (2020): 203–206.

2 Machine Learning in Healthcare
The Big Picture

Ananta Charan Ojha and C. Vinitha
School of CS & IT, Jain (Deemed to be University),
Bangalore, India

CONTENTS

DOI: 10.1201/9780367548445-3

2.1   OVERVIEW OF MACHINE LEARNING

Machine Learning is influencing our everyday life. Think of the Google search engine that gives you accurate autocomplete prompts in the search box as if it is reading your mind about the search context. Amazon is so good at recommending to you what you like to buy. Facebook and Twitter monitor their social media platforms using Machine Learning algorithms to check hate messages and fake news, just like a human moderator. Nowadays, Machine Learning has been very instrumental in checking the spreading of Covid-19 infection by alerting you if someone nearby is Covid-19 positive. There are workplaces and supermarkets integrated with Machine Learning systems to help ensure customers and employees are safe by detecting if they wear masks and maintain social distancing as per Covid-19 protocols.

Learning, in general, builds intelligence and the ability to perform a task better. It is an essential part of being intelligent. The ability to learn is naturally possessed by humans, and animals. It is the process of acquiring new knowledge, understanding, skills, and behavior through study, observations, experience, or being taught. With advancements in computer and information technologies, now computers and machines can learn and act as humans do. These intelligent systems take inputs in the form of observations and real-world interactions, improve their learning over time autonomously, and perform activities closer to that of humans. Thus, learning is a fundamental building block of an intelligent machine.

2.1.1   WHY SHOULD MACHINES LEARN?

Traditionally, a computer programmer gives step-by-step instructions to a machine to solve a task in a particular area of application. However, there are situations where the traditional programming approach is either difficult or undesirable and the machine learning approach becomes useful in instructing the computer.

Learning is essential for unknown environments. In such environments, a computer programmer or system designer lacks every knowledge that is required to give step-by-step instructions to the computer to perform a task. For example, if you are developing a program to navigate a robot, you may not know the layouts of different places where the robot may be used. So, the robot must learn to navigate in a new place. Similarly, environments that keep on changing require a programmer to use Machine Learning techniques to program a computer to perform tasks. In such environments, the programmer cannot anticipate all possible changes over time to provide explicit instructions to the computer.

Through learning, a computer program gains newfound information and alters its existing decision-making process, thereby performing a task better. Sometimes, learning can also be used as a software engineering approach. Instead of programming the computer every step of the way, the programmer uses a Machine Learning approach to expose the computer program to the real world. The computer program learns from the data without taking new instructions from the programmer. Thus, the essence of Machine Learning is that the learning algorithm creates new rules or steps from the data and solves a problem. By using different datasets, the same learning algorithm can be used to perform different tasks. For example, you can use the same learning algorithm to detect a spam email as well as to predict the price change in the stock market.

2.1.2 What Is Machine Learning?

Machine learning is a field of computer science that makes use of algorithms and statistical methods to learn from the data, drawing inferences and recognizing patterns without explicit programming. It is an approach to developing a software system that learns from data and adapts itself to improve its ability to perform tasks, such as prediction. While several authoritative definitions of Machine Learning exist involving different perspectives, Tom Mitchell, includes in his book (Mitchell, 1997) a formal and widely used definition of Machine Learning.

According to him, a computer program is said to learn from experience **E** with respect to some classes of tasks **T** and performance measure **P**, if its performance at tasks in **T**, as measured by **P**, improves with experience **E**.

The definition serves as a template to analyze Machine Learning problems with less ambiguity. The template could be used to design Machine Learning software systems by clearly identifying three features: what task or decision the software needs to make (T), what data inputs the software is provided with (E), and how to evaluate its results (P). For example, a computer program that checks spam emails might improve its performance of correctly classifying an email as spam and putting it in the correct folder. In this case, classifying an email as spam or not-spam is the task (T), a set of example emails consisting of spam and not-spam is the experience or data inputs (E), and classification accuracy as a percentage is the performance measure (P).

The set of example emails as input to the computer program is called training data. The program learns from the examples, and the learned program is called a model. The model has a state and persists to classify an incoming email, which is called test data. The training is performed once and can be rerun if needed with new examples to gain new insights.

Machine Learning programs are fundamentally different from traditional programs, as they can adapt themselves to the new data by recognizing new-found knowledge therein. Whereas a traditional algorithm is restricted by design to work for a particular task or domain, a Machine Learning algorithm works in different domains and tasks.

2.1.3 Types of Machine Learning

In artificial intelligence and computer science, intelligent machines may use two broad classes of learning techniques, based on how knowledge is represented: deductive learning and inductive learning.

2.1.3.1 Deductive Learning

Deductive learning requires a programmer or knowledge engineer to find a set of rules based on the data or facts known in the task domain, and feed the set of rules to the computer so that the computer can use these rules to make inferences or deductions. Deductive learning follows an approach where general rules are used to determine a specific outcome. Yesteryears' rule-based expert systems are classic examples of intelligent systems using deductive learning. Let's take the example of a medical expert system for heart disease that uses a set of rules to diagnose heart diseases that are available in the medical science domain. These rules are stored in its knowledgebase beforehand. When the symptoms and medical conditions of a person are fed to the system, it employs reasoning techniques such as forward chaining or backward chaining on the rules in its knowledgebase to conclude whether the person suffers from heart disease or not.

2.1.3.2 Inductive Learning

On the other hand, inductive learning does not require a programmer to find a set of rules from the data. Rather, the intelligent system accepts the data inputs and learns a set of rules or recognizes patterns in the data automatically, using a learning algorithm. Inductive learning follows an approach where specific examples are used to formulate general rules, and then these rules are used to determine a specific outcome. For example, if you want to develop an intelligent system using inductive learning to diagnose heart diseases, then you need to feed the system a set of training examples with various symptoms and medical conditions of people with and without heart diseases. The learning algorithm in the system then learns from these examples and forms a general rule, which the system uses to diagnose a person when given his symptoms and medical conditions.

Intelligent systems based on inductive learning are far superior to deductive learning. Today, Machine Learning primarily refers to inductive learning. Based on the training methods used to train the learning algorithms, there are three basic types of machine learning: supervised, unsupervised, and reinforcement learning.

   a. **Supervised Learning**

     Supervised learning is the most common type of Machine Learning used in various applications. The computer program is provided with a set of labeled data called a training set. In the training set, each input data is labeled with an output value, as per the prior knowledge, known as ground truth. The goal of the learning algorithm is to approximate a function that maps each input to the desired output as observed in the sample data. Once learned, the mapping function, called model is used to predict a label for unseen input data. Supervised learning can be of two types: classification and regression. The Machine Learning task where the output is a categorical value is called classification, and if the output is a continuous real value, it is called a regression task. Detecting emails using Machine Learning and categorizing them as spam or not-spam is a classification task. Similarly, diagnosing a person with heart disease and predicting disease or no disease is also a classification task. On the other hand, stock price prediction and predicting rainfall based on historical data are regression tasks.

b. Unsupervised Learning

There are many application areas of machine learning where the ground truth or prior knowledge on data is not available or difficult to obtain. In those situations, the available data is unlabeled, i.e., not in the form of input-output pairs. So, no training data is provided to the computer program, and the learning algorithm is made to learn by itself through searching hidden patterns in the input data. It then draws inferences that describe the hidden patterns. That is why unsupervised learning is called "learning from observation" and supervised learning is called "learning from examples." Unsupervised learning has the advantage of being able to work with unlabeled data, which removes the human effort of making the dataset machine-readable.

Unsupervised learning can be of two types: clustering and association rule mining. Clustering is the most common type of unsupervised learning. It involves organizing unlabeled data into similar groups called clusters by finding similarities among data objects. Within a cluster, the data objects are cohesive, i.e., the similarity among them is very high, whereas the clusters are very distinct from one another. Grouping documents based on the nature of their contents is an example of clustering. Similarly, segmenting customers of an enterprise based on their profiles to apply target marketing is another case of clustering. Association rule mining is a type of unsupervised learning in which the learning algorithm tries to learn the hidden relationship among data objects. It identifies interesting insights and represents them in the form of rules or a frequent itemset. Market basket analysis is one classic example of association rule mining application in the retail business where the algorithm operates on a large transactional data and generates which items are brought together by customers.

c. Reinforcement Learning

Reinforcement learning is completely different from supervised and unsupervised learning. In reinforcement learning, there is no training data provided to the computer program, rather the learning algorithm, called an agent, is put to interact with a work environment that employs a reward system. The agent learns to carry out its intended task by performing actions in the environment and looking at the outcomes. The reward system interprets the outcome of an action and decides if it is favorable or not. If the outcome is favorable, the reward system assigns a reward score, and the action is encouraged or reinforced. If the outcome is not favorable, the reward system assigns a punish score, and the agent iterates for a better action that provides a reward score. The agent learns through experience, following a trial-and-error approach, to maximize the expected cumulative score. Since the agent gathers examples in the form of good-action and bad-action, reinforcement learning is called "learning from actions." Primarily, reinforcement learning is used in autonomous robots, self-driving cars, and gaming applications. It can also be used in many other domains, such as recommender systems, real-time bidding, stock trading using bots, chatbots, and portfolio management. Reinforcement learning has potential applications in the healthcare domain (Yu et al., 2019). For example, patients can receive treatments for chronic diseases and critical care based on policies learned from the reinforcement learning system.

2.1.4 DEEP LEARNING

Deep Learning is a subset of Machine Learning that uses a multi-layered neural network to mimic the complex information processing of the human brain. Similar to biological neurons in the human brain, a perceptron is used as the basic building block of an artificial neural network and performs information processing. A neural network consists of a large number of neurons (or perceptrons) interconnected with each other in layers and is responsible for the massively parallel processing of input data to produce outputs. A neural network consists of several layers: one input layer, one output layer, and several hidden layers. In a deep learning system, the hidden layers decide the deepness of the network. The number of hidden layers depends on the nature of the problem at hand and the volume of data considered for processing. A deep learning network uses both supervised and unsupervised learning approaches for training the neural network. A variety of architectures and supervised learning algorithms are used in deep learning, the most common being Recurrent Neural Network (RNN) and Convolutional Neural Network (CNN).

A recurrent neural network is a feedback neural network architecture in which the outputs of neurons of one layer are used as inputs to the neurons of the previous layer. This feedback mechanism in RNN allows it to maintain its internal state and to be capable of modeling temporal problems. RNN can process an enormous amount of sequential data using its feedback connections. Due to its sequential data processing capability, RNN is widely used in process control automation, time-series prediction, text processing, speech recognition, gene sequence analysis, and very suitable bioinformatics applications.

A convolutional neural network is a fully connected feedforward neural network and popular in deep learning systems. CNN is presently the de facto standard for object recognition and image classification. It consists of several layers that perform feature extraction and classification tasks. Although a CNN architecture contains several hidden layers, it always performs two basic operations: convolution and pooling. The convolution operation is the primary operation of CNN and responsible for extracting higher-order features from input data using multiple filters. The pooling operation is used to reduce the dimensionality of the features generated from the convolution operation and keep the most significant features in the data, using max-pooling, min-pooling, or average pooling. The resultant features are then fed to a fully connected layer for necessary classification and prediction tasks.

Deep Learning is commonly used in medical imaging for quicker diagnosis. It has been successfully used to detect melanoma, the deadliest form of skin cancer, at an early stage in order to cure the disease. Deep Learning-based systems use MRI images to visualize and quantify blood flow to detect cardiovascular diseases. Diabetic retinopathy, which is the fastest-growing cause of blindness around the world, has been effectively diagnosed using Deep Learning methods. Deep

Learning-based biomedical analysis is popularly used in genomics and proteomics (Miotto et al., 2018).

2.2 DRIVING FORCES FOR ML IN HEALTHCARE

The global healthcare industry faces several challenges that hinder better services. These challenges open doors for the use of modern technologies in general and Machine Learning techniques, in particular, to improve the quality of services and population health outcomes while reducing the cost of healthcare. Machine Learning can improve the accuracy of diagnosis, prognosis, and risk prediction. It can automate the detection of relevant findings in pathological and medical imaging investigations. Further, it can prevent and reduce many medical errors. Some of the prominent driving factors that endorse the use of ML in healthcare are outlined here.

2.2.1 BIG DATA IN HEALTHCARE

Unlike other domains, the healthcare industry deals with massive data amassed from a large number of sources. The diversity of health data spans several areas, including demography, vital signs, lab tests, medications, notes and transcripts, medical imaging, payor records, bio-signal data from wearable and medical devices, genomics, proteomics, pharmaceutical research, clinical trials, etc. The data is highly variable in terms of its structure and nature. Big data in healthcare is growing rapidly and is projected to grow faster than other sectors, such as manufacturing, financial services, and media, in the next five years. Healthcare data will grow at a CAGR (compound annual growth rate) of 36 percent by 2025 according to IDC (International Data Corporation) reports. This massive leap in growth will pose challenges of managing, utilizing, and monetizing these data assets for organizations involved in healthcare.

It has been reported that about 80% of the medical data remains untapped after it is created. This happens because the data resides at the point of care in the form of free-text or semi-structured documents and remains out of the electronic medical records or hospital information systems. The data tends to be unmanaged and ignored but holds huge potential and value-creation if properly captured and processed (Kong, 2019). Despite several challenges in data aggregation and processing, the adoption of big data in healthcare has the potential to transform the sector fundamentally from a fee-for-service model toward a promise of value-based healthcare. Healthcare big data analysis will provide evidence-based information and help sharpen the understanding of diagnosis and patient care, increasing the efficiency of services to deliver superior patient experience and outcomes at a lower cost (NEJM, 2018).

2.2.2 DEMOGRAPHIC SHIFT

According to the UN report on World Population Ageing 2019, one in six people in the world will be over the age of sixty-five by 2050 (Un Report, 2019). It is a huge jump from the current state of one in eleven, as reported in 2019. The global population of people older than sixty is expected to reach two billion. It is a global phenomenon. Almost every country is experiencing a growing population of older people. Population aging in East and South-East Asia, Latin America, and the Caribbean region has been the fastest in the world. Life expectancy is increasing, with survival beyond age sixty-five. The dramatic increase in population aging will put significant pressures on healthcare systems to align care services with older people.

As older people suffer from chronic diseases such as diabetes, heart diseases, high cholesterol, cancer, arthritis, dementia, and hypertension, they require long-term care which will create an increasing burden on available resources, including healthcare professionals. Focusing on a single disease while addressing the comorbidity of an elderly patient remains a challenge for caregivers. Increasing chronic diseases among older people require healthcare service providers to think of a multidisciplinary approach to treatment. They need to look at not only reactive care but also preventive care for elderly people. Consequently, the need for technology-driven low-cost but high-impact healthcare services will be promoted.

2.2.3   GLOBAL PANDEMIC

Another rising concern in the healthcare domain is the global pandemic. Not only the present Covid-19, but the pandemics of the past decades, such as Ebola, SARS, MERS, and H1N1, have also established the speed at which infections spread throughout the world. Covid-19 took roughly three months to spread over the whole world after it was first reported in Wuhan, China, in December 2019. As of March 10, 2021, the virus has infected about 118,714,747 people and caused death to about 2,633,819 people globally, according to WHO (World Health Organization). The numbers are rising daily since the pandemic is continuing. Global pandemics have far-reaching economic consequences, evident from the present Covid-19 pandemic. Many countries, both poor and rich, have been forced to lock down their business and social systems, incurring a global loss of several trillion dollars. Global pandemics cause huge impacts on healthcare systems and necessitate solutions to manage them. As evident from the Covid-19 pandemic, a huge quantity of assets and infrastructure, healthcare professionals, and other supporting manpower is required to tackle new disease, including a quick diagnosis, treatment and patient management, drug discovery and manufacturing, clinical trials, and proper planning and coordination.

Pandemics are likely and will exert periodic and significant disruptive pressure on healthcare systems. Thus, coordinated, concerted, and technology-driven pandemic management is essential to contain the outbreaks quickly as and when they occur. Artificial Intelligence and Machine Learning techniques can be used in early screening and treatment of infected people, contact tracing of persons who are found positive for infection, predicting and forecasting the spread and impact of the pandemic, finding the choice of drugs for treatment, drug and vaccine discovery, etc. (Lalmuanawma et al., 2020).

2.2.4   PERVASIVE MEDICAL ERRORS

Unfortunately, healthcare is not as safe as one would expect. Physicians, nurses, and other caregivers are fallible, committing unintentional errors that cause injuries and death to patients, as well as impacting their families. Errors may also create adverse mental and emotional reactions in the involved caregivers. Medical errors cost billions of dollars every year. Reports suggest that more Americans die due to medical errors than deaths from automobile accidents (Bari et al., 2016). Medical errors are pervasive. Most of these errors committed are under-reported due to a lack of courage to face the consequences. Medical errors may happen due to a variety of reasons and at different places in the healthcare system. The cause may be inexperienced caregivers, fatigue on the part of the caregiver, inadequate staffing, improper communication and coordination, lack of established protocols and standards in the workplace, etc.

The most common forms of medical errors are diagnostic errors, delayed treatment, medication errors, hospital-acquired infections, surgical procedure-related errors, inadequate follow-up after treatment, unsatisfactory monitoring of the patient, improper precautions on the part of the caregivers, errors in test results, and errors related to faulty medical equipment and devices. Medical errors may vary from minor errors to serious errors that result in enormous impact on patients and other stakeholders. Minimizing avoidable medical errors to enhance patient safety and robust healthcare services is a challenging task before service providers. Identifying the risk factors for medical errors, proper coordination, and communications, adequate information flow, and learning from mistakes are some of the crucial steps toward the prevention of medical errors. Intelligent techniques and automated information systems may help in these directions (Choudhury and Asan, 2020).

2.2.5 PATIENT-CENTRIC HEALTHCARE

Patient expectations are high due to competition among healthcare providers. In an era of abundant information, patients are very choosy about their healthcare needs. Now, patients have more choices of where, when, how, and from whom to receive healthcare services. Tech-savvy patients with digital awareness want active participation and greater interaction with the service providers throughout the entire business process of the healthcare service. Surveys find that a vast majority of patients want access to their full medical history that rests with healthcare organizations. They strongly believe that information sharing is an essential aspect of high-quality services. Similarly, a survey reports that patients would switch care providers if they can get an appointment faster. Patients in a clinical trial are no longer subjects today. They are partners of the pharmaceutical company conducting the trial.

Understanding and managing patient's expectations can improve the patient's satisfaction and help achieve the fulfilment of the healthcare needs of a patient (El-Haddad et al., 2020). Healthcare organizations need to change their service models, accommodating patient expectations and engagement and working toward better healthcare delivery. They need to design and deliver their services in a personalized way, keeping patient's preferences in mind. The healthcare industry is transitioning from organization-driven healthcare to patient-centric healthcare that is based on patient outcomes and patient satisfaction. Understanding patient's needs and expectations is central to patient-centric healthcare, which is a necessity today. Machine Learning and data analytics can play a crucial role in generating actionable insights from patient data to deliver customized services.

2.3 OPPORTUNITIES FOR ML APPLICATIONS

Institutions in the healthcare ecosystem generate a massive amount of health data regularly. This data is diverse and poses difficulties for analysis using traditional data processing methods. Since the data is valuable for improving the quality of healthcare and population outcomes, use of Machine Learning algorithms presents a lot of opportunities in the healthcare domain (Ahmed et al., 2020). Some of the commonly used areas of ML applications are described here.

2.3.1 DISEASE DIAGNOSIS

Machine Learning algorithms play a major role in identifying diseases accurately. The errors in a disease diagnosis seriously affect a patient's treatment. ML helps doctors to diagnose and predict diseases at an early stage with greater accuracy. It also saves diagnostics time, cost, and the effort that accompanies pathological tests. There are several examples of fatal diseases which have been successfully diagnosed using ML.

Parkinson's disease is a neurological disorder commonly presenting with shaking movement in one or more parts of the body, slowness of movement, and balance issues. In addition to this symptom, there may be problems with speech, sensory disturbances, sleep issues, cognitive decline, and psychological problems. This disease will affect an individual's day-to-day activities. The Machine Learning algorithms like Artificial Neural Network (ANN) can be used for gait analysis and diagnose the disease accurately (Tahir and Manap, 2012).

Hepatitis is one of the liver infections caused by viruses, bacteria, fungi, or toxic substances like alcohol. The disease damages the liver cells and causes swelling or inflammation. One of the problems faced by liver patients is the early detection of the disease. Early detection of the disease is not easy because the liver can function normally even if it is partially damaged. The patient's survival rate can be increased by an early diagnosis of liver problems. By analyzing the enzymes of blood, liver disease can be diagnosed. Automated classification tools based on Machine Learning classification algorithms help doctors to identify and diagnose liver disease more accurately (Spann et al., 2020).

The common symptoms of dengue fever are high fever, headache, vomiting, muscle pain, joint pain, and skin rashes. The healthcare system must diagnose dengue fever early to save lives. Tools based on ML classification algorithms can be used by the doctor to diagnose dengue disease at an early stage with more accuracy (Mello-Román et al., 2019).

A large number of people die due to coronary heart disease (CHD). High blood pressure, diabetic conditions, smoking, and high cholesterol are some of the contributing factors to heart diseases. The early detection of heart disease can help to reduce the death rate. Machine Learning techniques have been successfully used for the early detection of coronary heart disease (Katarya and Meena, 2021).

Autism Spectrum Disorder is a neuro-disorder which has a lifelong effect on interaction and communication with others. It is very important to diagnose autism at an early stage. Early treatment will increase the IQ significantly and improve the quality of the life of the patient.

Haematological diagnosis is mainly based on the laboratory blood test. Doctors analyze the patterns and the deviations in the blood parameters. Machine Learning algorithms can handle large quantities of parameters simultaneously and easily identify the relationship between various parameters more accurately. Watson healthcare is one ML application developed by IBM that offers the treatment for blood cancer. QuantX is an ML application developed by the University of Chicago, which is used by radiologists for disease diagnosis. A company named PATHAI located in Cambridge, Massachusetts has developed ML-based tools to reduce the error in cancer diagnosis and provide methods for personalized treatment.

2.3.2 MEDICAL IMAGING ANALYSIS

Medical imaging is a technique to create the visual representation of the internal organs or tissues for clinical usage, for example, to monitor health, diagnose disease, and identify injuries. High-quality imaging techniques are available for data collection and can be used for better medical decision-making. Various parameters present in the medical images can be analyzed by the doctor to diagnose the disease and provide better treatment for the patient. The interpretation of the medical image purely depends on the subject knowledge of the medical experts. They have to consider various parameters to diagnose the disease, and the outcome depends on the observer. Machine Learning can handle more complex data with a large number of parameters. Machine Learning and Deep Learning can be used for medical image analysis so that medical professionals can diagnose the disease more accurately and reduce the errors committed by the doctors. Some of the common examples in medical image analysis are given here.

Radiologists analyze the x-ray image and measure the joint space-width to diagnose Rheumatoid Arthritis. This process is very time-consuming, and the accuracy depends on the observer. Machine Learning and Deep Learning algorithms can be used to automate this process so that the radiologist can diagnose the disease more accurately with less time.

Cardiovascular diseases can be easily identified from cardiac images. Doctors analyze various features of the image, find the abnormalities in the image and make a decision about the affected disease. The accuracy of the diagnosis is purely dependent on the knowledge of the expert, and his observations, which may not be reliable. By automating the diagnosis with Machine Learning, it is possible to detect the disease more accurately at an early stage, thus reducing the risk of erroneous diagnosis.

A large number of women suffer from breast cancer, which is a significant cause of death. Digital mammography is used to diagnose breast cancer. Using texture analysis and classification techniques on mammography images, it is possible to identify micro-calcification and diagnose the disease. The technique saves not only the time of diagnosis but also provides more accurate results. Only early and accurate prediction helps the doctor to suggest the proper treatment to save lives.

Diabetic retinopathy is a common cause of blindness in diabetic patients. If found at an early stage, necessary measures can be taken to control the progress of the disease. The aspects and stages of the disease are normally identified by analyzing retinal images. Machine Learning algorithms can be used to perform this analysis faster and more accurately.

There are several AI/ML applications in practical use today. Enlitic, a San Francisco-based company, has developed an ML/ DL medical tool that is used by radiologists to analyze medical images and create better insights into the requirements of the patient. Similarly, Freenome has also developed an AI/ML application for the early detection of cancer. Zebra Medical Vision provides AI/ML applications used by radiologists for disease diagnosis from medical images. Beth Israel Deaconess Medical Center, a teaching hospital of Harvard University, is using AI enhanced microscopes for the early detection of blood disease.

2.3.3 MEDICAL PROGNOSIS

One of the important areas in healthcare decision-making is to predict the future health of the patient. Necessary treatment can be started if it is possible to predict the future outcome of treatment and disease. Machine Learning can be used to predict the outcome of a disease or post-operative health issues. Prognosis focuses on treatment sensitivity, life expectancy, survivability, and disease progression. The various parameters of age, weight, sex, smoking and drinking habits, and psychological factors affect the disease and evolution of diseases. Medical prognosis improves the treatment process and also avoids unnecessary surgical procedures and excess treatments. It helps to prioritize the patients' care. ML models help medical prognosis (Severson et al., 2020; Zhu et al., 2020).

2.3.4 SMART RECORD AND PERSONALIZED MEDICINE

Increasing numbers of hospitals maintain Electronic Health records and patient's care information. Machine Learning techniques can be used by doctors to analyze the individual patient's data and can generate insights from this information so that personalized treatment can be given to the patients. The workload of the doctor is not only reduced, but better service is provided for the patient.

Ciox is a Machine Learning application developed by Ciox Health, a European health technology company, to manage and exchange health information efficiently. They have developed an application smart-chart that identifies and extracts data from various medical reports and creates medical history in a single file.

Healthcare organizations maintain electronic records of all patients and physicians. Machine Learning and AI can be used to analyze these records and match the patients with appropriate physicians for correct and better treatment. Behavioral modification is an important aspect of preventive medicine. It plays a major role in disease prevention and identification, and treatment. ML-based applications can be used to identify gestures we make, understand the unconscious behavior, and suggest necessary corrective measures to prevent disease.

Machine Learning applications like chatbots can be used by patients for self-diagnosis or obtaining suggestions about competent physicians/doctors. It is also possible to reduce prescription errors. Harvard Medical School used an AI-based symptom and cure checker named Buoy Health to diagnose and treat illness quickly. In this application, a chatbot listens to the symptoms and health concerns of the patients, then guides the patients on correct care based on the diagnosis.

2.3.5 ROBOTIC SURGERY

All robot-assisted surgical procedures do not require Machine Learning, but some systems use computer vision aided Machine Learning to identify the body parts, for example, in hair transplantation surgery to identify the hair follicles and the distance. Another example of ML is Automation of Suturing - the process of sewing up an open wound, which is an important, time-consuming process in surgery. In 2013, the University of California at Berkeley designed a robot for laparoscopic surgery. In 2016, Johns Hopkins University designed a surgical robot called STAR (Smart Tissue Autonomous Robot) which used sensors and 3D computer imaging to guide the surgical process. Since robotic surgery is usually performed using tiny incisions in the body as compared to traditional open surgical procedure, it provides shorter recovery time. Robot-assisted surgery allows doctors to perform complex procedures with more precision than traditional surgery. Thus, robotic surgery gives more performance than the traditional process. Mazor Robotics is a 3D tool used by surgeons to visualize surgical plans and to recognize anatomical features. MicroSure's robots help the surgeon to improve the precision and performance of the surgical procedure. Carnegie Mellon University designed a mobile robot named HeartLander for heart therapy.

2.3.6 GENOMICS AND PROTEOMICS

Genetics is a branch of biology which deals with the study of genes, variations in genetics, and heredity. A genome of an organism contains the complete set of genetic materials. Genomics is a field of biology which gives more focus on function, structure, evolution, or editing of genomes. Genetics is the study of individual genes whereas the genome focuses on the quantification of genes of an organism, its characteristics, their interrelations. Both the areas deal with more complex data. Machine learning can be used to analyze complex data and can generate hidden insights from it more accurately. It can also reduce the time and cost of the analysis. During this covid pandemic, many of the start-up companies were working to predict the protein structure of Covid-19 using Machine Learning. Early discovery of the covid vaccine is one of the important achievements of ML application in genetics and genomics. One important area in genetics is the diagnosis of leukemia using DNA coding. During the ongoing coronavirus pandemic, Deep Learning has been used to find new information about the structure of proteins associated with Covid-19, which in turn provides important clues to the coronavirus vaccine formula.

2.3.7 DRUG DISCOVERY

The process of developing new drugs is time-consuming and costly. To ensure the effectiveness of the drug and the safety of the patients, the drug has to undergo a long procedure. The drug development process starts with basic research or drug discovery. Then a pre-clinical test has to be conducted to assess the efficacy of the drug, as well as its side effects on the body. A clinical trial is conducted to identify the dose-toxicity, side effects, and kinetic relationships. After this, the performance of the drug has to be measured and a comparison of molecules to the standard of care is evaluated. Post drug marketing is carried out to monitor the long-lasting side effects and to know the combination of the drug with other treatments. It takes a minimum of 5-15 years to complete all these stages. Suppose the drug has been discovered and a pre-clinical test is passed, but it failed during the post-clinical test or during the post-marketing stage. Then it will lead to huge financial loss, and the time spent is also wasted.

To accelerate the development of drugs and to reduce human-related errors, pharmaceutical companies have started using artificial intelligence and Machine Learning algorithms. By integrating the Machine Learning techniques with drug development, the cost and time for drug development can be reduced. Machine Learning is used to predict the success rate of the drug compounds and help in screening them, based on different biological factors in the early stage of drug discovery. It also identifies mechanisms for multifactorial diseases and suggests alternate paths for therapy. Project Hanover is a Machine Learning-based system developed by Microsoft which is used for developing cancer treatment and personalizing drug combinations for patients. Bioxcel Therapeutics uses AI/ ML for developing medicines for neuroscience and immuno-oncology. Similarly, Berg has developed an AI-based biotech platform to accelerate the discovery and development of new drugs.

2.3.8 CLINICAL TRIAL

A clinical trial undergoes four phases to find the efficacy of a drug. Researchers check the appropriate dosage, side-effects of the drug, patient safety, and drug effectiveness. If the first three phases are successful, the regulator approves the drug for clinical use. Machine learning and predictive analytics can be used to analyze a wide variety of data, including doctor visits and social media, to find the target population and candidates for clinical trials. It can monitor participants remotely by collecting real-time data related to biological and other signals to evaluate the effect of the drug. Social media analysis can also be done to identify the influence of physicians with more meaning and value by quantifying the success of a campaign.

2.3.9 EPIDEMIC OUTBREAK AND CONTROL

ML and AI applications can directly access the data from satellites, the web, and social media platforms to predict and monitor epidemic outbreaks around the world. The ongoing Covid-19 pandemic is a standout case for use of ML. The complex nature of novel coronavirus and its rapid spread across the globe posed unprecedented challenges for government agencies. ML models have been used heavily to control the coronavirus pandemic. Government agencies use real-time location data and ML models to trace people's movement in an infected place to predict the number of infected people and track the spread of the novel coronavirus. Machine-learning models have been used to analyze data from social media posts and search engine queries to predict the size and speed of the outbreak in different parts of a geographical area. With the help of big data tools, researchers around the world have been able to analyze massive data to generate various analytics to manage the pandemic. For example, organizations use interactive data platforms to analyze bed capacity in a covid hospital, identify covid clusters, and predict risk levels of different geographic locations. Researchers at Johns Hopkins School of Public Health developed a Covid-19 mortality risk calculator that advises on the mass vaccination program. It suggests who are the right people to be vaccinated first. Similarly, Medical Home Network developed a Machine Learning-based system to identify people with high vulnerability and severe complications to prioritize Covid-19 care for them.

2.4 KEY CHALLENGES FOR ML IN HEALTHCARE

Machine Learning models demonstrate its efficiency that surpasses human capabilities in areas such as diagnostics in healthcare. However, it faces quite a few significant technical challenges and organizational barriers, which create hindrances to achieve the full potentials.

2.4.1 CAUSALITY PROBLEM

There are enough success stories that establish the efficacy of Machine Learning in healthcare. Most of the standout examples are related to learning patterns and predicting outcomes. Such applications are great at finding correlation (or association) in data, but not causation. Most of the ML problems in healthcare are about a causal relationship. For example, observational data finds that people who are Covid-19 positive get relief when they take acetaminophen (i.e., paracetamol). Acetaminophen is a medication commonly used to reduce fever, pain, and aches. Patients infected with Covid-19 experience fever, headaches, and body aches. It establishes a strong association of the drug with Covid-19 symptoms. But the observational data fails to answer the question, can acetaminophen cure the Covid-19 infection? It is a fact that correlation does not imply causation. However, there might be causal signatures hidden in data, but they are ambiguous and usually corrupted by missing variables and observations, noise, and bias, making it difficult to find the cause-and-effect relationships. Most of the ML models that are very effective in learning input-output predictions fail to reason about causal relations.

Judea Pearl, one of the pioneers of Bayesian networks and probabilistic reasoning in artificial intelligence, in his book "Book of WHY: The New Science of Cause and Effect" argues for building intelligent systems with the notion of causality to prove the efficacy of AI and ML (Pearl and Mackenzie, 2018). He suggests three levels of causal hierarchy and classification of causal information.

At the lowest level, Association involves a purely probabilistic relationship that exists in the observational data. Most of the ML methods work at this level to answer associated questions given the data. For example, what are the symptoms of Covid-19? How does paracetamol help relieve Covid-19 symptoms?

The second level, Intervention, is a higher level that deals with not only Association but also estimating one's action. It finds causal relationships. A typical question could be: What if I take paracetamol? Will it cure the Covid-19 infection? Questions at this level can also answer the questions raised at the first level of the hierarchy.

The highest level, Counterfactual, deals with retrospective reasoning. It deals with hypothetical situations not observed in the data. It tries to estimate the unobserved outcome and subsumes the questions raised at the first and second levels of the hierarchy. So, if we can have a model to reason about counterfactuals, then it can also deal with queries on association and intervention. For example, a counterfactual question, "What would have happened had I taken paracetamol when I suffered from Covid-19 infection?", can answer the intervention question, "What if I take paracetamol? Will it cure the Covid-19 infection?"

Different levels require different data and models to answer different questions. Data required at a level can be used to answer questions at that level or a lower level. Data at the lower level cannot answer questions at a higher level. Data at the association level is characterized by conditional probability. Such evidential data can be processed using Bayesian Network or Deep Learning. However, data required at intervention and counterfactual levels is more analytical and logical. It requires formalisms and models such as Causal Bayesian Networks and Structural Equation Models to answer questions. We need to develop effective Machine Learning models at these two higher levels of the hierarchy to address the reasoning requirements involving counterfactual and interventional logic in real-world healthcare applications.

2.4.2 DATA LIMITATIONS

Data is the backbone of any Machine Learning application, and healthcare is no exception to it. The universally accepted premise "garbage-in-garbage-out" is very significant in healthcare ML applications. How effective the ML algorithm may be, it will fail to deliver its promises if the data fed to it is of poor quality. To take full advantage of ML algorithms, healthcare data must be reliable, accurate, unbiased, and representative. However, there are skeptics about the quality of data available in healthcare, due to a variety of reasons.

Although electronic health records (EHRs) provide a wealth of data, it is not free from challenges when it comes to missing data (Wells et al., 2013). Data present in EHR is of two types: structured and textual. The structured data consists of numeric data and categorical data stored in well-defined databases for effective retrieval. Textual data is the free form data available in the form of reports, notes, and transcripts that are difficult to analyze due to the use of abbreviations and acronyms both standard and non-standard, different individual expressions, different contextual interpretations, and possibly language errors. Whether it is structured data or textual data, missing data is of great concern in healthcare data. Missing data in EHRs can be a particular data value not available or a variable not captured in the database. Missing data is attributed to different reasons, such as lack of collection or lack of documentation. Sometimes, the patient is not asked a question to answer. Even if the question is asked and the patient has answered it, the data may not be documented for several reasons. When data is missing of one or more variables or aspects of a diagnosis or disease, then the data becomes incomplete. Such incomplete data is also common in medical records.

Similarly, data in EHRs can be noisy. The noise in data can be due to faults in the devices used for collecting the data or it can be due to the disturbances occurring in the environment. For example, while taking an MRI, a slight movement of the patient's head or body may result in unwanted artifacts recorded in the resultant image, leading to the risk of misdiagnosis. There is also a possibility of noise in the form of erroneous data recording by unqualified or inexperienced healthcare workers while collecting data from the subject. Sometimes, data noise may result when the ground truth is an ambiguous or improper annotation of the data sample (Qayyum et al., 2021). Further, human biases cannot be ruled out when it comes to data noise in healthcare data.

Often, the data available is small in size and not adequate to train the ML model. Although there is no definite size of a dataset required to train a model, Machine Learning and Deep Learning algorithms cannot be effectively used for healthcare tasks with small datasets. Further, the class imbalance problem in healthcare data arises when the distribution of samples among classes is not uniform. Normally, ML models tend to be biased in their outcomes when trained using an imbalanced dataset. Developing ML models using an imbalanced dataset may create significant life-threatening consequences. Healthcare data should be representative of the population at hand to predict trustworthy outcomes. Arguments exist that the data on a disease representing a particular population, say Asian, may be different from another population data, say African. Findings from one population data cannot be used to generalize another.

Despite several strategies available to address data quality in EHRs, missing data, incomplete data, noise, class imbalance, and outliers are prevalent. The presence of such issues reduces the data quality and significantly affects the outcomes. Preprocessing of healthcare data in ML models remains an open challenge.

2.4.3 Model Interpretability Problem

Advancements in Machine Learning and Deep Learning algorithms have resulted in their performance improvement, which is evident from several standout applications in different domains, including healthcare. Despite the recognition, ML algorithms are often criticized for a major drawback, which is known as the interpretability problem. The black-box nature of advanced ML models such as CNN and RNN fails to provide explanations about their learning behavior and decision-making in an understandable manner. Often it is difficult or even impossible to query as to why and how the model has arrived at a particular conclusion. Healthcare applications require not only well-performing models but also interpretable models (Ahmad et al., 2018). Model interpretability is essential in healthcare for algorithm fairness, robustness, and generalization. Healthcare applications necessitate ML models that are trustworthy, transparent, and explainable. Explainable models enable system verification when data is incomplete, imbalanced, noisy, contains missing values, and involves unqualified biases which are difficult to identify beforehand. It enables the ML professionals to recognize model errors and help remove model bias, thereby optimizing the model (Kelly et al., 2019). When a model is trained using a large dataset, it may unravel hidden patterns in the data. An interpretable model can provide clear visibility of this new-found knowledge and make it easily accessible. Further, an interpretable model provides decision traceability when needed since model outcomes may significantly affect users in healthcare (Holzinger et al., 2017). While interpretable ML model is imperative for healthcare applications, model interpretability remains an impediment of the advanced Machine Learning algorithms.

2.4.4 ADOPTION BARRIERS

Innovations in AI and ML have a lot of potentials in healthcare. It can result in smarter and faster healthcare processes, lower cost, enhancement of clinical quality and provision of better care. Despite its recognition, ML adoption is slow and faces several organizational barriers. The costs of Machine Learning projects are fairly high. Projects are time-consuming with relatively complicated processes, as compared to traditional software development projects. Small and medium healthcare providers may not be able to afford the projects. Convincing the business stakeholders about a positive return on investment is not so easy.

Skilled manpower to implement ML projects in healthcare is limited due to the difficulty of acquiring complex inter-disciplinary knowledge. Healthcare organizations do not have the required AI and ML talent pool to initiate internal experimentations and innovations. Usually, they rely on third-party vendors and costly solutions. The physicians and healthcare workers need to be AI and ML-reliant. However, the current medical science curriculum is not aligned to build the skillsets required. Thus, the present preparedness allowing growth toward an AI and ML competent healthcare workforce is not encouraging (Singh et al., 2020).

Healthcare is human-driven. Patients believe in doctors, not machines. Handing over the critical illness diagnosis and care to a machine is not acceptable to a large number of people. While a human can provide gentleness and compassion to a patient, a machine cannot do so. Hence, there is a natural reluctance to accept the machine's precision diagnosis and care. How a machine could give recommendations that will form the basis for patient treatment is hard to believe, not only by the patients but also by many physicians. Thus, the black-box nature of the ML models creates a trust deficit in the minds of the caregiver and the care receiver.

Patient data is incredibly valuable and raises concern about its ethical use for years. There are regulations such as the Health Insurance Portability and Accountability Act of 1996 (HIPAA) that mandate the insurance companies and healthcare providers to inform patients about the use of their data. Even data sharing for ML system development can be considered a violation of these laws. Further, several data breaches in recent years create concerns about the security and privacy of electronic patient records. While multi-institutional healthcare data is necessary for generalization and effective model development, data sharing remains a major issue. Healthcare data across the care institutions needs to be shared and used following ethical practices. However, data breaches and unethical practices pose security and privacy issues as another major barrier to the adoption of ML models in healthcare (Gerke et al., 2020). Approaches to Machine Learning systems with federated data need to be developed, keeping data privacy in mind.

Believe it or not, there are many healthcare service providers who do not have the right electronic health record systems in place. They operate in silos and use disparate technological standards, resulting in poor quality digitization of health records. When a patient shifts from one care provider to another, the patient data becomes fragmented and distributed in different organizational systems. Consequently, data sharing and consolidation remain a big challenge due to interoperability issues. Even after efforts and pushes by governments, many organizations still operate on the age-old method of paper and fax machines. When there are no proper digital records, the availability of patients' data for use in ML models raises eyebrows. Healthcare organizations must improve digitization and work on ways to improve data consolidation and sharing so that consolidated patient data can be available for processing and analysis using ML models.

2.5 CONCLUSIONS

Several breakthroughs in Machine Learning have been noticed. Machine Learning applications tend to be ubiquitous and are trusted across the industries gradually. There is a lot of excitement surrounding Machine Learning in healthcare applications. Machine Learning has the potential to transform the healthcare sector in terms of clinical decision-making and R&D capabilities. While there are powerful use cases of AI and ML in healthcare, several open issues exist, both technical and non-technical. To realize the full benefits of Machine Learning, the issues and barriers need to be addressed. People, processes, and technology, all should be integrated and goals should be aligned to create an AI and ML-reliant healthcare ecosystem. Only then will a healthcare ecosystem with the integration of Machine Learning models be the rule rather than the exception.

REFERENCES

Ahmad, M., Teredesai, A., and Eckert, C. 2018. Interpretable Machine Learning in Healthcare, in *2018 IEEE International Conference on Healthcare Informatics (ICHI)*, New York City, NY, 2018: pp. 447–447. doi: 10.1109/ICHI.2018.00095

Ahmed, Z., Mohamed, K., Zeeshan, S., and Dong, X. 2020. Artificial Intelligence with Multi-Functional Machine Learning Platform Development for Better Healthcare and Precision Medicine, *Database*, 2020. doi:10.1093/database/baaa010

Bari, A, Khan, R.A., and Rathore, A.W. 2016. Medical Errors: Causes, Consequences, Emotional Response and Resulting Behavioral Change, *Pakistan Journal of Medical Sciences*, 32(3): 523–528. doi: 10.12669/pjms.323.9701

Choudhury, A., Asan, O. 2020. Role of Artificial Intelligence in Patient Safety Outcomes: Systematic Literature Review, *JMIR Medical Informatics*; 8(7): e18599. doi: 10.2196/18599

El-Haddad, C., Hegazi, I., Hu, W. 2020. Understanding Patient Expectations of Health Care: A Qualitative Study, *Journal of Patient Experience*. doi: 10.1177/2374373520921692

Gerke, S., Minssen, T., and Cohen, G. 2020. Ethical and Legal Challenges of Artificial Intelligence-Driven Healthcare. *Artificial Intelligence in Healthcare*, 295–336. doi: 10.1016/B978-0-12-818438-7.00012-5

Holzinger, A., Biemann, C., Pattichis, C.S. 2017. What do we need to build explainable AI systems for the medical domain? Available at arXiv; https://arxiv.org/abs/1712.09923. Accessed 15 March 2021.

Katarya, R., Meena, S.K. 2021. Machine Learning Techniques for Heart Disease Prediction: A Comparative Study and Analysis. *Health and Technology*; 11:87–97. doi: 10.1007/s12553-020-00505-7

Kelly, C.J., Karthikesalingam, A., Suleyman, M., et al. 2019. Key Challenges for Delivering Clinical Impact with Artificial Intelligence. *BMC Medicine*, 17, 195. doi: 10.1186/s12916-019-1426-2

Kong, H.J. 2019. Managing Unstructured Big Data in Healthcare System, *Healthcare Informatics Research*, 25(1):1–2. doi: 10.4258/hir.2019.25.1.1

Lalmuanawma, S., Hussain, J., and Chhakchhuakb, L. 2020. Applications of machine learning and artificial intelligence for Covid-19 (SARS-CoV-2) pandemic: A review. *Chaos, Solitons & Fractals*, 139. doi: 10.1016/j.chaos.2020.110059

Mello-Román, J.D., Mello-Román, J. C., Gómez-Guerrero, S., and García-Torres, M. 2019. Predictive Models for the Medical Diagnosis of Dengue: A Case Study in Paraguay, *Computational and Mathematical Methods in Medicine*, 2019. doi: 10.1155/2019/7307803

Miotto, R., Wang, F., Wang, S., Jiang, X., and Dudley, J.T. 2018. Deep Learning for Healthcare: Review, Opportunities and Challenges, *Briefings in Bioinformatics*, 19(6):1236–1246. doi: 10.1093/bib/bbx044

Mitchell, T.M. 1997. *Machine Learning*, McGraw-Hill, United States; ISBN: 0070428077.

NEJM Catalyst, 2018. Healthcare Big Data and the Promise of Value-Based Care, *NEJM Catalyst Innovations in Care Delivery*, https://catalyst.nejm.org/doi/full/10.1056/CAT.18.0290.

Pearl, J., and Mackenzie, D. 2018. *The Book of Why: The Nepw Science of Cause and Effect*, New York: Basic Books.

Qayyum, A., Qadir, J., Bilal, M., and Al-Fuqaha, A. 2021. Secure and Robust Machine Learning for Healthcare: A Survey, *IEEE Reviews in Biomedical Engineering*, 14, 156–180. doi: 10.1109/RBME.2020.3013489.

Severson, K.A., Chahine, L.M., Smolensky, L., Ng, K., Hu, J. and Ghosh, S. 2020. Personalized Input-Output Hidden Markov Models for Disease Progression Modeling. *Proceedings of the 5th Machine Learning for Healthcare Conference*, in PMLR 126:309–330.

Singh, R. P., Hom, G. L., Abramoff, M. D., Campbell, J. P., Chiang, M. F., and AAO Task Force on Artificial Intelligence. 2020. Current Challenges and Barriers to Real-World Artificial Intelligence Adoption for the Healthcare System, Provider, and the Patient. *Translational Vision Science & Technology*, 9(2), 45. doi: 10.1167/tvst.9.2.45

Spann, A., Yasodhara, A., Kang, J., Watt, K., Wang, B., Goldenberg, A., and Bhat, M. 2020. Applying Machine Learning in Liver Disease and Transplantation: A Comprehensive Review, *Hepatology*: 71(3):1093–1105. doi: 10.1002/hep.31103

Tahir, N. Md., and Manap, H. H. 2012. Parkinson Disease Gait Classification based on Machine Learning Approach, *Journal of Applied Science*: 12(2):180–185. doi: 10.3923/jas.2012.

UN Report, 2019. World Population Ageing 2019, available at https://www.un.org/en/development/desa/population/publications/pdf/ageing/WorldPopulationAgeing2019-Highlights.pdf, Accessed on 10 March 2021.

Wells, B.J., Nowacki, A.S., Chagin, K., Kattan, M.W. 2013. Strategies for Handling Missing Data in Electronic Health Record Derived Data. *eGEMs* (Generating Evidence & Methods to improve patient outcomes):1(3):7. doi: 10.13063/2327-9214.1035

Yu, C., Liu, J., and Nemati, S. 2019. Reinforcement Learning in Healthcare: A Survey, available at https://arxiv.org/abs/1908.08796, Accessed on 5 March 2021.

Zhu, W., Xie, L., Han, J., and Guo, X. 2020. The Application of Deep Learning in Cancer Prognosis Prediction, *Cancers*:12(3), 603. doi: 10.3390/cancers12030603

3 Heart Disease Assessment Using Advanced Machine Learning Techniques

Vasantham Vijay Kumar, D. Haritha, and Durga Bhavani Dasari
K L Deemed to be University, Vaddeswaram, India

Venkata Rao Maddumala
Vignan's Nirula Institute of Technology & Science for Women, Guntur, Andhra Pradesh, India

CONTENTS

3.1 INTRODUCTION

Today, the chance of developing heart disease is great. It is urgent to be able to predict whether a patient is likely to have major symptoms [1]. Heart disease will be predicted by the different techniques explored in this chapter, such as K-Nearest Neighbor, Random forest classifier, Naive Bayes, Support Vector Machines, etc. [2].

A data mining technique is used to find better solutions. Now, these techniques can be used to check which technique performs best, because a certain algorithm might not work well for prediction [3]. The most innovative field in computer science is the study of various statistical techniques. Also, we are using data extraction from

DOI: 10.1201/9780367548445-4

a data set in this model, like data mining techniques to search valuable data from a large database, as many data mining methods are utilized to predict heart diseases and nowhere is diagnosis using Machine Learning techniques used in order to have an accurate result [4].

Currently, more literature studies have been done on the expectation of coronary illness. Each time, we are drawing one stage nearer to our goal. With AI calculations, immense progress has been made in various fields [5], individuals can do things effectively which were recently considered to be impossible. With the progression of these AI calculations, calculating the expectation of illness in a patient has become a typical practice [6].

The other danger factors incorporated are gender classification, age, sexual orientation, and based on work stress and unfortunate choices in eating habits. The accuracy of predicting heart disease improves when an individual is older. Men have a greater threat of heart disease. Also, ladies have a similar danger after menopause. Focusing on survival can in the same way harm the blood vessels and add to the possibility of heart disease.

Thus, using the list above, we attempt to anticipate the danger of heart disease. Work has been completed with heart forecast structures through using various methods and calculations by a number of creators. These strategies might have profound potential on data mining, Machine Learning concepts, and deep learning methods. A goal of all these studies is to be precise in finding a way to predict the variations of a heart attack [7].

The hazard factors of heart disease are: High circulatory strain, Cholesterol levels, Chest torment, Obesity, and so on. Some of the best ML techniques are used for prediction using the following methods.

1. K-Nearest Neighbor
2. Random Forest classifier
3. Support Vector machines
4. Naïve Bayes
5. Logistic regression
6. Decision Tree

These techniques are used to predict different analyses, have the potential to check the relationship of the data, and can be applied for various tasks. The dataset using in this study is taken from the UCI Machine Learning repository. It contains 303 instances.

3.2 LITERATURE SURVEY

M. Gandhi and S. N. Singh [8], utilized Naïve Bayes conditions, Decision Tree and Neural networks methods and examined the clinical or health or medical information dataset. Also there are an immense number of highlights included. With this process, there is a need to diminish the number of highlights. This should be possible using choice. According to Gandhi and Singh, time is also decreased. They utilized choice tree and neural network process.

J. Thomas and R. Theresa Princy [9] utilized K closest Neighbor algorithm, neural method, credulous Bayes, and choice tree for heart disease prediction. They utilized information mining strategies to recognize the heart disease hazard rate.

Sana Bharti and Shailendra Narayan Singh [10] utilized Particle Swarm Optimization, Artificial neural method, and Genetic calculation for forecasting. Cooperative grouping is another productive strategy which incorporates affiliation, rules, and conditions for data mining and characterization to a model for forecast. This strategy was the most efficient.

Purushottam, Kanak Saxena, and Richa Sharma [11] proposed

A computerized framework in medical determination would improve clinical consideration to take information from data set, and it can like decrease costs. In this investigation, we have organized a framework that can proficiently information find the standards to anticipate the danger position of patient's dependent on the given boundary about their wellbeing. The standards can be organized dependent on the patient's prerequisite. The exhibition of the framework is assessed regarding grouping precision, and the result shows that the report has incredible potential in anticipating the heart disease hazard phases all the more precisely.

Sellappan Palaniyappan and Rafiah Awang [12] utilized choice conditional of Naïve Bayes, Decision tree and Artificial Neural Networks to fabricate Intelligent Heart Disease expect or Predict systems. This system shows upgrade representation and simplicity of understanding; it shows the outcomes pair in plain and graphical structures. By giving powerful medicines, it likewise assists with lessening treatment costs. Revelation of concealed examples and connections has gone unexploited. Progressed information mining methods helped fix the present circumstance.

Himanshu Sharma and M. A. Rizvi [13] utilized Decision support system or tree uphold vector machine, profound learning, and K closest neighbor calculations. Also, because the datasets contain clamor, they attempted to diminish the commotion by noise remove and pre-handling the dataset, then attempted to decrease the dimensionality and data information storage locations in the dataset. They found that great precision can be accomplished with neural methods.

H. Animesh, S. Mandal, and Gupta [14] talked in detail about information for cardiovascular illness and various manifestations of coronary failure. Various sorts of arrangement and bunching calculations and instruments were utilized.

Krishnaiah, Narsimha, and Subhash [15] introduced an investigation utilizing information mining. The study indicated that utilizing various procedures and getting a distinctive number of properties gives varying correctness for anticipating heart illnesses.

K. Ramandeep, Er. P. Kaur [16] have indicated some information for heart disease information represents pointless copy data. This data also must be pre prepared. Likewise, they state that include choice must be done on the dataset to get better outcomes.

J. Vijayashree and N Ch S. Narayana Iyenger, [17] utilized information mining. An enormous measure of information is created consistently. All things considered, it can't be deciphered physically. Information mining can be adequately used to anticipate illnesses from these datasets. In this paper, distinct information mining procedures are examined on heart disease data set. Taking everything into account, this paper examines and analyzes how unique grouping calculations work on a heart disease information base.

3.3   METHODOLOGY

The dataset was taken from the UCI repository and executed in Kaggle Kernel. The dataset contains fourteen attributes and 303 instances (Table 3.1).

3.3.1   PROPOSED METHODS

Here our main agenda is to analysis the heart disease. From the data set, we check how many members develop heart disease and how many are not; then, how many males and how many females develop heart disease as a percentage as well as by numbers. [5]. Next, we check every aspect (like Frequency of Age, Cholesterol, Blood Pressure, Chest pain) and analyze the heart disease [6]. In Logistic regression, we use sigmoid function to perform better calculations. By taking all the values, we are performing different algorithms [18] to measure and to analyze the accuracy to predict the heart disease [7].

3.3.1.1   K-Nearest Neighbors (KNN)

Figure 3.1 represents a data set stored particular information and pre-processing supervised algorithm because we are trying to portray that a point depends on the known classifier of other points. A supervised learning algorithm [19] makes it easy to understand. KNN implementation also helps to solve the classification problem and regression problems [20]. Learn to label other points. It takes a batch of labeled points closest to the new point, checks the nearest neighbors, and has its votes. So where the label has most of the neighbors having a label for the new points, it takes "k" value as the number of neighbors of the new points.

The data set is divided into pre-processing, and post processed and test sets when the training set is used for a particular model structure and preparation. In k, rank is concluded methodically on the square foundation of the quantity of perceptions.

TABLE 3.1
UCI Dataset Attributes Detailed Information

Attribute	Description	Type
age	Patient's age in years	Numeric
sex	(1 = male; 0 = female)	Nominal
cp	Chest pain type	Nominal
trestbps	resting blood pressure (in mm Hg on admission to the hospital)	Numeric
chol	serum cholestoral in mg/dl	Numeric
fbs	(fasting blood sugar > 120 mg/dl) (1 = true; 0 = false)	Nominal
restecg	resting electrocardiographic results	Nominal
thalach	maximum heart rate achieved.	Numeric
exang	exercise induced angina (1 = yes; 0 = no).	Nominal
oldpeak	ST depression induced by exercise relative to rest.	Numeric
slope	the slope of the peak exercise ST segment.	Nominal
ca	number of major vessels (0-3) colored by flourosopy.	Numeric
thal	3 = normal; 6 = fixed defect; 7 = reversible defect.	Nominal
target	have disease or not (1=yes, 0=no).	Nominal

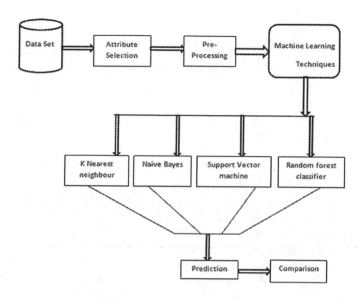

FIGURE 3.1 Experimental approach.

Currently, our method is that test information is anticipated on the fabricated model [21]. There are distinctive distance measures. For endless factors, Euclidean Distance, the alternative Manhattan distance, and Minkowski distance measures can be utilized.

The working process of KNN under the following steps is:

Step1: Finding the K value
Step2: Computing the distance between all the training data and test data (the new points). To compute the distance, the common technique uses Euclidean, Minkowski, and Manhattan alternatives.
Step3: To find the distance and find the KNN based on minimum distance values.
Step4: Separate the grouping of those neighbors and distribute the characterization for the test data reliant on bigger part vote
Step5: Place the predicted class.

The maximum accuracy percentage of KNN is 88.52%, which has a larger percentage compared to other algorithms. Here, the Shortest path is 3-5-8.

3.3.1.2 Random Forest Classification

In this classification, Random Forest is a best supervised learning technique used in both techniques for classification and regression, as it may be. In any case, it is mostly utilized for order challenges. As we know that a forest is composed of decision trees, and more trees indicate another technique [22] is Robust Forest. Essentially, Random forest algorithms pick trees for knowledge checking and later get the prediction.

The greatest one is the frail connection between models. Similarly, as assets with low relationships, (for example, stocks and securities) structure a portfolio that is bigger than the aggregate of its parts, uncorrelated models may produce expectations [23] that are more exact than any of the individual predictions. We understand the functioning of the Random Forest algorithm with the benefit of the following steps:

Step 1:   First, we start with a random sample chosen from a given dataset.
Step 2:   In this algorithm create a decision tree for each sample. So every decision tree process will have the predicted result.
Step 3:   Voting for any expect or predict outcome will be carried out in this step.
Step 4:   Eventually, pick the most voted outcome as the final outcome of the prediction.

3.3.1.3   Support Vector Machines

Supervised classification algorithm for learning Machine Learning, which has become increasingly popular nowadays, yields the best results. This can be used to detect Category, Regression, and Out layer. This is a discriminative classifier [24] explicitly constructed by a separate hyperplane.

The Support Vector Machines (SVM) calculation is utilized and SVM is foresee this disease by plotting the training data on the dataset where a hyperplane groups [25] the focuses into divided, one is presence and another is nonattendance of heart disease. SVM works by distinguishing the hyperplane which expands the edge between two classes.

Here, punished SVM is utilized to deal with class lopsidedness. Class irregularity is an issue in AI when the absolute number of positive and negative class isn't the equivalent. In the event that the class awkwardness isn't taken care of, at that point the classifier won't perform well. The accompanying plot shows the class irregularity.

The fundamental target of our SVM is to isolate the given information in the most ideal manner. SVM for directly distinguishable double sets is to plan a hyperplane (Figure 3.2) that orders all preparation vectors in two classes. The most ideal decision will be a hyperplane that parts the greatest edge structure between the two classes (Figure 3.3).

3.3.1.4   Naive Bayes Classification

This classification is a grouping calculation, utilized when the data dimensionality of the information is extremely very high. A Naive Bayes classifier accepts to conditional properties that the presence of a specific component in a class is disconnected to the presence of some other element. It depends on Bayes hypothesis.

Naive Bayes classifiers are a group of classification techniques based on Bayes' Theorem. This is not a particular specific algorithm but an algorithm family in which that information shares a general concept, i.e., each pair of features described is independent of each other.

$$\text{Equation: } P(A) = P(B)P(A)/P(B).$$

Using Bayes theorem, we can consider the likelihood of A occurring, since B happened. Here, proof is B, and theory is A. Here, the presumption is that the predictors/

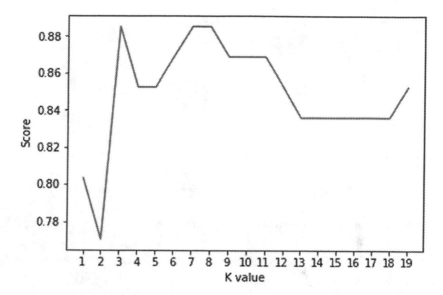

FIGURE 3.2 Performance analysis chart.

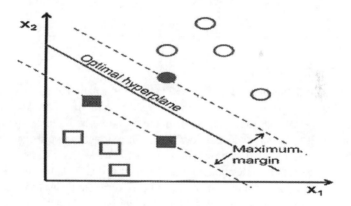

FIGURE 3.3 Graph between two classes.

functions are independent (Figure 3.4). That is the potentiality of the theorem, and this one particular trait does not affect another method, so is called naïve.

- P(h): a particular Probability of hypothesis h and its being true (output regardless data).
- P(D): In a Probability of data, and this can be said Prior or basic probability.
- P(h|D): Both Probability of h and D, and this can be said Posterior Probability.
- P(D|h): final Probability of D and if h was true, and it is said to be posterior probability.

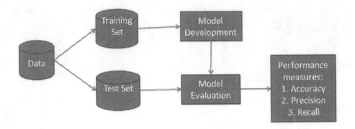

FIGURE 3.4   Classification of Naïve Bayes process on data set.

TABLE 3.2
Result of Various Models with Machine Learning Model

Models	Accuracy	Classification	Precision	F-score measure	Sensitivity	Specificity
K-Nearest Neighbors	88.52%	21.4	93.5	87.6	98.8	33.3
Support Vector Machine	86.89%	13.0	85.1	91.5	98.9	0.0
Random Forest	88.52%	13.3	86.7	92.1	97.8	10.2
Naive Bayes	86.89%	24.2	90.1	84.4	78.9	50.0
Logistic Regression	86.89%	16.1	87.9	92.3	93.6	22.0
Machine Learning	87.4%	12.8	90.7	93.5	98.5	34.3

3.3.1.5   Logistic Regression

Logistic represents a logistic regression model to the information to discover the capacity that best fits the preparation set by utilizing Support Vector Machine as a method process and assessing the slope of the exact risk or the variations between the model and the preparation set [26]. It is otherwise called KNN and is a straightforward characterization method that chooses the class estimation of another occasion by taking a gander at a bunch of the K nearest examples in the preparation set and picking the most successive class particular value among them. In Table 3.2 I used different types: models, accuracy of data, classification of attributes of data, precision, and F-Measure, Sensitivity, and Specificity information for heart attacks and different types of patients on data set.

3.4   RESULTS

After analysis we see that the accuracy of KNN and Random Forest is better when compared to Logistic, Naïve Bayes, and Decision Tree. In this chapter, we discuss the different AI calculations, for example: uphold vector machine, Naïve Bayes, choice tree, and k-closest neighbor, which were applied to the informational index. It uses the information [27, 28] (for example: circulatory strain, cholesterol, and diabetes) and afterward attempts to anticipate the possible heart disease for the next ten years.

A family background of heart disease can likewise be an explanation behind the approach of heart disease, as referenced before. Thus, this information about the patient can likewise be incorporated for expanding the precision of the model.

This work will be helpful in distinguishing the potential patients, who may experience the ill effects of heart disease in the following ten years. This work may encourage taking preventive measures and, consequently, ward off the chance of heart attack for the patient [29]. So, when a patient is anticipated to be a certain heart disease victim, then the clinical information or data attributes for the patient can be shown a particular status that is firmly broken down by the specialists. A model would assume the particular patient has diabetes, which might be the reason for heart disease in the future. The patient can be offered treatment to control the diabetes, which may forestall the heart disease.

The heart disease prediction should be possible, utilizing other AI calculations. Calculated relapse can likewise perform well if there should be an occurrence of paired arrangement issues, for example, heart disease expectation [30]. Irregular timberlands can perform better than choice trees. Likewise, the troupe techniques and counterfeit neural methods can be applied to the informational collection. The outcomes can be thought about and ad-libbed (Figures 3.5–3.7).

We are checking the frequency of ages shown for the age a person is affected with heart disease and at what age the frequency is high (Figures 3.8–3.10).

Above all, analysis of different types of algorithms gives particular reports. In forecasting heart disease utilizing AI strategies, we have taken a dataset with fourteen characteristics and 303 lines. With these insights, we finished analyzing interesting systems like KNN Algorithm, Naive Bayes, Neural system, Support Vector Machine, and Random Forest [31]. We had better outcomes at 88.52% by KNN and Random Forest of prediction of heart infections. Later, Random Forest, SVM, and Naïve Bayes [32] matched this percentage. We found less accuracy (78.69%) with Decision Tree.

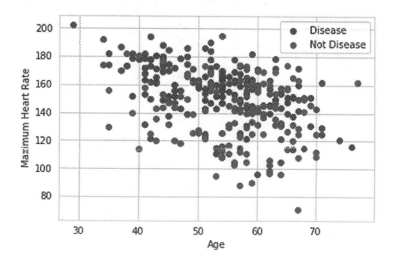

FIGURE 3.5 Classification of heart disease.

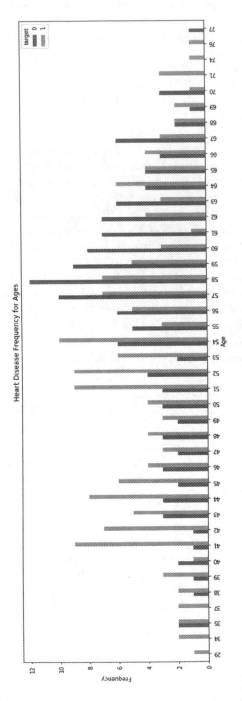

FIGURE 3.6 Performance comparison with various ages.

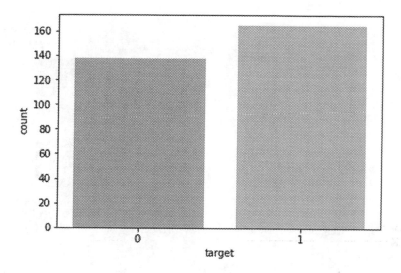

FIGURE 3.7 Bar chart representation for heart disease accuracy.

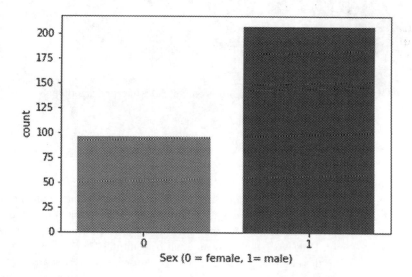

FIGURE 3.8 Bar chart representation for heart disease based on gender.

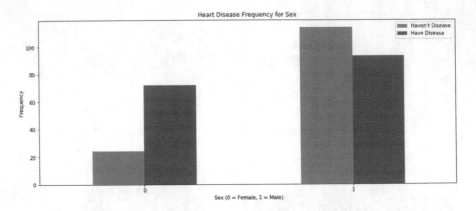

FIGURE 3.9　Heart disease classification based on gender.

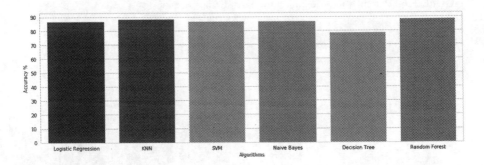

FIGURE 3.10　Comparison of results for heart disease classification algorithms.

3.5　CONCLUSION

Identifying how best to utilize the health domain information of heart data will help save human life with early identification of deviations in individual hearts. ML methods were utilized in this work to handle crude information and give novel insight into heart problems. Testing for the potential of a heart attack is a vital issue in the medical field. However, the death rate can be definitely controlled if the illness is identified in the beginning phases and precautionary measures are taken at the earliest opportunity. Basic highlights among these data sets are removed, then utilized in a later examination for a similar illness in any data set. The outcomes show that the grouping exactness of the gathered dataset is higher than normal of the characterization precision of all different datasets. This work can be useful for heart attack classification and its precision can be improved. The entire cycle of classification can be utilized with different sorts of datasets(for example, stress and clinical datasets).

REFERENCES

1. Prakash, K.B., Rangaswamy, M.A.D., Raman, A.R., (2012). Statistical interpretation for mining hybrid regional web documents, *Communications in Computer and Information Science*, Vol. 292, CCIS, pp. 503–512.
2. Ismail, M., Prakash, K.B., Rao, M.N., (2018). Collaborative filtering-based recommendation of online social voting, *International Journal of Engineering and Technology (UAE)*, Vol. 7, No. 3, pp. 1504–1507.
3. Prakash, K.B., Rajaraman, A., (2016). Mining of bilingual Indian web documents, *Procedia Computer Science*, Vol. 89, pp. 514–520.
4. Prakash, K.B., (2016). Content extraction studies using total distance algorithm, 2017, *Proceedings of the 2016 2nd International Conference on Applied and Theoretical Computing and Communication Technology*, iCATccT, 7912085, 673–679.
5. Prakash, K.B., (2015). Mining issues in traditional indian web documents, *Indian Journal of Science and Technology*, Vol. 8, No. 32, pp. 1–11.
6. Prakash, K.B., DoraiRangaswamy, M.A., Ananthan, T.V., Rajavarman, V.N., (2015). Information extraction in unstructured multilingual web documents, *Indian Journal of Science and Technology*, Vol. 8, No. 16. doi: 10.17485/ijst/2015/v8i16/54252
7. Prakash, K.B., Rangaswamy, M.A.D., Raja Raman, A., (2012). ANN for multi-lingual regional web communication, Lecture Notes in Computer Science (including subseries Lecture Notes in Artificial Intelligence and Lecture Notes in Bioinformatics), 7667, LNCS, PART 5, pp. 473–478.
8. Gandhi, M., Singh, S.N. 2015. Predictions in heart disease using techniques of data mining, *2015 International Conference on Futuristic Trends on Computational Analysis and Knowledge Management (ABLAZE)*, Noida, pp. 520–525, doi: 10.1109/ABLAZE.2015.7154917.
9. Thomas, J., Princy, R. T. (2016). Human heart disease prediction system using data mining techniques, *2016 International Conference on Circuit, Power and Computing Technologies (ICCPCT)*, Nagercoil, pp. 1–5, doi: 10.1109/ICCPCT.2016.7530265.
10. Bharti, S., Singh, S. N. (2015). Analytical study of heart disease prediction comparing with different algorithms, *International Conference on Computing, Communication & Automation*, Noida, pp. 78–82, doi: 10.1109/CCAA.2015.7148347.
11. Purushottam, Kanak Saxena, Sharma, Richa. (2016). Efficient heart disease prediction system, *Procedia Computer Science*, Vol. 85, pp. 962–969, ISSN 1877-0509, doi:10.1016/j.procs.2016.05.288.
12. Palaniyappan, Sellappan, Awang, Rafiah. (2008). Intelligent heart disease prediction using data mining technique, *IJCSNS International Journal of Computer Science and Network Security*, Vol. 8, No. 8.
13. Sharma, Himanshu, Rizvi, M. A. (2017). Prediction of heart disease using machine learning algorithms: A survey, *International Journal on Recent and Innovation Trends in Computing and Communication*, Vol. 5, No. 8, pp. 99–104. doi: 10.17762/ijritcc. v5i8.1175.
14. Hazra, Animesh, Mandal, Subrata, Gupta, Amit, Mukherjee, Arkomita, Mukherjee, Asmita. (2017). Heart disease diagnosis and prediction using machine learning and data mining techniques: A review, *Advances in Computational Sciences and Technology*. doi: 10.2137-2159.
15. Krishnaiah, V., Narsimha, G., and Chandra, Subhash N. (2016). Article: Heart disease prediction system using data mining techniques and intelligent fuzzy approach: A review, *International Journal of Computer Applications*, Vol. 136, No. 2, pp. 43–51.
16. Kaur, Ramandeep, Kaur, Prabhsharn. (2016). A review – heart disease forecasting pattern using various data mining techniques, *IJCSMC*, Vol. 5, No. 6, pp. 350–354.

17. Vijayashree, J., Iyenger, N. Ch. Sriman Narayana. (2016). Heart disease prediction system using data mining and hybrid intelligent techniques: A review, *International Journal of Bio-Science and Bio-Technology*, Vol. 8. 139–148. doi: 10.14257/ijbsbt.2016.8.4.16.
18. Maddumala, V.R., Arunkumar, R. (2020). Big datadriven feature extraction and clustering based on statistical methods, *Traitement du Signal*, Vol. 37, No. 3, 387–394. doi: 10.18280/ts.370305
19. Jagajeevan Rao, L., Venkata Rao, M., Vijaya Saradhi, T. (2016). How the smartcard makes the certification verification easy, *Journal of Theoretical and Applied Information Technology*, Vol. 83, No. 2, pp. 180–186.
20. Kumar, Singamaneni Kranthi, Reddy, Pallela Dileep Kumar, Ramesh, Gajula, Maddumala, Venkata Rao. (2019). Image transformation technique using steganography methods using LWT technique, *Traitement du Signal*, Vol. 36, No. 3, pp. 233–237.
21. Mounika, Banavathu, Khadherbhi, S. Reshmi, Maddumala, Venkata Rao, Lakshmi, R.S.M. (2020). Data distribution method with text extraction from big data, *Journal of Critical Reviews*, Vol. 7, No. 6, pp. 376–380.
22. Lakshman Narayana, V., Naga Sudheer, B., Maddumala, Venkata Rao, Anusha, P. (2020). Fuzzy base artificial neural network model for text extraction from images, *Journal of Critical Reviews*, Vol. 7, No. 6, pp. 350–354.
23. Lakshmi Patibandla, R.S.M., Tarakeswara Rao, B., Sandhya Krishna, P., Maddumala, Venkata Rao. (2020). Medical data clustering using particle swarm optimization method, *Journal of Critical Reviews*, Vol 7, No. 6, pp. 363–367.
24. Tejaswini, J., Mohana Kavya, T., Ramya, R. Devi Naga, Sai Triveni, P., Maddumala, Venkata Rao. (2020). Accurate loan approval prediction based on machine learning approach, *Journal of Engineering Science*, Vol 11, No 4, pp. 523–532.
25. Maddumala, Venkata Rao, Gundabattini, Sanjay Gandhi, Anusha, P., Sandhya Krishna, P. (2020). Classification of cancer cells detection using machine learning concepts, *International Journal of Advanced Science and Technology*, Vol. 29, No. 3, pp. 9177–9190.
26. Maddumala, Venkata Rao, Maha Lakshmi, K., Anusha, P., Lakshman Narayana, V. (2020). Enhanced morphological operations for improving the pixel intensity level, *International Journal of Advanced Science and Technology*, Vol. 29, No. 3, pp. 9191–9201.
27. Anusha, P., Ravikiran, Aala, Narayana, V. Lakshman, Maddumala, Venkata Rao. (2020). Energy priority with link aware mechanism for on-demand multipath routing in manets, *International Journal of Advanced Science and Technology*, Vol. 29, No. 3, pp. 8979–8991.
28. Kalava, Jagadish, Maddumala, Venkata Rao, Venkata Ranga Rao, K., Reddy, Chavva Ravi Kishore. (2020). Credit score based on bank loan prediction system using machine learning techniques, *International Journal of Advanced Science and Technology*, Vol. 29, No. 3, pp. 9418–9429.
29. Sandhya Krishna, P., Khadherbhi, Reshmi, Pavani, V. (2019). Unsupervised or supervised feature finding for study of products sentiment, *International Journal of Advanced Science and Technology*, Vol. 28 No.16.
30. Patibandla, R.S.M.L., Veeranjaneyulu, N. (2018). Performance analysis of partition and evolutionary clustering methods on various cluster validation criteria, *Arabian Journal for Science and Engineering*, Vol. 43, pp. 4379–4390.
31. Lakshmi Patibandla, R. S. M., Kurra, Santhi Sri, and Veeranjaneyulu, N. (2015). A study on real-time business intelligence and big data, *Information Engineering*, Vol. 4, pp. 1–6.
32. Maddumala, V.R. (2020). A weight based feature extraction model on multifaceted multimedia bigdata using convolutional neural network, *Ingénierie des Systèmes d'Information*, Vol. 25, No. 6, pp. 729–735.

4 Classification of Pima Indian Diabetes Dataset Using Support Vector Machine with Polynomial Kernel

P. Pujari

Guru Ghasidas Vishwavidyalya, Central University, Bilaspur, India

CONTENTS

DOI: 10.1201/9780367548445-5

4.1 INTRODUCTION

Classification is one of the supervised machines learning techniques, which includes two phases. The first phase is to construct a classifier to describe a set of predetermined data classes or labels. This phase is known as learning phase or training phase. In this phase classifier is constructed using classification algorithm by analyzing or examining a set of database samples and their associated class category. The n measurements of the sample X from n database attributes (Attribute_1, Attribute_2,..,Attribute_n respectively) is represented as an attribute vector with n-dimension as X = (x1, x2, ..,xn). There also exists an attribute in the database to determine the class category of each sample. The class level or category attributes are discrete and not ordered. It is categorical because each value is used as a category or class. The samples that make up the training set are called training samples, and they are taken from the database samples being analyzed. This phase is called supervised learning, as class category or label is assigned to each sample. The first phase can predict the associated class category Y of a given sample X and can be considered as a learning function or mapping Y=f(X). The learning or mapping function can be expressed in various forms like a decision tree, mathematical formula, or classification rule. The future data samples can be classified using these rules. The second phase is used for classification purposes, using the model. In order to test the new samples and their associated class category the test samples are used. From ordinary samples the test samples are selected randomly. The performance of the classifier in terms of accuracy on new samples called test samples is determined from the percentage of test samples that are correctly classified by the classifier. For each test sample, comparison is made between the associated class category and the predicted category, predicted by the learned classifier for that sample. If desired accuracy level is obtained by the learned classifier, the learned classifier can be used effectively for classification of future data samples whose category is unknown. In this chapter, data mining classification techniques Support Vector Machine (SVM) with polynomial, Radial Basis Function (RBF) kernel, sigmoid kernel, linear kernel, ANN, and C5.0 decision tree are applied on the Pima Indian Diabetes dataset. The statistical indicators classification accuracy, specificity and sensitivity are used for evaluating the performance of SVM classifier with different kernel. Performance comparison is carried out among the SVM classifier with different kernels, ANN, and the C5.0 decision tree algorithm.

4.2 BACKGROUND DETAILS

Following are descriptions of the research work carried out by different authors in the related fields.

In [1] Stavros et.al have introduced e *Class* architecture for evolving fuzzy rule-based systems by arranging the order of incoming data. The proposed model is applied on the Pima Indian Diabetes dataset for classification. A novel method Recursive-Rule eXtraction (Re-RX) algorithm with J48graft, combined with sampling selection techniques has been proposed in [2] for classification of (sampling Re-RX with J48graft) Pima Indian Diabetes (PID) dataset and the proposed model has achieved an accuracy of 83.83%. In [3] the authors have employed the SVM technique for classification of homogeneous data. The proposed method has been applied on the VidTIMIT database and achieved a good result in the classification of homogeneous data. A novel methodology using SVM classifier and Discrete Wavelet Transform (DWT) has been presented in [4] for detection of heart arrhythmias. The classification accuracy of the proposed model is found to be 95.92%. In [5] Zayrit Soumay et al. have introduced discrete wavelet transform, Genetic Algorithm (GA), and SVM classifier for the detection of Parkinson disease and achieved an accuracy of 91.18%. A Decision Tree SVM (DTSVM) method has been presented in [6] and applied on a large scale dataset to reduce computational costs. In [7] the authors have employed the SVM model for making correct decisions in the stock market for prediction of investor sentiment and achieved a high level of accuracy. In [8] Sidheswar Routray et.al have proposed a hybrid denoising technique based on SVM. The proposed model is applied on standard noisy images and achieved a good denoising performance in preserving edges and textures of noisy images. In [9] the authors have presented three feature extraction methods (convolution neural network of aspect ratio, HOG, and Hu invariant moment) for extraction of important features from peanuts images. The extracted features have been applied to SVM classifier and obtained an accuracy of 96.72% with the aspect ratio feature extraction method. E. A. Zanaty has proposed a combination of Gaussian Radial Basis Function (RBF) and Polynomial (POLY) kernels called Gaussian Radial Basis Polynomials Function (GRPF) and applied the proposed model to a variety of non-separable data sets with different number of attributes in [10]. The proposed model yielded better performance as compared to other kernels. In [11] the authors have proposed SVM and principal component analysis (PCA) for multiclass classification in industrial processes. The proposed model has been applied on the Tennessee Eastman (TE) challenging benchmark, which contains twenty-one abnormalities from the real world for showing the effectiveness of the approaches.

From the literature, data mining classification technique support vector machines have been applied successfully for classification of homogeneous data [2], detection of heart arrhythmias [4], detection of Parkinson disease [5], investor sentiment prediction in the stock market [7], classification of peanuts image [9] and many more. On the other hand, classification of the Pima Indian Diabetes Dataset has been less explored in literature. In this chapter the classification technique Support Vector Machine (SVM) with polynomial kernel has been proposed and applied on the Pima Indian Diabetes dataset. The performance of SVM classifier with polynomial kernel is compared with the performance of SVM with RBF kernel, sigmoid and liner kernel, ANN, and the C5.0 decision tree classifier. From experimental results the SVM with polynomial kernel is found to provide a better performance as compared with other classifiers.

4.3   DATASET DESCRIPTION

The dataset Pima Indian Diabetes used in this proposed work is taken from the UCI machine learning repository [12]. The dataset contains data of female patients with Pima Indian Heritage who are at least twenty-one years old. Each sample of the dataset is categorized under two categories: "has diabetes" and "no diabetes."

The category is a binary valued attribute, which shows whether a patient shows signs of diabetes or not, as per world health organization criteria. The class label 'class 0' is interpreted as "no diabetes" case, and the class label 'class 1' is interpreted as "has diabetes" case. The dataset contains 768 numbers of samples in total. Of these, 500 samples belong to category 'class 0,' and 268 samples belong to category 'class 1'. The dataset contains nine numbers of attributes, of which the 9th attribute represents the target output, and the remaining eight attributes are used as the input attributes. Through the concept of balanced nodes, there are 768 samples in the Pima Indian Diabetes Dataset, of which 611 samples are used as the training samples and 157 samples are used as the test samples.

4.4 METHODOLOGY

In order to meet the goal of the proposed work data mining classification technique, SVM with a number of kernels (like polynomial, RBF, sigmoid linear, ANN and C5.0 decision tree) are used as the classifier for classification of the Pima Indian Diabetes dataset. The dataset is divided into 80% of training samples and 20% of testing samples. The performance of the individual classifier is measured by using statistical indicators like accuracy, specificity, and sensitivity. Gain graph and response graph are also used to analyze the performance of the models.

4.4.1 SVM (Support Vector Machine)

Support Vector Machine [13] (SVM) is a powerful classification and regression technique which increases the predication accuracy of a model to a maximum extent, maximizes without over fitting the training samples. SVM is found to be suitable for the analysis of input data when the numbers of attributes are large in the input data. SVM maps the input data into a higher dimension space, where the data points are easily separated, even if the data cannot be linearly separated in other ways. SVM finds the separator between categories, and then transforms the data into a way that the separator finds acts as a hyperplane. Using this, behaviors of new samples can be analyzed for predication of category of the new samples. The Support Vector Machines (SVM) is a kind of general learning architecture, inspired by statistical learning theory, which implements structural risk minimization for the nested set structure of separating hyperplanes. Based on the generalization error, SVM generates a hyperplane that best separates the data points. Given a training data, the SVM learning technique will generate the best separation hyperplane. Support vector machine is one of the popular classification techniques in many areas of classification as a high-performance classifier. A set of support vector are generated by SVM, which performs the classification of data with less complexity. For maximizing the separation between positive and negative cases, SVM constructs a hyperplane as the decision surface. For this purpose SVM uses the structural risk minimization principle. Here, the error rate of the learning machine is considered to be limited by the sum of the training error rate and Vapnik Chervonenkis (VC) 1 dimension term. For N training samples (X_i, Y_i) which are labeled, where $X_i \in R^n$ and $Y_i \in \{-1, 1\}$, the hyper-plane that best separates the data points is represented as:

$$f\left(X_q\right) = \sum_{i=1}^{N} Y_i \propto_i K\left(X_q, X_i\right) + b \qquad (4.1)$$

Where the kernel function is represents as K (.) and determines the membership of query sample, X_q is determined from the sign of $f(X_q)$. Constructing an optimal hyperplane is equivalent to determining all nonzero $\alpha_i^{'s}$, which corresponds to the support vectors and the bias b. The expected loss of making decision is the minimum.

4.4.2 SVM KERNEL

Kernel is a method to solve nonlinear problems with the help of linear classifiers. This is called kernel trick, used to bridge linearity and non-linearity. The kernel is used as parameters in SVM. A kernel helps to form a hyperplane in higher dimensions without increasing complexity. It determines the shape of the hyperplane and decision boundary. The kernel function can transform the training data, so that the nonlinear decision surface can be converted into a linear decision surface in higher dimension. Kernel function reduces the complex calculation into a simpler one. It provides a more convenient and inexpensive way to transform data from a lower dimension into a higher dimension. Kernel can be used to scale a number of higher dimensions. A number of mathematical functions are used by SVM classifier known as kernel to transform the input data into a suitable form. There are different mathematical functions for kernels like sigmoid, polynomial, radial basis function (RBF) and linear.

4.4.2.1 Polynomial Kernel Function

The polynomial kernel is a general representation of kernels with a degree of more than one. It's useful for image processing. A polynomial kernel is a more generalized form of the linear kernel. The curved or nonlinear input spaces are easily distinguishable by the polynomial kernel. There are two types of this:

4.4.2.1.1 Homogenous Polynomial Kernel Function

$$K\left(X_i, X_j\right) = \left(X_i . X_j\right)^d \qquad (4.2)$$

where '.' is the dot product of X_i and X_j and d represents is the degree of the polynomial kernel.

4.4.2.1.2 Inhomogeneous Polynomial Kernel Function

$$K\left(X_i, X_j\right) = \left(X_i . X_j + c\right)^d \qquad (4.3)$$

where c is a constant.

4.4.2.2 Gaussian RBF Kernel Function

Gaussian RBF kernel function (Radial Basis Function) is a very well-known Kernel function used in SVM models. RBF kernel is a mathematical function whose value depends on the distance from the origin or from some point. RBF is the radial basis function. RBF kernel is more suitable when no prior knowledge is given about the data. The format of Gaussian kernel is as follows

$$K\left(X_i, X_j\right) = \exp\left(-\gamma X_i - X_j\right)^2 \tag{4.4}$$

Where $\|X_i - X_j\|$ is the Euclidean distance between X_i and X_j. The dot product of X_i and X_j i can be calculated using the distance in the original dimension. The value of width parameter γ ranges from 0 to 1.

4.4.2.3 Sigmoid Kernel Function

Sigmoid kernel is represented as

$$K\left(X_i, X_j\right) = \tanh\left(\sigma X_i^T X_j + c\right) \tag{4.5}$$

Where σ represents a scaling parameter of the input samples, and c represents a shifting parameter which controls the threshold of mapping.

4.4.2.4 Linear Kernel Function

This kernel is one-dimensional and is the most basic form of kernel in SVM. Normal dot product of any two given observations can be used for linear kernel. The product between two vectors is the sum of the products of each pair of input values. The equation is:

$$K\left(X_i, X_j\right) = X_i . X_j + c \tag{4.6}$$

The sum of the multiplication of each pair of input values represents the product between two vectors X_i and X_j is and c is a constant factor.

4.5 PERFORMANCE MEASURES

The performance of an individual classifier is obtained by using the statistical indicators accuracy, sensitivity, and specificity. These indicators are defined by positive cases that are true (TPcase), negative cases that are true (TNcase), positive cases that are false (FPcase), and negative cases that are false (FNcase). Table 4.1 represents a matrix showing number of True Positive case (TPcase), True Negative case (TNcase), False Positive case(FPcase), and False Negative case (FNcase).

For N number of case using the above table the following statistical performance indicators are evaluated.

TABLE 4.1

Matrix for Actual and Predicted Cases

	P′(Predicted case)	N′(Predicted case)
P(Actual case)	TPcase	FNcase
N(Actual case)	FPcase	TNcase

4.5.1 CLASSIFICATION ACCURACY

It considers positive and negative inputs and measures the proportion of correct prediction. The distribution of data plays an important role in measuring classification accuracy of a system which sometimes produces erroneous system performance. The calculation is represented as:

$$\text{Accuracy} = \text{Total no of samples correctly classified/}$$
$$\text{Total number of samples}$$
$$= \left(\text{True positive}\left(\text{TPcase}\right) + \text{True Negative}\left(\text{TNcase}\right)\right)/$$
$$\left(\text{positive}\left(\text{P}\right) + \text{Negative}\left(\text{N}\right)\right) \quad (4.7)$$

4.5.2 SENSITIVITY

Sensitivity measures the proportion of positive cases that are correctly predicted. It is the ability of the system to predict the positive cases. The calculation is represented as:

$$\text{Sensitivity} = \text{Positive cases that are correctly classified/}$$
$$\text{Total number of positive cases}$$
$$= \text{True Positive}\left(\text{TPcase}\right)$$
$$/\left(\text{True Positive}\left(\text{TPcase}\right) + \text{False Negative}\left(\text{FNcase}\right)\right) \quad (4.8)$$

4.5.3 SPECIFICITY

Specificity measures the proportion of negative cases that are correctly predicted. It is the ability of the system to predict the negative cases. The calculation is represented as:

$$\text{Specificity} = \text{Negative cases that are correctly classified/}$$
$$\text{Total number of negative cases}$$
$$= \text{True Negative}\left(\text{TN}\right)/$$
$$\left(\text{True Negative}\left(\text{TN}\right) + \text{False Positive}\left(\text{FP}\right)\right) \quad (4.9)$$

4.6 SIMULATION AND EXPERIMENTAL RESULT

For simulation purpose the Pima Indian Diabetes Dataset collected from the UCI Machine Learning repository is explored in this chapter. The development of the model involves two stages: training and testing. Training refers to the use of historical data to build a new model, while testing refers to the trial of the model on totally new data for determining performance characteristics of the model. Training is usually conducted on a large portion of the dataset available, while testing is conducted on a small portion of the dataset. Using the training dataset, the model is build. Once the model is developed on the training dataset, the performance of the model can be found on new dataset known as test dataset. Two mutually exclusive data sets are created by using partition node and balanced node segmentation methods for balancing the training and testing datasets. A training data set accounts for 80% of the total Pima Indian Diabetes Dataset and test dataset accounts for 20%. The training sample is applied on SVM with polynomial, sigmoid, RBF, and linear kernels, ANN and C5.0 decision tree classifiers. The parameter of SVM classifier with a polynomial kernel is set as regularization parameter = 20, degree = 5. A fivefold cross validation technique is used for testing the performance characteristics of the models. Statistical indicators accuracy, specificity, and sensitivity are used to evaluate the performance of the models. The performance characteristics of each model are obtained by using the training and testing data sets. A confusion matrix is obtained after applying the training and testing dataset to the classifier to identify positive cases that are true (TP), negative cases that are true (TN), positive cases that are false (FP), and negative cases that are false (FN). Table 4.2 shows the confusion matrices of different models for training and testing data sets. Tables 4.3 and 4.4 show correctly classified instances with accuracy and comparative statistical measures of different models for training and testing datasets respectively.

TABLE 4.2

Confusion Matrices of Different Models for Training and Testing Dataset

Models	Class	Training Set		Testing Set	
		Class 0	Class 1	Class 0	Class 1
SVM with RBF Kernel	Class 0	357	47	79	17
	Class 1	64	143	24	37
SVM with Polynomial Kernel	Class 0	394	10	91	5
	Class 1	10	197	1	60
SVM with Sigmoid Kernel	Class 0	348	56	80	16
	Class 1	94	113	31	30
SVM with Linear Kernel	Class 0	359	45	79	17
	Class 1	68	139	24	37
ANN	Class 0	349	55	77	19
	Class 1	50	157	15	46
C.5 Decision Tree	Class 0	394	10	83	13
	Class 1	42	165	24	37

TABLE 4.3
Correctly Classified Instances of Different Models for Training and Testing Dataset

Models	Cases	Training Set		Testing Set	
		Number of Instances	Accuracy	Number of Instances	Accuracy
SVM with RBF Kernel	Correct	500	81.83%	116	73.89%
	Wrong	111	18.17%	41	26.11%
SVM with Polynomial Kernel	Correct	591	96.73%	151	96.18%
	Wrong	20	3.27%	6	3.82%
SVM with Sigmoid Kernel	Correct	461	75.45%	110	70.06%
	Wrong	150	24.55%	47	29.94%
SVM with Linear Kernel	Correct	498	81.51%	116	73.89%
	Wrong	113	18.49%	41	26.11%
ANN	Correct	506	82.82%	123	78.34%
	Wrong	105	17.18%	34	21.66%
C.5 Decision Tree	Correct	559	91.49%	120	76.43%
	Wrong	52	8.51%	37	23.57%

TABLE 4.4
Statistical Indicators of Different Models for Training and Testing Dataset

Models	Phase	Accuracy	Sensitivity	Specificity
SVM with RBF Kernel	Training	81.83%	88.86%	69.08%
	Testing	73.81%	82.29%	60.66%
SVM with Polynomial Kernel	Training	96.73%	97.52%	95.17%
	Testing	96.18%	94.79%	98.36%
SVM with Sigmoid Kernel	Training	75.45%	86.14%	54.59%
	Testing	70.06%	83.33%	50.82%
SVM with Linear Kernel	Training	81.51%	88.86%	67.15%
	Testing	73.89%	82.29%	60.66%
ANN	Training	82.82%	86.39%	75.85%
	Testing	78.34%	80.21%	75.41%
C.5 Decision Tree	Training	91.49%	97.52%	79.71%
	Testing	76.43%	86.46%	60.66%

4.6.1 Gain Graph

Gain graphs [14, 15] are used for measuring performance of a model. Gain is represented as the ratio of the number of correctly classified cases in each increment to the total number of correctly classified cases in the tree, using the following formula:

$$\left(\text{Hit increment/total hits}\right) \times 100\% \qquad (4.10)$$

The values in the "Gain Percentage" column of the table are represented with the gain graph. From left to right the cumulative gain graph always starts at 0% and ends

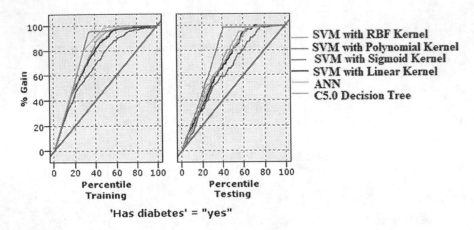

FIGURE 4.1   Gain Graph of all models for 'Has diabetes' = "yes" case.

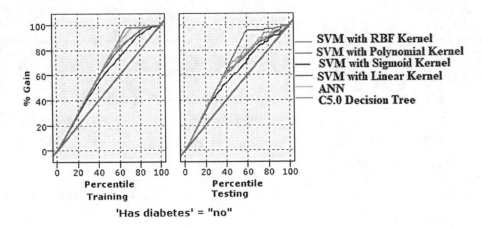

FIGURE 4.2   Gain Graph of all models for 'Has diabetes' = "no" case.

at 100%. For a good model the gain graph will rise sharply to 100% and then level off. A model that does not provide any information will stop along a diagonal line from bottom to top right. The gain is higher if the curve is steeper. Figure 4.1 shows the cumulative gain plots of all models used in the training and testing data sets for "has diabetes' = "yes" case. Figure 4.2 shows the cumulative gain plots of all models used in the training and testing data sets for "has diabetes' = "no" case.

4.6.2 Response Graph

For comparison of classification methods, response graph [14, 15] is a very benefi-cial visual tool. The Response graph shows the trade-off between the positive cases that are truly classified (TPcase) rate and the positive cases that are classified false (FPcase) rate for a given model. Response graphs plot the values in the Response percentage column of the table. The response is the percentage of records in the hit increment, using the equation:

$$(\text{Incremental Response/Incremental Records}) \times 100\% \qquad (4.11)$$

The response chart usually starts at 100% and gradually decreases at the right edge of the chart until it reaches the overall response rate (total number of cases correctly classified / total number of cases). For better model the line starts at 100 percent on the left or near and stays on a higher platform as movement is taken towards right, then on the right hand side of the graph, the line drops sharply towards the overall response rate. For models that do not provide information, the line will remain around the overall response rate of the entire graph. Figure 4.3 shows the response graph of all models used in the training and testing data sets for "has diabetes' = "yes" case. Figure 4.4 shows the response graph of all models used in the training and testing data sets for "has diabetes' = "no" case.

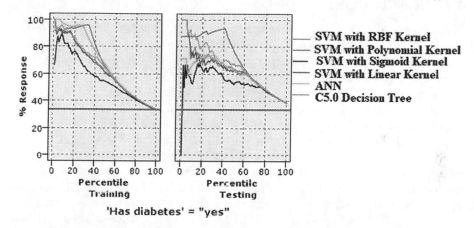

FIGURE 4.3   Response Graph of all models for 'Has diabetes' = "yes" case.

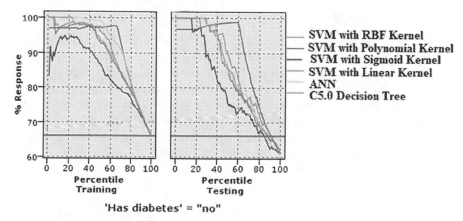

FIGURE 4.4   Response Graph of all models for 'Has diabetes' = "no" case.

From the experimental study the accuracy of the model SVM with RBF kernel, SVM with polynomial kernel, SVM with sigmoid kernel, SVM with linear kernel, ANN, and C5.0 decision tree is found to be 81.83%, 96.73 %, 75.45%, 81.51%, 82.82%, and 91.49% respectively on the training data set. The accuracy of the model SVM with RBF kernel, SVM with polynomial kernel, SVM with sigmoid kernel, SVM with linear kernel, ANN, and C5.0 decision tree is found to be 73.81%, 96.18%, 70.06%, 73.89%, 78.34%, and 76.43% respectively on the test data set. The results obtained are also compared for analyzing the performance of the classifier. It has been found that the SVM with polynomial kernel yields the highest result as compared to other models. The performance of the model is also visualized using Gain graph and Response graph. From the analysis of both gain and Response graphs, the SVM model with polynomial kernel is found to provide very good results as compared to other models.

4.7 CONCLUSION

In this chapter the SVM data mining classifier with polynomial kernel is proposed for classification of the Pima Indian Diabetes dataset taken from the UCI Machine Learning repository. The performance of the SVM classifier with polynomial kernel is compared with the SVM classifier with kernels RBF, Sigmoid, and linear. The performance of the SVM classifier with polynomial kernel is also compared with ANN and Decision Trees data mining techniques. The statistical performance indicators, such as accuracy, specificity, and sensitivity, are used for analyzing the performance of all models. In addition, with the help of Gain graph and Response graph, the performance of each model is studied on the training and test datasets. From experimental study, SVM classifier with polynomial kernel is found to provide very good results, with an accuracy of 96.18 % as compared to other models. The Gain graph and Response graph also show the high performance of the SVM model as compared to other models. Therefore, the proposed SVM model with polynomial kernel can be effectively used for the prediction of diabetes.

4.8 FUTURE SCOPE

In this chapter, few data mining techniques like SVM, ANN, and Decision Tree have been explored for the classification of the Pima Indian diabetes dataset. Other classification techniques like Convolutional Neural Network (CNN) and Deep Learning techniques can be used to improve the performance of the system. Various feature extraction techniques can be used before the classification phase to reduce the dimensionality of the dataset to make the classification phase less complex. Different optimization techniques can be integrated with individual classifiers to develop optimization models for classification of the Pima Indian diabetes dataset.

REFERENCES

1. Lekkas, Stavros, and Mikhailov, Ludmil. 2010. Evolving fuzzy medical diagnosis of Pima Indians diabetes and of dermatological diseases, *Artificial Intelligence in Medicine*, 50(2):117–126.

2. Hayashi, Yoichi, and Yukita, Shonosuke. 2016. Rule extraction using Recursive-Rule extraction algorithm with J48graft combined with sampling selection techniques for the diagnosis of type 2 diabetes mellitus in the Pima Indian dataset, *Information in Medicine Unlocked*, 2:92–104.
3. Li, Huan, Chung, Fu-Lai, Wang, Shitong. 2015. A SVM based classification method for homogeneous data, *Applied Softcomputing*, 36: 228–235.
4. Usha Kumri, Ch., Murthy, A. Sampath Dakshina, Prasanna, B. Lakshmi, Reddy, M. Pala Prasad, Panigrahy, Asisa Kumar. 2020. An automated detection of heart arrhythmias using machine learning technique: SVM, In proceedings of Materialstoday (In Press).
5. Soumaya, Zayrit, Taoufiq, Belhoussine Drissi, Benayad, Nsiri, Yunus, Korkmaz, Abdelkrim, Ammoumou. 2020. The detection of Parkinson disease using the genetic algorithm and SVM classifier, *Applied Acoustics*, 171:107528.
6. Nie, Feiping, Zhu, Wei, Li, Xuelong. 2020. Decision tree SVM: An extension of linear SVM for non-linear classification, *Neurocomputing*, 401:153–159.
7. Wang, Diya and Zhao, Yixi. 2020. Using news to predict investor sentiment: Based on SVM model, *Procedia Computer Science*, 174: 191–199.
8. Routray, Sidheswar, Ray, Arun Kumar, Mishra, Chandrabhanu, Palai, G. 2018. Efficient hybrid image denoising scheme based on SVM classification, *Optik*, 157:503–511.
9. Li, Zhenbo, Niu, Bingshan, Peng, Fang, Li, Guangyao, Yang, Zhaolu, Wu, Jing. 2018. Classification of peanut images based on multi-features and SVM, *IFAC PapersOnLine*, 51(17):726–731.
10. Zanaty, E. A. 2012. Support Vector Machines (SVMs) versus Multilayer Perception (MLP) in data classification, *Egyptian Informatics Journal*, 13(3):177–183.
11. Jing, Chen, and Hou, Jian. 2015. SVM and PCA based fault classification approaches for complicated industrial process, *Neuro Computing*, 167:636–642.
12. UCI Machine Learning Repository of machine learning databases. University of California, School of Information and Computer Science, Irvine. C.A. http://www.ics.uci.edu/~mlram,?ML.Repositary.html
13 Mitra, S., Acharya, T. 2004. *Data mining multimedia, soft computing and bioinformatics*, Wiley-Interscience Publication, Hoboken, United States, ISBN: 9780471460541.
14. Elsayad, Alaa M. 2010. Predicting the severity of breast masses with ensemble of Bayesian classifiers, *Journal of Computer Science* 6(5): 576–584.
15. Elsayad, Alaa M. 2010. Diagnosis of Erythemato-Squamous diseases using ensemble of data mining method, *ICGST-BIME Journal* 10(1):13–23.

5 Analysis and Prediction of COVID-19 Pandemic

Bichitrananda Patra and Santosini Bhutia

Siksha O Anusandhan (Deemed to be) University,
Bhubaneswar, India

Sujata Dash

Maharaja Srirama Chandra Bhanjs Deo University,
Baripada, India

Lambodar Jena and Trilok Nath Pandey

Siksha O Anusandhan (Deemed to be) University,
Bhubaneswar, India

CONTENTS

5.1 INTRODUCTION

HCoV-229E and HCoV-OC43, which initially caused duct and tract infections, was discovered in the nasal cavities and respiratory tracts of human patients. A form of the virus was initially introduced in the sixties by the human corona virus (HCoV). SARSCoV-2(2019) is another coronavirus that has caused serious coronavirus infections (2019). A breathing disease occurrence was reported in China at the end of 2019, and a couple of weeks later, on January 11, 2020, a new virus was announced that leads to severe human infection. This infection may spread from a bat to people through an intermediate host and cause an extreme respiratory condition [1]. It is human-to-human transmittable through the air [2, 3]. This virus has propagated slowly and steadily across the world to cause this worldwide pandemic threatening the world's health and economy. This disease is transmitted by cough droplets from one infected person to another. Since there are no initial signs of infection, a

DOI: 10.1201/9780367548445-6

person could not be aware that his infection spreads to another person, causing a human propagation chain to become dangerous. It is an extremely dangerous disease because it can cause death for people who are elderly, have poor immunity, or are already suffering from another illness.

This virus has the ability to double consistently. So on the first day, there is one infection, on the second day there are two, on the third day there are four, and the fourth day eight, and so on, as shown in Figure 5.1. This implies that the infection has the property called exponential growth. This is how the pandemic works. In the beginning the outbreak is unnoticeable, but when it reaches a significant value, the development to the maxima is very fast.

As we know, the Corona virus epidemic started in December and spread across the world. It has had three stages of highly infectious disorder: local propagation, community spread, and broad transmission. In studying the dissemination of the virus from those initially exposed to the larger regional population, scientists pinpointed the Wuhan fish market as the first place from which local spread began, since the majority of the first cases of exposed people went to to the market. The second phase of transmission is community spread. The virus spreads from one city in China to other cities and countries during the third stage of large-scale transmission to cause a worldwide pandemic. It is common knowledge that this disease is spreading, damaging the health and economy of the world. Research is needed to study the spread chain and learn from visualization and prediction to plan ahead so that less harm may be done in the near future, as all of us know this disease cannot be stopped because there is no medication for it. This study analyzes, visualizes, and forecasts the spread of coronavirus diseases worldwide, using as a basis the previous results. The primary goal is to understand trends of health data so that the symptoms of this disease can be minimized.

First reported in China, the epidemic of coronavirus has spread throughout the world. Many countries, including Italy, the USA, Spain, and Germany, are in the worst situation and are confronted with major public health issues because of the restrictions imposed because of this pandemic. Many questions are presented because of the growth pattern of the disease. "How quickly is the virus spreading?" "How do government policies enhance the control of the situation?" This study will evaluate, imagine, and forecast reported cases of COVID-19 by means of predictive modeling to interpret health information trends. This constant expansion can also vary. This study presents the spread pattern of the coronavirus through visualization from

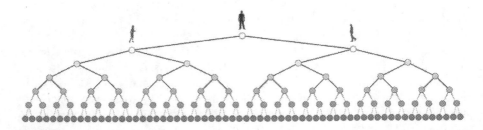

FIGURE 5.1   Exponential growth of the pandemic.

January 22, 2020 to July 28, 2020, followed by a prediction using prophet models, which are considered the best way to present time series-related data.

5.2 LITERATURE SURVEY

The on-going epidemic of COVID-19 (coronavirus disease, 2019), which was reported in December 2019, is an international concern. This epidemic, sparked in Wuhan, the capital city of Hubei, China, has spread across the World. According to the Chinese Center for Disease Control and Prevention (China CDC) there were 70,641 confirmed cases and 1,772 deaths due to coronavirus by February 16, 2020. In this public health emergency, 15% of the cases are severely affected [4]. Cardiac, hypertension, or diabetes patients are more affected by COVID-19. The number of infected patients is increasing, but there is a deficiency in healthcare resources, giving a positive correlation to the mortality rate. The Chinese government realized this problem and immediately provided new medical facilities to control the epidemic. The National Health Commission of China, on January 20, 2020, confirmed that COVID-19 is transmitted from human-to-human. Because of this knowledge, other countries controlled their borders to prevent transmission of the disease.

In January 2020, the World Health Organization (WHO) provided a provisional guideline to test the new virus outbreak in laboratories, along with prevention and control guidance [5, 6]. COVID19 is transmitted to human beings from an unknown animal. The symptoms of the disease are fever, malaise, dry cough, shortness of breath and respiratory distress. According to WHO, seven different strains of human coronaviruses (HCoVs) have been reported up to now [7, 8]. Among them 229E and NL63 are two strains of HCoVs belonging to Alpha coronaviruses, while OC43, HKU1, SARS, MERS, and COVID19 are five strains of HCoVs belonging to Beta coronaviruses [9, 10]. WHO also reported that SARS and MERS HCoVs are more destructive strains of coronaviruses where the SARS HCoVs has a 10% mortality rate and the MERS HCoVs has a 36% mortality rate [11].

This section focuses on the quality of the framework for the study and the prediction for the coronavirus. The proposed system and its feasibility will also be described. As we all know, coronavirus has spread worldwide. Scientists around the world aim to identify the trend of spreading and predicting the impacts of this disease so that governments and people can respond as required to reduce the incidence and spread of death. WHO itself works on forecasts to provide countries throughout the world with the necessary guidelines.

In the late Gregorian calendar month of 2019, the first case of coronavirus occurred in China. The White House Science and Technology Policy Workplace (STP) predicts a global virus that could be a worldwide pandemic and is pushing Alliance Scientific Teams and Companies (including Kaggle) to coordinate an Open Research Data Set (CORD-19) to arrange predictive models to better predict the event in new ways.

On January 30, 2020, the first COVID-19 case was diagnosed in India when a student traveled from Wuhan, China to Kerala, India. Two days later, Kerala announced two more cases. For almost thirty days, no new cases were found in India.

Then on March 2, 2020, five new coronavirus cases were confirmed in Kerala and cases have continued to rise.

The John Hopkins University dataset was considered for the proposed solution on the understanding that it is updated daily. Cumulative data from reported, dead, and recovered cases have been analyzed over time after the pre-processing of the datasets. This analysis examines the COVID-19 trend across countries. For each country, cases confirmed, recovered cases, deaths, and active cases were plotted for analysis and visualization. Prophet models are used as simplified models to provide accurate time series foresight to forecast reported cases worldwide. This helps to understand the spread and increase of the virus. Prediction is now important in planning ahead to reduce the impact of the crisis.

5.3 PROPOSED SYSTEM

Due to the exceptionally complex nature of the COVID-19 outbreak and the variety in its behavior from country to country, this study recommends Machine Learning as an effective tool to model the outbreak. The main focus for this study is to evaluate the earlier coronavirus spread set and imagine the spread on the graph, then predict the next spread using the dataset. It is necessary to speak to the forecasting of confirmed cases as well as analysis of deaths and recoveries. Forecasting requires adequate historical data. Simultaneously, no forecast is sure, as the future rarely repeats itself, but is affected by the reliability of the data and what factors are being predicted. It also affected by the psychological factor: how people see and respond to the danger of the disease.

In this study, we are going to produce a forty days ahead forecast of confirmed cases of COVID-19 by using Prophet. Prophet is open source software released by Facebook's Core Data Science group. It is accessible for download in CRAN and PyPI. We use Prophet, a method for forecasting time series data dependent on an additive model, where non-linear trends are fit. Our data were collected from World Health Organization. We considered the cases from January 22, 2020, to July 28, 2020. Our experiment requires special resources like Anaconda software and their different libraries in order to meet these functional requirements. Since the data were open to the public, it was not necessary to agree. Facebook's Prophet was used in GC (a free Jupyter Notebook environment) to forecast the virus development. This section focuses on various criteria for function and non-function, data flows, testing steps, and algorithm design steps. We will explicitly define the criteria for our study before we start any research. Our project should meet non-functional requirements, for instance if the data set changes, as we know that the information is not consistent. This experiment also works accurately according to the changed data set. This section also describes how the model works and that the dataset can be used with Prophet Model for visualization and prediction. The flow chart of the algorithm is shown in Figure 5.2.

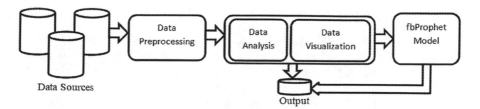

Data Sources

Output

FIGURE 5.2   Flowchart of the application.

Conception:
Along with the related plots, maps and graphs are shown inline.

Algorithm
1. Import Libraries
2. Dataset import
3. Dataset Preprocessing
    a. Active Case Calculations
    b. Change the country names to pycountry_convertlib as needed
    c. Disable data values for nan
4. Analysis and visualization of data
    a. World Confirmed Case Map
    b. World Death Case Map
    c. Country-wide Cases Reported
    d. Worldwide Confirmed Cases Graph
    e. Worldwide Death Cases Graph
    f. Worldwide Recovered Cases Graph
    g. Bar Graph of confirmed, deaths and revered in top 10 countries
    h. Pie Chart of confirmed, deaths and revered in top 10 countries
    i. Folium Map encircling the states and union territories affected in India
    j. Geo-scatter Plot to show timeline-spread progression in India
5. Spread Analysis
    a. The spread patterns across countries
    b. The spread patterns in India
6. Forecast
    a. Calculation of the mean factor of growth
    b. Create Timestamps from data from time series
    c. Plot Prophet Model forecasted data

5.4 THE DATA FROM THE WORLD

The data from the World Health Organization (WHO) were collected. WHO COVID-19 data were used to assess global and regional patterns of new confirmed COVID-19 and deaths worldwide. Cases from January 22, 2020, through July 28, 2020, were given. Day, year, year, area, death, cumulative death, confirmations, cumulative confirmations were the input variables found in the dataset. Since the data were open to the public, it was not necessary to agree. Facebook's Prophet was used in GC (a free Jupyter Notebook environment) to forecast the virus development. In Jupyter Notebook, the dataset and the corresponding library were imported. Pandas is an extremely fast and flexible data analysis and manipulation tool. By using Pandas, we can store and manipulate the tabular data. We also import visualization libraries like matplotlib, seaborn, and plotly. More data have been processed so that missing values are omitted and the data is in the right format. A model that is easy to build, fast to train, and provides interpretable designs has been introduced. Over the last six months of the COVID-19 events, we used the Facebook Prophet to estimate a base model using the seasonality parameters. Notice that seasonality has not been established annually and daily, since COVID-19 is not a seasonal disease. The estimate of our model has been compared to the official data on the official COVID-19 Updates website of the World Health Organization (WHO). The WHO official comparison dashboard can be downloaded at https://covid19.who.int/. Data were collected from "https://raw.githubusercontent.com/datasets/COVID-19/master/data/time-series-19-covid-combined.csv" for 188 countries, as shown in Table 5.1. This database is used for analysis and prediction.

5.5 RESULTS

At the time of data extraction from the official website of the WHO (28.07.2020), the cohort study included 188 countries. The reported Confirmed, Death, Active, Recovered, and Mortality rate for each country is shown in Figure 5.3 and Figure 5.4 shows the total number of confirmed, death, recovered and active cases throughout the world as on Dt: 28.07.2020.

Figures 5.5 and 5.6 represent the worldwide spread in the map.

This virus is a major threat and challenge to the health and economy of the world and has spread to more than 150 countries across the world as represented in maps, graphs, and pie charts. We also focus on the COVID-19 spread across the world in Figure 5.7. In the following Figures 5.8–5.10, the number of cases increased all over the world in the recent months. Also these figures show how this spread has advanced over time frame; we plotted the Confirmed cases, Deaths, and Recovered cases globally using plotly.graph_objects. The sharp exponential curve that can be seen on the right side of the graph shows the staggering rate at which the pandemic is spreading around the world.

The bar graph of the top ten countries with the rest of the countries for confirmed, death, and recovered cases are shown in Figures 5.11–5.13. Also the coronavirus diseases' spread across the world are represented in the pie chart in Figures 5.14–5.17.

TABLE 5.1

Dataset for 188 Countries

	Country_Region	Last_Update	Lat	Long_	Confirmed	Deaths	Recovered	Active	Incident_Rate	People_Tested	People_Hospitalized	Mortality
0	Australia	2020-07-29 05:35:02	-25.000000	133.000000	15582.0	176.0	9617.0	5789.0	61.202606	NaN	NaN	1.1
1	Austria	2020-07-29 05:35:02	47.516200	14.550100	20677.0	713.0	18379.0	1585.0	229.581187	NaN	NaN	3.4
2	Canada	2020-07-29 05:35:02	60.001000	-95.001000	116871.0	8957.0	101686.0	6229.0	308.727599	NaN	NaN	7.6
3	China	2020-07-29 05 35:02	30.592800	114.305500	86990.0	4657.0	80517.0	1816.0	6.192886	NaN	NaN	5.3
4	Denmark	2020-07-29 05 35:02	56.263900	9.501800	13811.0	613.0	12652.0	546.0	238.441229	NaN	NaN	4.4
...	...	...	...	...	...	...	...	...	...	...	...	...
183	West Bank and Gaza	2020-07-29 05:35:02	31.952200	35.233200	10938.0	79.0	3752.0	7107.0	214.411058	NaN	NaN	0.7
184	Western Sahara	2020-07-29 05:35:02	24.215500	-12.885800	10.0	1.0	8.0	1.0	1.674116	NaN	NaN	10.0
185	Yemen	2020-07-29 05:35:02	15.552727	48.516388	1703.0	484.0	840.0	379.0	5.709790	NaN	NaN	28.4
186	Zambia	2020-07-29 05:35:02	-13.133897	27.849332	5002.0	142.0	3195.0	1665.0	27.208507	NaN	NaN	2.8
187	Zimbabwe	2020-07-29 05:35:02	-19.015438	29.154857	2817.0	40.0	604.0	2173.0	18.953198	NaN	NaN	1.4

188 rows × 14 columns

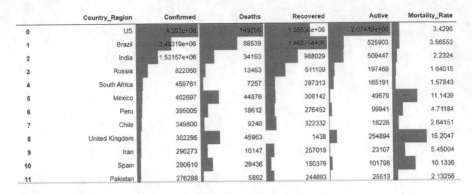

	Country_Region	Confirmed	Deaths	Recovered	Active	Mortality_Rate
0	US	4.352e+06	149256	1.35508e+06	2.07448e+06	3.4296
1	Brazil	2.48319e+06	88539	1.86875e+06	525903	3.56553
2	India	1.53167e+06	34193	988029	509447	2.2324
3	Russia	822060	13483	611109	197468	1.64015
4	South Africa	459761	7257	287313	165191	1.57843
5	Mexico	402697	44876	308142	49679	11.1439
6	Peru	395005	18612	276452	99941	4.71184
7	Chile	349800	9240	322332	18228	2.64151
8	United Kingdom	302295	45963	1438	254894	15.2047
9	Iran	296273	16147	257019	23107	5.45004
10	Spain	280610	28436	150376	101798	10.1336
11	Pakistan	276288	5892	244883	25513	2.13256

FIGURE 5.3   Country Wise Reported Cases (till July 28, 2020).

Confirmed	Deaths	Recovered	Active
16,737,842	660,383	9,749,159	5,555,415

FIGURE 5.4   Globally case count summary as on July 28, 2020.

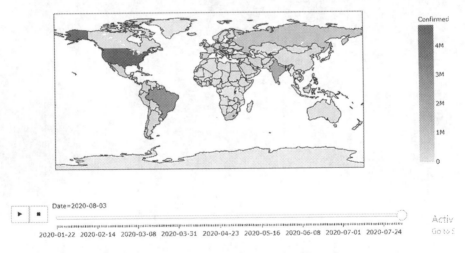

FIGURE 5.5   Map plot of Worldwide Confirmed Cases till July 2020.

According to the statistics Brazil, US, India, Russia, and South Africa are the top five infected countries. The infection rate is still growing across the world. But this can be controlled by social distancing. The number of infections depends on the weather condition and control strategy, which differ from nation to nation.

Figure 5.18 shows the progression of confirmed, recovered, deaths, and active cases in India from January 22, 2020 to July 28, 2020. We also focus on COVID-19 spread in different states and territories across India in Figure 5.19.

FbProphet is an open source library of Facebook [12], which utilizes the technique of time series decomposition and Machine Learning fitting with yearly, weekly, daily, seasonality, and also holiday effects. It not only predicts the future but also fills

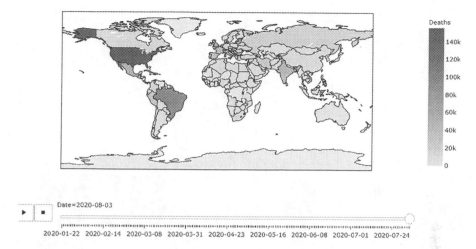

FIGURE 5.6 Map plot of Worldwide Death Cases till July 28, 2020.

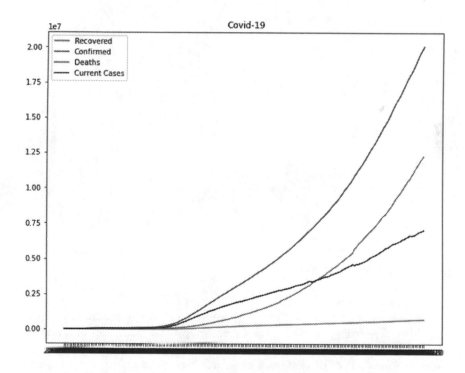

FIGURE 5.7 Graph for COVID19 spread across the World (till July 28, 2020).

the missing values and identifies anomalies. It can forecast time series with high accuracy. The study of this chapter is to construct a reliable forecast model for the usefulness of the people. The input to the Prophet model is a time series with two features, the date as 'ds' and the value as 'y'. In this study 'ds' is the date of the day

Total Coronavirus Confirmed Cases:Globally

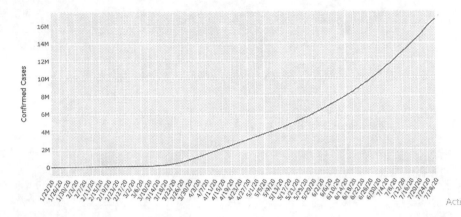

FIGURE 5.8   Worldwide Confirmed cases (till July 28).

Total Coronavirus Death Cases:Globally

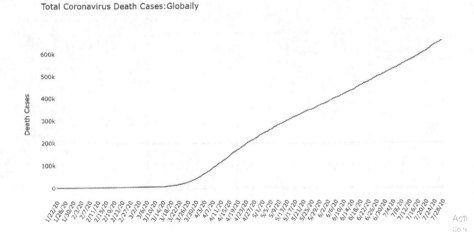

FIGURE 5.9   Worldwide Death cases (till July 28).

and 'y' is number of confirmed cases in India. Using this prediction model, the prediction of the reported confirmed case for the data is shown in Figure 5.19, which gives a good forecasting accuracy. The Figure shows that the model's accuracy in prediction for the confirmed instances was 95%, and it is rising exponentially.

Although the number of confirmed cases as observed in recent months did not increase significantly, the predictions of the pandemic COVID-19 in the next forty days (till September 6, 2020) showed a rise in the number of COVID-19 confirmed cases in India in Figure 5.20.The subsequent forecast data frame is made of 19 features. Further the analysis is done by decomposing the dataset components, which are shown in Figures 5.21–5.23.

Total Coronavirus Recovered:Globally

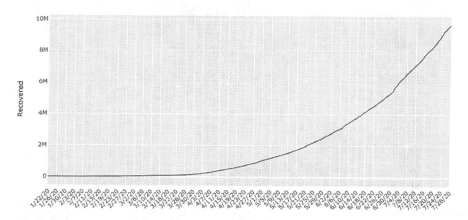

FIGURE 5.10 Worldwide Recovered cases (till July 28).

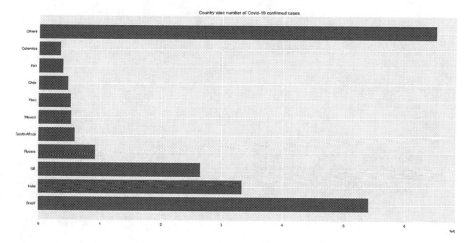

FIGURE 5.11 Bar Graph of Confirmed Cases with Top 10 countries.

In addition to lockdowns, six essential requirements have been established for the World Health Organization, which include proof of controlled COVID-19 transmission, successful contract tracking system, quarantine and testing, reduced vulnerability in care home outbreaks, continued adherence to preventive measures (physical drifting, hand-washing, etc.), In several countries, preliminary proposals to ease the lockdown of the pandemic COVID-19 have been released. As our predictive model shows, restarting the economies that had been severely affected by the sudden shutdowns should be done with caution to curb the destructive effects of COVID-19 [13]. It is important to prevent the risk of a second wave of the virus with more harmful effects [13].

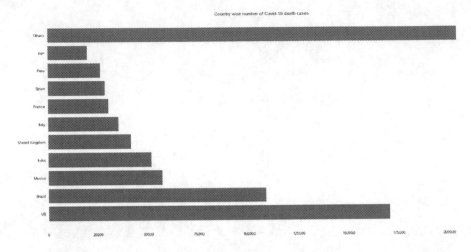

FIGURE 5.12   Bar Graph of Death Cases with Top 10 countries.

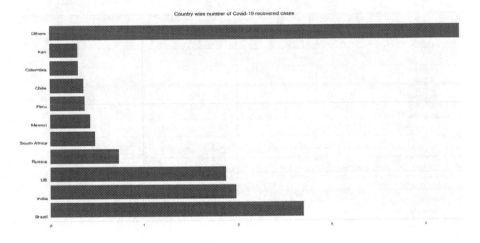

FIGURE 5.13   Bar Graph of Recovered Cases with Top 10 countries.

5.6   DISCUSSION

We used a time series prediction approach in this study for the prevision of COVID-19 trends in India from January 22, 2020, to July 28, 2020 (Figure 5.2) and the prediction of trends for the next forty days (September 6, 2020). A Facebook prophet used additive model was a forecast model that we developed. A precise, fast, fully-automatic Facebook prophet has been used, provides a tunable prediction and is Python or R compatible. It should be noted that because this is not a seasonal phenomenon and the absence of approved vaccinations at this stage, the pattern persists in modeling indefinitely. Our model averaged its predictable accuracy of 95%. There could be potential mistakes, delays, or inadequate documentation of different countries to cause variations between our forecast model and the WHO official case published.

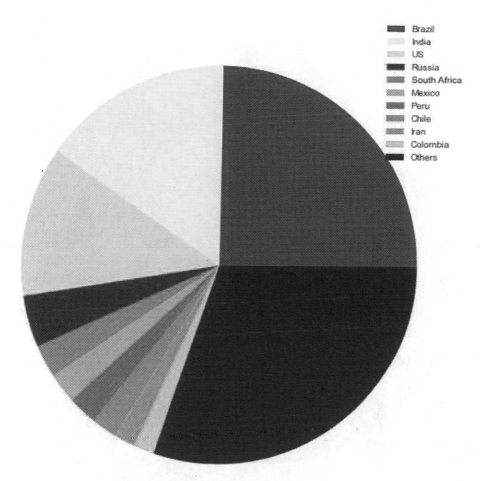

FIGURE 5.14 Pie chart of Confirmed Cases country wise.

Despite this change, our prediction model still showed reasonable accuracy in predicting confirmed cases for the target dates.

The number of confirmed cases is increasing at a high rate. This means that the pandemic in various countries may have started to peak. However, our model foresees an increase in the number of cases confirmed in India for the period forty days (Figures 5.19). Therefore, care should be taken with undue optimism about locking down in the country. The explanation is that the burden on our health centers is actually immense, both in terms of life and death. There are currently minimal to no services available for the care of COVID-19 patients in research locations to hospitals. Changes should be carried out regularly in measures that loosen lock-out and social isolation, or there could be a rise in the number of people in hospital. This can cause difficulty in treating COVID-19 patients adequately in the existing environments of the individual hospitals in the country. In the coming year, global infections

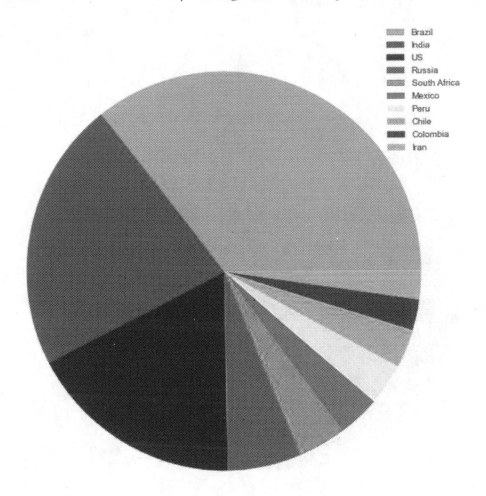

FIGURE 5.15   Pie chart of Confirmed Cases of Top 10 countries.

are projected to grow to 40%–70% [14]. The Governor of Massachusetts also expressed his concern about the rise in the number of reported cases when he estimated that coronavirus is expected to infect between 0.7 and 2.5% of the state's population [15]. The key question is how long the locking, quarantine, monitoring contracts, and social distance steps will flatten the curve? Of course, these initiatives have a huge global economic influence. Consequently, the authorities concerned (government, politicians, and other stakeholders) started to loosen lockout policies to mitigate economic consequences. As our predicted model indicates, however, a potential rise in reported cases stresses the need for broad-based testing and that hospitals have the resources to monitor COVID-19 patients in order to ensure control of exposures before switching gears [16]. The government and policymakers must therefore correct the trend towards virus spread. Comprehension of the COVID-19 roadmap means that the correct decision is made with respect to public policies [17].

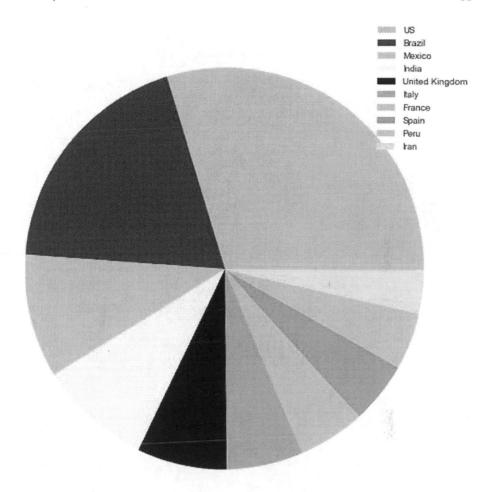

FIGURE 5.16 Pie chart of Death Cases of Top 10 countries.

Therefore, the projection that only a possible COVID-19 vaccine may bring back normalcy is realistic with an increase in the number of confirmations, as set out in our model, and informing citizens of the possibility of spending a year or more at home before developing a possible COVID-19 vaccine. Since COVID-19 prescription drugs are yet to be identified, the modality of citizens' exposure to the virus must be controlled. Vulnerable people slip into a state of rescue with herd immunity (exposure to the virus). The biggest problem with this approach is how this can be achieved ethically and morally. As our model has forecast, the increase in the number of cases means that people's exposure to herd immunity to COVID-19 must be made reasonable, sensitive, and controllable, while the world is waiting for the breakthrough of a COVID-19 vaccine [18]. Therefore, the government must strike a balance between this pandemic's health and economic impact. When not adequately treated, there is a high risk of viral re-entering (negative after initial exposure checked after exposure),

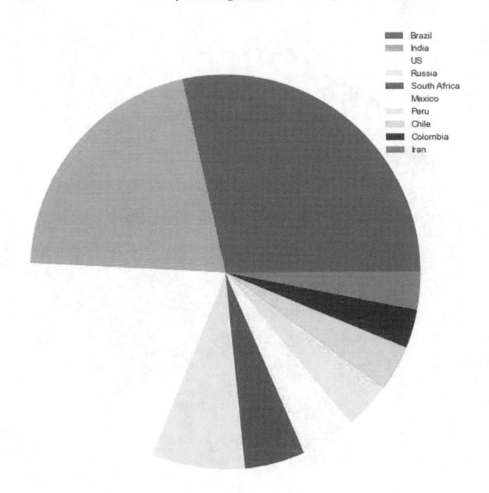

FIGURE 5.17   Pie chart of Recovered Cases of Top 10 countries.

intense healthcare stresses, and increased COVID-19 fatalities. While various countries, such as France, start reducing their lockdown rates, in order to avoid a potential
second wave of the pandemic, essential steps such as the compulsory enforcement
with face masks, increased security in the capital regions, a reduction in the number
of public gatherings and activities, and the temporary closure of schools and venues
could also take place. To conclude, the government and policymakers in various
countries should be very cautious in promoting lockdown steps, given the economic
effect of COVID-19. Systemic lockdown facilitation not only ensures that the curve
is slackened, but is crushed with managed virus exposure before potential therapeutic
treatment is available. It should not trigger too much panic, [18] but consider the

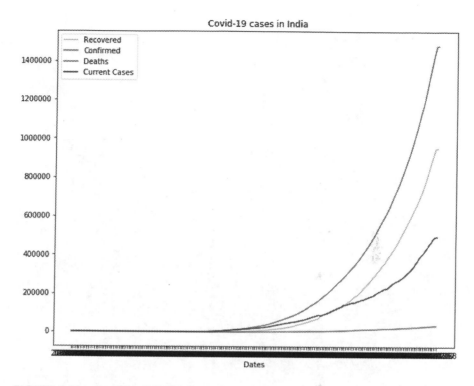

FIGURE 5.18 Graph for COVID-19 spread in India (till July 28).

trend towards the right decision-making, especially in high-risk situations where our model predicted a spike. It ensures that we not only flatten the curve but also smash the curve through the world's protected herd immunity.

5.7 CONCLUSION

This chapter visualizes the current spread of the ongoing epidemic COVID-19 from January 22, 2020, to July 28, 2020. Here the graph of the outbreak is forecasted by the FbProphet model through August 2020, for the time series data. The graph represents that the number of confirmed cases is increasing at a high rate in India, like USA, and Brazil. We have already crossed the number of 1,900,000 cases. However, this speed of spreading is dependent on every citizen doing a social duty and on government strategies that are taken in regard to this circumstance. This representation and forecast can help the residents and government to make appropriate decisions and plan accordingly to stop this pandemic. At present, India is at a dangerous place. Hence, we can battle the pandemic by isolating ourselves, by staying inside and ensuring the safety of ourselves as well as other people around us.

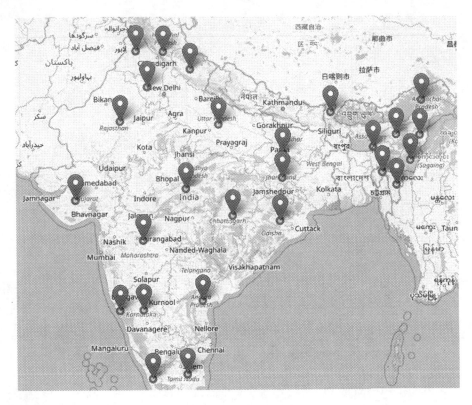

FIGURE 5.19   States and union territories affected in India (till July 2020).

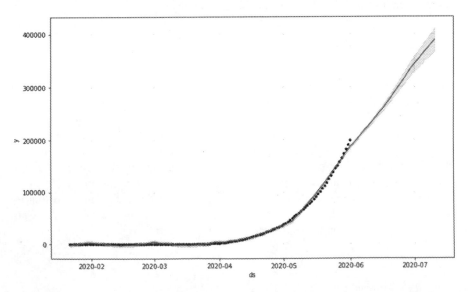

FIGURE 5.20   Possible future scenario for COVID-19 in INDIA.

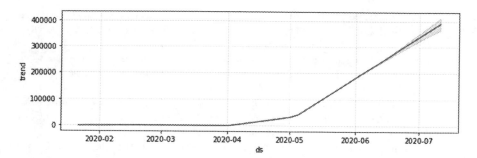

FIGURE 5.21 COVID-19 trend.

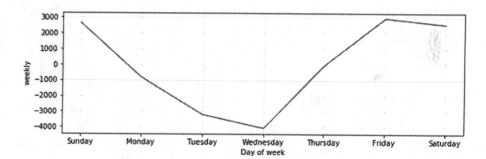

FIGURE 5.22 COVID-19 weekly analysis.

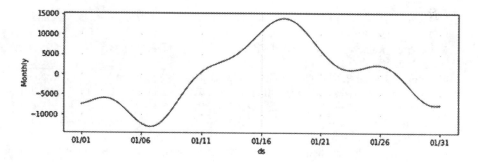

FIGURE 5.23 COVID-19 monthly analysis.

REFERENCES

1. Li, Xingguang, Junjie Zai, Qiang Zhao, Qing Nie, Yi Li, Brian T. Foley, and Antoine Chaillon. "Evolutionary history, potential intermediate animal host, and cross-species analyses of SARS-CoV-2." *Journal of Medical Virology* 92, no. 6 (2020): 602–611.
2. Alhazzani, W., M. H. Møller, Y. M. Arabi, M. Loeb, M. N. Gong, B. Du, and E. Fan. "Surviving Sepsis Campaign: guidelines on the management of critically ill adults with Coronavirus Disease 2019 (COVID-19)". *Intensive Care Medicine* 48 (2020): 1–34.
3. Bai, Yan, Lingsheng Yao, Tao Wei, Dong-Yan Jin Fei Tian, Lijuan Chen, and Meiyun Wang. "Presumed asymptomatic carrier transmission of COVID-19." *JAMA* 323, no. 14 (2020): 1406–1407.

4. Flegal, Katherine M., Barry I. Graubard, David F. Williamson, and Mitchell H. Gail. "Cause-specific excess deaths associated with underweight, overweight, and obesity." *JAMA* 298, no. 17 (2007): 2028–2037.

5. World Health Organization. "Infection prevention and control during health care when novel coronavirus (nCoV) infection is suspected: interim guidance, 25 January 2020." (2020).

6. World Health Organization. *Laboratory testing of human suspected cases of novel coronavirus (nCoV) infection: interim guidance, 10 January 2020.* No. WHO/2019-nCoV/ laboratory/2020.1. World Health Organization, 2020.

7. Hui, David S., Esam I. Azhar, Tariq A. Madani, Francine Ntoumi, Richard Kock, Osman Dar, Giuseppe Ippolito et al. "The continuing 2019-nCoV epidemic threat of novel coronaviruses to global health—The latest 2019 novel coronavirus outbreak in Wuhan, China." *International Journal of Infectious Diseases* 91 (2020): 264–266.

8. Bleibtreu, Alexandre, S. Jaureguiberry, N. Houhou, D. Boutolleau, H. Guillot, D. Vallois, J. C. Lucet et al. "Clinical management of respiratory syndrome in patients hospitalized for suspected Middle East respiratory syndrome coronavirus infection in the Paris area from 2013 to 2016." *BMC Infectious Diseases* 18, no. 1 (2018): 1–9.

9. Elfiky, Abdo A., Samah M. Mahdy, and Wael M. Elshemey. "Quantitative structure-activity relationship and molecular docking revealed a potency of anti-hepatitis C virus drugs against human corona viruses." *Journal of Medical Virology* 89, no. 6 (2017): 1040–1047.

10. World Health Organization. *Clinical management of severe acute respiratory infection (SARI) when COVID-19 disease is suspected: interim guidance, 13 March 2020.* No. WHO/2019-nCoV/clinical/2020.4. World Health Organization, 2020.

11. Miller, Meg. "2019 Novel Coronavirus COVID-19 (2019-nCoV) Data Repository." *Bulletin-Association of Canadian Map Libraries and Archives (ACMLA)* 164 (2020): 47–51.

12. Taylor, Sean J., and Benjamin Letham. "Forecasting at scale." *The American Statistician* 72, no. 1 (2018): 37–45.

13. Jacobsen, Kathryn H. "Will COVID-19 generate global preparedness?." *The Lancet* 395, no. 10229 (2020): 1013–1014.

14. Jernigan, Daniel B. CDC COVID, and Response Team. "Update: public health response to the coronavirus disease 2019 outbreak—United States, February 24, 2020." *Morbidity and Mortality Weekly Report* 69, no. 8 (2020): 216.

15. Myint, S. H. "Human coronaviruses: a brief review." *Reviews in Medical Virology* 4, no. 1 (1994): 35.

16. Fehr, Anthony R., and Stanley Perlman. "Coronaviruses: An overview of their replication and pathogenesis." *Coronaviruses* 1282, (2015): 1–23.

17. World Health Organization. *Water, sanitation, hygiene and waste management for COVID-19: technical brief, 03 March 2020.* No. WHO/2019-NcOV/IPC_WASH/2020.1. World Health Organization, 2020. https://www.who.int/emergencies/diseases/novel-coronavirus-2019 [Accessed on March 30, 2020].

18. Zhu, Na, Dingyu Zhang, Wenling Wang, Xingwang Li, Bo Yang, Jingdong Song, Xiang Zhao et al. "A novel coronavirus from patients with pneumonia in China, 2019." *New England Journal of Medicine* 382, (2020).

6 Variational Mode Decomposition Based Automated Diagnosis Method for Epilepsy Using EEG Signals

Akshith Ullal
Department of EECS, Vanderbilt University, Nashville, TN, USA

Ram Bilas Pachori
Discipline of Electrical Engineering, Indian Institute of Technology Indore, India

CONTENTS

DOI: 10.1201/9780367548445-7

6.1 INTRODUCTION: BACKGROUND AND DRIVING FORCES

It is estimated around fifty million people in the world are diagnosed with epilepsy, making it a very common neurological disorder. The number of new epilepsy diagnoses per year has also been steadily increasing due to increasing life expectancy and people surviving head injury accidents, which generally trigger epileptic seizures [1]. People diagnosed with epilepsy not only suffer from the physical aspects of the disease, but also face wide-spread discrimination and stigma both socially and economically. Exacerbating this fact is that 80% of the diagnosed cases occur in low or middle income countries where access to healthcare is constrained [1]. However, with proper treatment epilepsy patients can lead normal lives. Hence, it is essential that patients are diagnosed correctly and given the proper medications to control the seizures caused due to epilepsy.

Although magnetic resonance imaging (MRI) based techniques have shown to be effective to a certain degree in epilepsy diagnosis [2], the equipment involved in the process is very expensive and not available in low income population areas. Hence, electroencephalogram (EEG) is the most commonly used method for diagnosis. An important part of the diagnosis is analyzing the seizure affected locations of the signal. Generally, seizures only form a small part of the total signal, and manual inspection is impractical. Hence, an automated diagnosis method for detecting the seizure affected parts of the signal will be helpful in the accurate diagnosis of epilepsy. Seizure signals have a few peculiar characteristics. During the seizure phase there are a lot of uncontrolled electrical disturbances taking place between the neurons of the brain. Generally, normal activity of the brain consists of non-synchronous signals, as the neurons are firing in different directions [3]. However, during seizure the electrical activity can become highly synchronous, which gives rise to the more uniform patterns when compared to normal electrical activity, which can be used as a distinguishing characteristic for classification. Although seizures have distinct characteristics, not all of them have the same cause. Not all seizures are caused by epilepsy [4]. Some other causes include psychogenic non-epileptic seizures (PNES), brain infections that include meningitis and encephalitis, and head injuries [5]. It is quite common that Psychogenic nonepileptic seizures (PNES), which are also paroxysmal behaviors, are misdiagnosed as epileptic seizures; This is because both these disorders share similar traits, such as alterations in behavior and consciousness. However, PNES does not involve changes in the cortical state of wakefulness that is commonly associated with epileptic seizures. Hence, the treatment approach to both these disorders is different. Patients with PNES are often asked to consult neurologists, who give them the wrong medication, including antiepileptic drugs (AEDs) which worsen PNES, and vice versa. Hence, it is important that misdiagnosis of epileptic seizure is avoided. Part of the goal of the proposed system is to be able to classify epileptic from non-epileptic seizures.

The standard method used for classification between epileptic seizure versus non-seizure signals is to first identity certain unique characteristics related to seizure EEG signals, then use these as features into a neural or deep network for classification [6]. In the earlier days, EEG signal analysis was conducted by assuming the signal to be stationary. Hence, time and frequency domain analyses were conducted independently [7]. Some of the frequency domain features used were normalized spectral entropy, average power spectral density, and average frequency, while time domain features were signal rhythmicity and relative signal amplitude [8]. The problem with this approach was that a time domain analysis would lose resolution in frequency and vice versa, which meant some information would be lost. In order to tackle this problem, the time-frequency domain analysis was used by considering the EEG signal to be non-stationary and multicomponent [9, 10]. Some of the commonly used time-frequency (TF) domain techniques are, the short time Fourier transform (STFT), where the Fourier transform is taken of a pre sized window, which is slid across every instant of time [10], wavelet transform in combination with chaos theory for modeling non-linear dynamics of the EEG signal [11], the multiwavelet transform which have two or more scaling function and wavelets [12], and the smoothed pseudo-Wigner–Ville distribution [10]. Many other features from TF techniques have also been used for classification of normal and seizure EEG signals, which include the decomposed sub-band frequencies from the sub-band signals and the line length feature of the discrete wavelet transform (DWT) [13].

The TF techniques such as STFT also have drawbacks. A major one includes the size of the selected window. In STFT, the size of the window is indirectly proportional to the frequency resolution, i.e., smaller windows limit lower frequencies. More recent signal decomposition techniques, such as empirical mode decomposition (EMD) [14], have been proposed for analyzing non-linear and non-stationary signals and have been applied to EEG signal classification [15], [16], [17]. EMD is similar to fast Fourier transform (FFT). Just as FFT decomposes the input signal into frequency components, EMD decomposes them into intrinsic mode functions (IMF). However, the advantage with EMD is that, unlike FFT which assumes the input signal as periodic and the decomposed signal consists of only sine wave components, the IMFs are derived signal components which have inherent characteristics of the original signals, i.e., the characteristics of the IMFs are dependent on the input signal, in other words empirically derived. They mainly have to satisfy two conditions: 1.) in the entire signal, the difference between the number of maxima, minima, and zero crossings should be zero or at most differ by one; 2.) the average value of the IMF should be zero. These conditions define the IMFs as amplitude and frequency modulated (AM-FM) signals. The ability to obtain the decomposed signal in time domain itself has allowed EMD to be used in a variety of applications, such as, signal decomposition in audio analysis [18], climate and weather prediction [19], and the study of respiratory and neuromuscular signals in biomedical related fields [20, 21, 22]. The problem with EMD is that there is a lack of mathematical theory behind it as the algorithm is empirical in nature. Hence, the algorithm's results are highly dependent on the methods of finding the maxima and minima, interpolation of the maxima and minima points to generate the signal envelopes, and the termination criteria used for the iteration, all which can introduce subjectivity into the algorithm results [23].

In order to overcome some of the drawbacks of EMD, a new technique was developed called variational mode decomposition (VMD) in [23]. One of the main drawbacks of EMD is that it is recursive, which means a small error in the initial steps will propagate or even tends to magnify in the subsequent steps. In contrast, VMD is an intrinsic, and variational method that determines the frequency band of each mode adaptively and the modes concurrently. Hence, if there is an error in one of the iterations while calculating the frequency bands, it can be quickly balanced while calculating the modes. The noise characteristics obtained during VMD have a close relation to Wiener filtering [23]. Our main aim in this chapter is to apply the VMD algorithm in order to classify EEG signals.

6.2 LITERATURE REVIEW

EEG signal classification is a well-studied topic, and there have been many methods used. The literature can be mainly divided into two subcategories, i.e., signal processing techniques used and different signal models and architectures used for the classification.

6.2.1 EEG Signal Processing Techniques

In addition to the TF techniques mentioned in the introduction, additional TF techniques have been applied [24]. Previous studies have demonstrated that the EEG signal, although non-linear and complex, can be characterised by four frequency bands, namely *delta* (0.4–4 Hz), *theta* (4–8 Hz), *alpha* (8–12 Hz), *and beta* (12–30 Hz). Investigations have been conducted to check the power content and amplitude of these bands during seizure, and it was shown that there is a change when compared to normal times [25]. Building on this, in [26] it was shown that during the pre-ictal period the power of the *delta* had decreased when compared to other spectral bands. Chu et al. have investigated the transition process from a seizure free state to seizure affected state in patients, using the mathematical technique of attractor analysis to predict the occurrence of seizures. This is done by calculating the Fourier coefficients of the frequency bands of the normal EEG measure, using a twenty-second window with an overlap of ten seconds for both seizure and seizure free parts of the signal. Based on these coefficient values the attractor analysis is applied and the warning signals are classified. A sensitivity of 86.7% was obtained [27].

Many studies using a combination of signal decomposition methods and techniques from system dynamics, including chaos theory, have also been applied for epileptic seizure classification. In [15], EEG signals were first deconstructed into their IMFs using EMD technique, from which entropy features were extracted. The features included different types of entropies, such as Shannon entropy, average approximate entropy, average Renyi entropy, average sample entropy, and average phase entropy. Using all these entropy features for classification between focal and non-focal EEG signals, an accuracy of 87% was obtained. Building on the work in [15], in [28], a new index called focal and non-focal index (FNFI) was developed to classify between focal and non-focal seizure signals. For this index, the discrete wavelet transform (DWT) was used to obtain the coefficients and sub band signals from which entropy features were derived. These features were permutation, phase, and fuzzy entropies. Although a slightly lower accuracy of 84% was obtained in comparison [15], here a single number FNFI could classify between focal and non-focal signals. The advantage of using wavelets is that custom wavelets can be used, depending on the amount of time and frequency localization required. In [29], the authors utilized the tunable-Q wavelet transform (TQWT) from which entropy features were extracted. TQWT has much less frequency localization, but a high time resolution feature. This aspect of TQWT was used in [29] for classifying between focal and non-focal EEG signals, and an accuracy of 95% was obtained. An even higher accuracy of 99%, with a sensitivity of 98% and specificity of 100%, was obtained for the study conducted in [30] where a dual-tree complex wavelet transform (DTCWT) method was applied to the EEG signals, with the mean and standard deviation of the wavelet transform coefficients used as features for classification. Compared to DTCWT, DWT and TQWT have a disadvantage in that they cause large changes in the wavelet coefficients for small changes in the input.

Integrating the advantages of both wavelet and time series signal decomposition methods, in [31], EMD and DWT technique were combined and applied in EEG signal classification. The entropy features were extracted, and an accuracy of 89% was obtained. EEG signals have also been classified using Fourier Bessel series expansion (FBSE-EWT) method [32] and two band biorthogonal wavelet filter banks [33].

6.2.2 MODELS AND METHODS OF CLASSIFICATION USED

The parametric model, as the name suggests, assumes that the EEG signal satisfies a particular generating model and its behavior can be described by using a formula with a number of parameters to be estimated. Some of the widely used parametric methods are auto-regressive (AR) model, auto-regressive moving average (ARMA) model and moving average (MA) model [34]. The AR model is very good at representing signals with high and narrow peaks. The MA model is better at representing broad peaks. Non-parametric models are mainly dependent on the power spectral density (PSD) in order to represent spectral resolution. There are two main types, which are periodogram and correlogram. These two techniques provide reasonable resolution for sufficient data lengths. However they suffer with the problem of having a high variance, even for longer data lengths, and methods have been tried to reduce the variance by making a compromise on the resolution [35]. Eigen vector methods are used for estimating frequencies when the signals have been corrupted by noise [36]. For classification of EEG signals, older methods such as independent component analysis (ICA) [37], support vector machines (SVM) [38], artificial neural networks (ANN) [10, 39] and newer deep learning based approaches have been used [40].

6.3 DATASET

The dataset used for seizure classification is available publicly online and has been taken from the University of Bonn website [41]. The dataset consists of five subgroups which are named as Z, O, N, F, and S. It consists of a mixture of signal samples from both healthy and seizure affected patients that also includes intracranial measurements. Each of this set has got One hundred single channel EEG signal recordings, that were obtained by sampling at a rate of 173.61 Hz using a 12-bit resolution for a duration of 23.6 seconds. The spectral bandwidths of the recorded signals was dependent on the acquisition device and are between 0.5 to 85 Hz. All the recorded EEG signals were done with the same 128-channel amplifier system with an average common ground reference. The signals in the groups Z and O have been recorded from healthy subjects extracranially, i.e., with the electrodes placed on the surface of the scalp, that follows the international ten-twenty system of electrode placement. Subset Z signals are with eyes open and O is with eyes closed. The subsets F and N contain signals from epilepsy patients when they are seizure free, with F being recorded from the epileptogenic zone and N from the hippocampus of the other hemisphere of the brain. Both of these sets were intracranial recordings. The subset S also recorded intracranially contains signals from all parts of the brain that were showing seizure activity. The subset S contains signals selected from all recording sites exhibiting ictal activity. For the intracranial measurements, depth electrodes were used for most of the contacts, with strip electrodes used for recording signals from the lateral and basal regions of the brain neocortex.

We have tested the proposed variational mode-based decomposition (VMD) method on three main types of classification used in medical diagnosis. These are classifying signals between healthy individuals and seizure signals from patients, between seizure- free interval signals and seizure signals from patients and between healthy and all signals from patients. For the healthy/seizure classification, the datasets Z, O, and S were used. For the seizure/seizure-free classification the datasets N, F and S were used, and finally, for the seizure/non-seizure classification, all the five datasets of Z, O, N, F, and S were used. A sample of healthy, seizure-free, and ictal signals is shown in Figure 6.1.

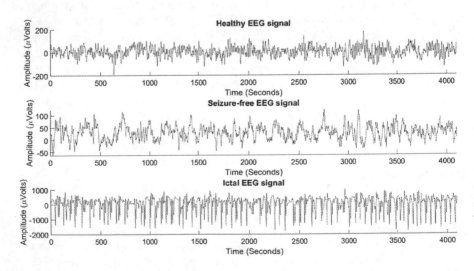

FIGURE 6.1 A sample of a healthy signal (top), seizure-free signal (middle), ictal signal (bottom) used in the proposed method.

6.4　VARIATIONAL MODE DECOMPOSITION

VMD uses variational techniques, and the main objective is to decompose the original signal into its discrete number of principal modes or sub modes that have specific sparse properties, while also keeping in mind the reconstruction of the original signal from these modes. In the VMD algorithm, we have chosen the bandwidth as the sparse property for each decomposed mode [23]. Hence, our final objective is to decompose the signal into its respective modes, which oscillate around a frequency called center frequency that we have to calculate. This means each of the sub-modes will have a center frequency ω_k which will be calculated during the decomposition. The calculation of bandwidth (center frequency) for each mode is done as follows: 1) the analytical signal of the mode is calculated by taking the Hilbert transform of the signal, which gives us the unilateral frequency spectrum of the mode. 2) Each mode signal is then shifted to baseband by mixing it with another signal, which is tuned to the respective estimated center frequency. 3) The bandwidth is now estimated by taking the square of the L-2 norm gradient. The variational problem with the reconstruction constraint can be written as:

$$min_{u_k,w_k}\left\{\sum_k\left|\partial_t\left[\left(\delta\left(t\right)+\frac{j}{\pi t}\right)*u_k\left(t\right)\right]e^{-j\omega_k t}\right|^2_2\right\} \tag{6.1}$$

where u_k represents all of the decomposed modes of the signal and ω_k represents their center frequencies respectively. The reconstruction constraint can be enforced in different ways. In this algorithm, two methods are used, namely, the Lagrange multiplier and the quadratic penalty method. The Lagrangian multiplier is used for enforcing the reconstruction constraint and the quadratic penalty method is used to increase the convergence of the result. Applying this condition, the above equation can be rewritten as follows:

$$L\left(u_k,\omega_k,\lambda\right)=\alpha\sum_k\left|\partial_t\left[\left(\delta\left(t\right)+\frac{j}{\pi t}\right)*u_k\left(t\right)\right]e^{j\omega_k t}\right|^2_2+\left|f-\sum u_k\right|+\left(\lambda,f-\sum u_k\right) \tag{6.2}$$

To find the optimal center frequency, we need to minimize the above Lagrangian $L(u_k,\omega_k,\lambda)$, in other words to find the saddle point. This is done in iterations using the alternate direction method of multipliers (ADMM), as shown in [29]. The three main equations of the iteration are shown below.

$$\hat{u}_k^n+1=\frac{\hat{f}-\sum_{i<k}\hat{u}_i^{n+1}-\sum_{i<k}\hat{u}_i^n+\frac{\hat{\lambda}^n}{2}}{1+2\alpha\left(\omega-\omega_k^n\right)^2} \tag{6.3}$$

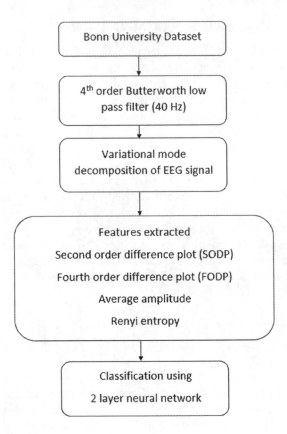

FIGURE 6.2   Block diagram of the proposed method.

$$\omega_k^{n+1} = \frac{\int_0^\infty \omega \left| \hat{u}_k^{n+1} (\omega) \right|^2 d\omega}{\int_0^\infty \left| u_k^{n+1} (\omega) \right|^2 d\omega} \qquad (6.4)$$

$$\hat{\lambda}^{n+1} = \hat{\lambda}^n + \tau (\hat{f} - \sum_k \hat{u}_k^{n+1}) \qquad (6.5)$$

The theory of how the results of equations shown above are obtained is also shown in [29]. The flowchart of the proposed method using VMD is shown in Figure 6.2.

6.5   FEATURES

Once we obtain the principal modes of the input signal after applying VMD, we can extract features from these modes. The principal modes obtained from VMD for seizure-free and ictal signals are shown in Figure 6.3 and 6.4 respectively. Most of

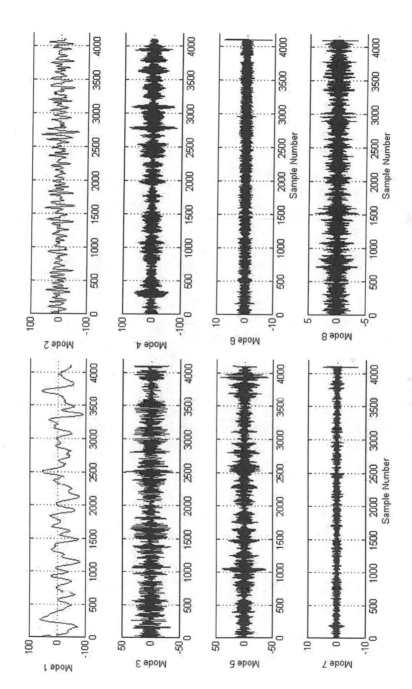

FIGURE 6.3 Principal modes obtained by VMD of a sample seizure free EEG signal.

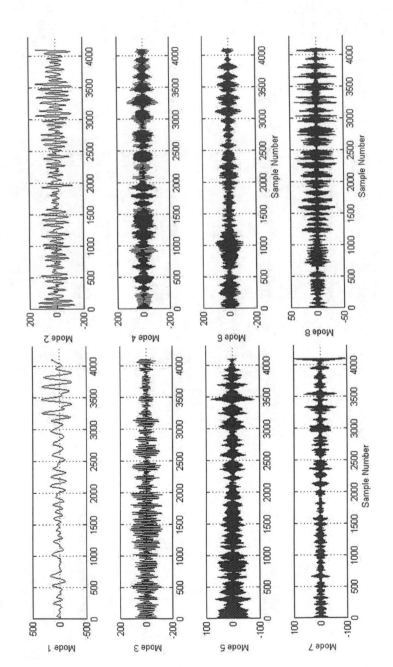

FIGURE 6.4 Principal modes obtained by VMD of a sample ictal EEG signal.

the features selected in the proposed method were used in previous signal processing techniques like EMD, wavelet transform, etc., and have shown to be accurate in classifying seizure affected EEG signals.

6.5.1 SECOND AND FOURTH ORDER DIFFERENCE PLOT AND COMPUTATION OF ELLIPSE AREA

Complex phenomenon with a multitude of factors can be modeled using chaos theory [42]. Difference plots can be used as a measure of chaos in the system [43, 44]. The second order difference plot (SODP) and fourth order difference plot (FODP) of the decomposed VMD modes have been shown to be a helpful tool in classifying between seizure and seizure-free affected signals [45]. The SODP intuitively tells us the rate of variations of successive samples of the signal.

Suppose we have a signal $a(n)$, its SODP can be calculated by plotting $A(n)$ against $B(n)$ which are defined as [45]:

$$A(n) = a(n+1) - a(n) \tag{6.6}$$

$$B(n) = a(n+2) - a(n+1) \tag{6.7}$$

Recently, SODP has been used in variability analysis for EEG signals and Center of pressure (COP) signals, which are important in balance and postural control [46]. The 95 % area of confidence has been used to calculate the SODP area [47]. The area is calculated for each of the decomposed signal modes, for which we first calculate the mean values of signals $A(n)$ and $B(n)$ as follows:

$$R_A = \sqrt{\frac{1}{N} \sum_{n=0}^{N-1} A(n)^2} \tag{6.8}$$

$$R_B = \sqrt{\frac{1}{N} \sum_{n=0}^{N-1} B(n)^2} \tag{6.9}$$

$$R_{AB} = \frac{1}{N} \sum A(n) B(n) \tag{6.10}$$

Compute the F parameter as;

$$F = \sqrt{\left(R_A^2 + R_B^2\right) - 4\left(R_A^2 R_B^2 - R_{AB}^2\right)} \tag{6.11}$$

$$x = 1.7321 \sqrt{R_A^2 + R_B^2 + F} \tag{6.12}$$

$$y = 1.7321 \sqrt{R_A^2 + R_B^2 - F} \tag{6.13}$$

From the calculated values of x and y, the ellipse area is calculated as:

$$A_{ellipse} = \pi xy \tag{6.14}$$

The SODP for the principal modes for ictal and seizure-free signals are shown in Figure 6.5 and 6.6 respectively. For the fourth order difference plot (FODP), the procedure is the same as SODP except that we shift the signal sample by two samples in both the A and B axis. Hence, the $A(n)$ and $B(n)$ in Equations (6.6) and (6.7) change to the following:

$$A(n) = a(n+2) - a(n) \tag{6.15}$$

$$B(n) = a(n+4) - a(n+2) \tag{6.16}$$

The rest of process remains the same as in SOPD calculations.

6.6 FEATURES: RENYI ENTROPY

The idea of using entropy to classify signals stems from the fact that entropy characterizes the disorderliness or randomness of the signal. This gives it the ability to characterize many non-linear chaotic systems and differentiate between them, like EEG which is non-stationary in nature. Generally normal signals tend to have more random impulses compared to ictal signals. This can be seen in the spectrum of seizure-free signals (Figure 6.7), which are more broadly distributed across a wider range of frequencies when compared to ictal signals (Figure 6.8). There are different types of entropies, but we are specifically interested in spectral entropies which include Shannon, log energy, and Renyi entropy [15]. The Shannon entropy is used to measure the average information contained in the signals, whereas Renyi entropy measures the mutual information between two signals. Central to the idea of both the entropy values is measuring the PSD, which gives the distribution of the power of the signal across different frequencies, quantifying the spectral complexity of the time series of the signal. The PSD is obtained by taking the Fourier transformation of the time series of the signal. All three entropies use the normalized power frequency metric p_f as defined below [15]:

$$p_f = \frac{P_f}{\sum P_f} \tag{6.17}$$

The Shannon entropy (SE), log energy entropy (LE) and Renyi entropy (RE) are defined as:

$$SE = -\sum_{f=1}^{N} p_f^2 \log_2\left(p_f^2\right) \tag{6.18}$$

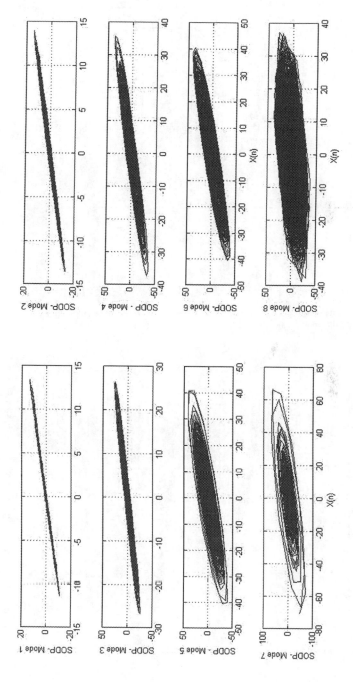

FIGURE 6.5 SODP of the principal modes of an ictal EEG signal.

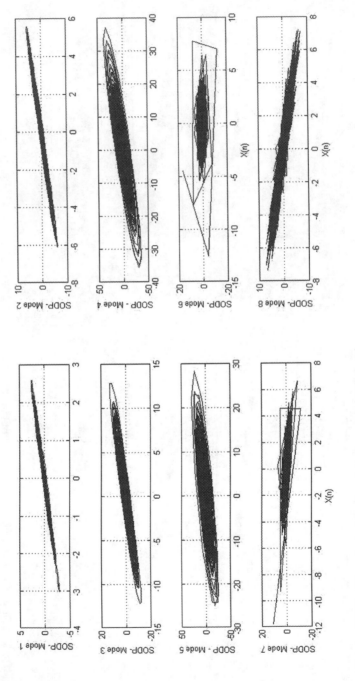

FIGURE 6.6   SODP of the principal modes of a seizure free EEG signal.

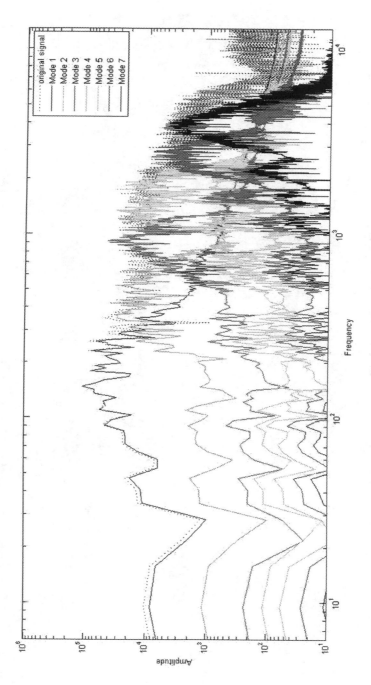

FIGURE 6.7 Frequency spectrum of the principal modes of a seizure free signal. The frequency unit is in Hz and the Amplitude is the FFT magnitude.

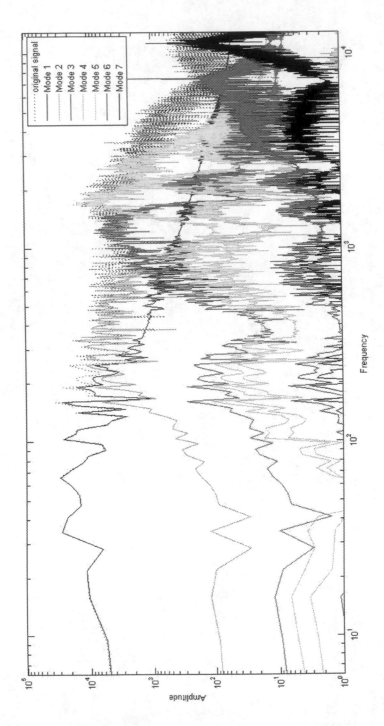

FIGURE 6.8 Frequency spectrum of the principal modes of an ictal signal. The frequency unit is in Hz and the Amplitude is the FFT magnitude.

$$LE = \sum_{f=1}^{N} \log_2\left(p_f^2\right) \tag{6.19}$$

$$RE = \frac{1}{1-\alpha} \log \sum_{f=1}^{N} \left(p_f\right)^{\alpha} \tag{6.20}$$

where N is the length of the time series signal and α is the entropy parameter. Among the entropies, we have selected the Renyi entropy with $\alpha = 2$ as it emphasises both the sub Gaussian and super Gaussian components, which has shown to be effective in EEG signal classification [28].

6.7   FEATURES USED: AVERAGE AMPLITUDE

During seizure, since the neurons fire in a synchronous manner, their amplitudes are superimposed, giving rise to an EEG signal with larger amplitude when compared to non-seizure signals. This is also seen in the principal modes after VMD has been applied. Hence, this can be used as a distinguishing feature. It is calculated by taking the sum of the absolute value of all the samples and dividing it by 4097, i.e., the total number of samples. We calculate the average amplitude for all the decomposed principal modes and use it as a feature vector.

6.8   MULTILAYER PERCEPTRON (MLP) BASED CLASSIFICATION

The classifier used for this analysis is the multilayer perceptron (MLP) which is a feed forward based artificial neural network (ANN). ANNs are inspired to emulate the working of the human brain and contain different numbers of nodes which can be distributed across many layers, with all the elements of one layer connected to the next one. They have been widely used in EEG signal classification [13, 48]. Each layer has an activation function and a weight that connects it to another node of the next layer. In our classifier we used a sigmoid activation function for each node. The configuration that yielded the best results were two hidden layers, each consisting of ten nodes each, and an input and output layer.

6.9   RESULTS AND DISCUSSION

VMD decomposes a given signal into its principal modes. The analysis was done using four features extracted from the principal nodes, which are the Renyi entropy, SODP, FODP, and average amplitude.

The features were tested both individually and also combining all the features, i.e., all the above features were used for classification, with a ranking methodology used between them. In VMD, the first principal mode has the lowest frequency. The consecutive modes constitute the higher frequencies as seen in Figure 6.3 and 6.4, unlike in EMD where the first few modes constitute the higher frequencies. Among

TABLE 6.1
Accuracy (%) with Features Used Individually

Feature	Normal/Seizure Accuracy (%)	Seizure/Seizure-Free Accuracy (%)	Seizure/Non-seizure Accuracy (%)
SODP	93.3	96.6	96
FODP	91.6	100	96
Average amplitude	98.3	98.3	98
Renyi entropy	98.3	88.3	90

the four features individually investigated, the Renyi entropy has the least accuracy, the results of which are shown in Table 6.1. This table shows the results for which the first eighty samples were used as the training data, and the remaining twenty samples (each set Z, O, N, F, S contains one hundred samples) were used as classification test data. The entropy was calculated for each of the decomposed modes and fed as separate inputs to the MLP classifier. The results for Normal/Seizure classification (Z, O and S) have the highest accuracy, as this set consists of signals from healthy patients and confirmed seizure patients. The seizure/seizure-free (N, F and S) and seizure/non-seizure (Z, O, N, F and S) classifications consist of the N and F datasets, which are seizure free, but may have other brain conditions, which increases the risk of misclassification and reduces the accuracy.

The SODP of the decomposed modes is plotted in Figures 6.5 and 6.6. We can observe the elliptical structure of the modes. Hence, we can calculate the area of the ellipse. We have used 95 % area of confidence of the ellipse as a feature for differentiating ictal EEG signals from the seizure-free EEG signals. The results obtained are shown in Table 6.1. The SODP concept can be extended to the fourth order by delaying the signal samples twice the amount. This is called the fourth order difference plot (FODP). The calculation of the ellipse area remains the same as in the case of SODP. The final feature used is the average amplitude of each of the decomposed modes. Compared to EMD, the difference in amplitude using VMD is more pronounced between the decomposed modes of seizure free and ictal signals, which gives a high degree of accuracy as shown in Table 6.1.

We have also tested using all three features in the classification logic. Renyi entropy was left out due to its lower accuracy. The ranking methodology used for this is as follows: The classification is initially done using each feature individually. Once the classification results for all three features are obtained, they are compared. If all three features output the same result, then that result will be the output. If the results differ, then the majority classification is taken as the result (Table 6.2). In order to get rid of the bias of selecting a particular set of training and testing samples, for this classification eighty samples were randomly selected without replacement, and the remaining twenty samples, which were not in the training set, were used as the test set. The same feature ranking method was used as the previous method. Twenty-four iterations were run, where in each iteration the training and test set is random [17]. The results are shown in Table 6.3:

TABLE 6.2
Accuracy (%) with Features Ranked

Feature	Normal/Seizure Accuracy (%)	Seizure/Seizure-Free Accuracy (%)	Seizure/Non-seizure Accuracy (%)
All features (except entropy)	98.3	100	99

TABLE 6.3
Accuracy (%) with Training and Testing Data Randomized

Iteration	Normal/Seizure Accuracy (%)	Seizure/Seizure-Free Accuracy (%)	Seizure/Non-seizure Accuracy (%)
1	93.3	96.3	96
2	100	98.3	97
3	100	95	98
4	98.3	98.3	97
5	98.3	96.3	99
6	96.3	96.3	97
7	100	98.3	99
8	100	98.3	97
9	100	95	100
10	96.3	93.6	98
11	100	93.6	98
12	100	100	95
13	96.3	93.6	99
14	95	98.3	98
15	98.3	100	98
16	98.3	96.3	98
17	96.3	93.6	98
18	95	96.3	97
19	96.3	96.3	100
20	98.3	98.6	95
21	96.3	93.6	96
22	100	93.6	99
23	98.3	95	100
24	98.3	98.6	100
Average	**98.2**	**96.4**	**97.9**

6.10 CONCLUSION

In this chapter, the VMD method was applied to classify EEG signals as seizure-free/ seizure. The VMD had many improvements over previous methods, like EMD and wavelet transforms. It has also given us promising results, with an average accuracy of 98.2%, 96.4%, and 97.9% for the classification of normal/seizure, seizure/seizure-free, and seizure/non-seizure cases respectively. It is a slight improvement on the accuracy of 97.7% obtained in [45], and a significant improvement over the accuracy of 94% and 95% obtained in [7, 49] respectively. However, we would like to try our algorithm over a larger data set and also are looking at applying new features like Lypanouv exponent, Shannon entropy, etc., in our future work.

REFERENCES

1. May, W., and H. Pdf. "World Health Organization. *Epilepsy: a public health imperative.* World Health Organization, 2019," vol. 1, 2014. doi: 10.1525/sp.2007.54.1.23.
2. Hong, S. J., B. C. Bernhardt, D. S. Schrader, N. Bernasconi, and A. Bernasconi, "Whole-brain MRI phenotyping in dysplasia-related frontal lobe epilepsy," *Neurology*, vol. 86, no. 7, pp. 643–650, 2016. doi: 10.1212/WNL.0000000000002374.
3. Pati, S., and A. V. Alexopoulos, "Pharmacoresistant epilepsy: From pathogenesis to current and emerging therapies," *Cleve. Clin. J. Med.*, vol. 77, no. 7, pp. 457–467, 2010. doi: 10.3949/ccjm.77a.09061.
4. Devinsky, O., D. Gazzola, and W. C. Lafrance, "Differentiating between nonepileptic and epileptic seizures," *Nat. Rev. Neurol.*, vol. 7, no. 4, pp. 210–220, 2011. doi: 10.1038/nrneurol.2011.24.
5. Delanty, N., C. J. Vaughan, and J. A. French, "Medical causes of seizures," *Lancet*, vol. 352, no. 9125, pp. 383–390, 1998. doi: 10.1016/S0140-6736(98)02158-8.
6. Kuhlmann, L., K. Lehnertz, M. P. Richardson, B. Schelter, and H. P. Zaveri, "Seizure prediction — ready for a new era," *Nat. Rev. Neurol.*, vol. 14, no. 10, pp. 618–630, 2018, doi: 10.1038/s41582-018-0055-2.
7. Altunay, S., Z. Telatar, and O. Erogul, "Epileptic EEG detection using the linear prediction error energy," *Expert Syst. Appl.*, vol. 37, no. 8, pp. 5661–5665, 2010. doi: 10.1016/j.eswa.2010.02.045.
8. Srinivasan, V., C. Eswaran, and A. N. Sriraam, "Artificial neural network based epileptic detection using time-domain and frequency-domain features," *J. Med. Syst.*, vol. 29, no. 6, pp. 647–660, 2005. doi: 10.1007/s10916-005-6133-1.
9. Tzallas, A.T., M. G. Tsipouras, and D. I. Fotiadis, "Automatic seizure detection based on time-frequency analysis and artificial neural networks," *Comput. Intell. Neurosci.*, vol. 2007, 2007. doi: 10.1155/2007/80510.
10. Tzallas, A. T., M. G. Tsipouras, and D. I. Fotiadis, "Epileptic seizure detection in EEGs using time-frequency analysis," *IEEE Trans. Inf. Technol. Biomed.*, vol. 13, no. 5, pp. 703–710, 2009, doi: 10.1109/TITB.2009.2017939.
11. Ghosh-Dastidar, S., H. Adeli, and N. Dadmehr, "Mixed-band wavelet-chaos-neural network methodology for epilepsy and epileptic seizure detection," *IEEE Trans. Biomed. Eng.*, vol. 54, no. 9, pp. 1545–1551, 2007. doi: 10.1109/TBME.2007.891945.
12. Guo, L., D. Rivero, and A. Pazos, "Epileptic seizure detection using multiwavelet transform based approximate entropy and artificial neural networks," *J. Neurosci. Methods*, vol. 193, no. 1, pp. 156–163, 2010. doi: 10.1016/j.jneumeth.2010.08.030.
13. Guo, L., D. Rivero, J. Dorado, J. R. Rabuñal, and A. Pazos, "Automatic epileptic seizure detection in EEGs based on line length feature and artificial neural networks," *J. Neurosci. Methods*, vol. 191, no. 1, pp. 101–109, 2010. doi: 10.1016/j.jneumeth.2010.05.020.
14. Huang, N. E. et al., "The empirical mode decomposition and the Hubert spectrum for nonlinear and non-stationary time series analysis," *Proc. R. Soc. A Math. Phys. Eng. Sci.*, vol. 454, no. 1971, pp. 903–995, 1998. doi: 10.1098/rspa.1998.0193.
15. Sharma, R., R. B. Pachori, and U. R. Acharya, "Application of entropy measures on intrinsic mode functions for the automated identification of focal electroencephalogram signals," *Entropy*, vol. 17, no. 2, pp. 669–691, 2015. doi: 10.3390/e17020669.

16. Flandrin, P., G. Rilling, and P. Gonçalvés, "Empirical mode decomposition as a filter bank," *IEEE Signal Process. Lett.*, vol. 11, no. 2 PART I, pp. 112–114, 2004. doi: 10.1109/LSP.2003.821662.

17. Bajaj, V., and R. B. Pachori, "Classification of seizure and nonseizure EEG signals using empirical mode decomposition," *IEEE Trans. Inf. Technol. Biomed.*, vol. 16, no. 6, pp. 1135–1142, 2012. doi: 10.1109/TITB.2011.2181403.

18. Klügel, Niklas. "Practical empirical mode decomposition for audio synthesis." In *Proc. Int. Conf. Digital Audio Effects (DAFx-12)*, no. 2, pp. 15–18. 2012.

19. Barnhart, B. L., and W. E. Eichinger, "Empirical mode decomposition applied to solar irradiance, global temperature, sunspot number, and CO2 concentration data," *J. Atmos. Solar-Terrestrial Phys.*, vol. 73, no. 13, pp. 1771–1779, 2011, doi: 10.1016/j.jastp.2011.04.012.

20. Assous, S., A. Humeau, and J. P. L'Huillier, "Empirical mode decomposition applied to laser doppler flowmetry signals: Diagnosis approach," *Annu. Int. Conf. IEEE Eng. Med. Biol. - Proc.*, vol. 7, pp. 1232–1235, 2005. doi: 10.1109/iembs.2005.1616647.

21. Andrade, A. O., S. Nasuto, P. Kyberd, C. M. Sweeney-Reed, and F. R. Van Kanijn, "EMG signal filtering based on empirical mode decomposition," *Biomed. Signal Process. Control*, vol. 1, no. 1, pp. 44–55, 2006. doi: 10.1016/j.bspc.2006.03.003.

22. Liu, S., Q. He, R. X. Gao, and P. Freedson, "Empirical mode decomposition applied to tissue artifact removal from respiratory signal," *Proc. 30th Annu. Int. Conf. IEEE Eng. Med. Biol. Soc. EMBS'08 - "Personalized Healthc. through Technol.*, Vancouver, BC, Canada. pp. 3624–3627, 2008. doi: 10.1109/iembs.2008.4649991.

23. Dragomiretskiy, K., and D. Zosso, "Variational mode decomposition," *IEEE Trans. Signal Process.*, vol. 62, no. 3, pp. 531–544, 2014. doi: 10.1109/TSP.2013.2288675.

24. Pachori, R. B., and V. Gupta, Biomedical engineering fundamentals. In *Intelligent Internet of Things*, pp. 547–605. Springer, Cham, 2020.

25. Park, Y., L. Luo, K. K. Parhi, and T. Netoff, "Seizure prediction with spectral power of EEG using cost-sensitive support vector machines," *Epilepsia*, vol. 52, no. 10, pp. 1761–1770, 2011. doi: 10.1111/j.1528-1167.2011.03138.x.

26. Mormann, F. et al., "On the predictability of epileptic seizures," *Clin. Neurophysiol.*, vol. 116, no. 3, pp. 569–587, 2005. doi: 10.1016/j.clinph.2004.08.025.

27. Chu, H., C. K. Chung, W. Jeong, and K. H. Cho, "Predicting epileptic seizures from scalp EEG based on attractor state analysis," *Comput. Methods Programs Biomed.*, vol. 143, pp. 75–87, 2017. doi: 10.1016/j.cmpb.2017.03.002.

28. Sharma, R., R. B. Pachori, and U. Rajendra Acharya, "An integrated index for the identification of focal electroencephalogram signals using discrete wavelet transform and entropy measures," *Entropy*, vol. 17, no. 8, pp. 5218–5240, 2015. doi: 10.3390/e17085218.

29. Sharma, M., A. Dhere, R. B. Pachori, and U. R. Acharya, "An automatic detection of focal EEG signals using new class of time–frequency localized orthogonal wavelet filter banks," *Knowledge-Based Syst.*, vol. 118, pp. 217–227, 2017. doi: 10.1016/j.knosys.2016.11.024.

30. Deivasigamani, S., C. Senthilpari, and W. H. Yong, "Classification of focal and nonfocal EEG signals using ANFIS classifier for epilepsy detection," *Int. J. Imaging Syst. Technol.*, vol. 26, no. 4, pp. 277–283, 2016. doi: 10.1002/ima.22199.

31. Das, A. B., and M. I. H. Bhuiyan, "Discrimination and classification of focal and non-focal EEG signals using entropy-based features in the EMD-DWT domain," *Biomed. Signal Process. Control*, vol. 29, pp. 11–21, 2016. doi: 10.1016/j.bspc.2016.05.004.

32. Gupta, V., and R. B. Pachori, "Epileptic seizure identification using entropy of FBSE based EEG rhythms," *Biomed. Signal Process. Control*, vol. 53, p. 101569, 2019. doi: 10.1016/j.bspc.2019.101569.

33. Bhati, D., R. B. Pachori, M. Sharma, and V. M. Gadre, *Automated Detection of Seizure and Nonseizure EEG Signals Using Two Band Biorthogonal Wavelet Filter Banks.* Springer, Singapore, 2020.

34. Patil, A., and K. Behele, "Classification of human emotions using multiclass support vector machine," *2017 Int. Conf. Comput. Commun. Control Autom. ICCUBEA 2017*, Pune, India. pp. 3–6, 2018, doi: 10.1109/ICCUBEA.2017.8463656.

35. Übeyli, E. D. "Decision support systems for time-varying biomedical signals: EEG signals classification," *Expert Syst. Appl.*, vol. 36, no. 2 PART 1, pp. 2275–2284, 2009. doi: 10.1016/j.eswa.2007.12.025.

36. Übeyli, E. D., and I. Güler, "Comparison of eigenvector methods with classical and model-based methods in analysis of internal carotid arterial Doppler signals," *Comput. Biol. Med.*, vol. 33, no. 6, pp. 473–493, 2003. doi: 10.1016/S0010-4825(03)00021-0.

37. Oveisi, F. "EEG signal classification using nonlinear independent component analysis," *ICASSP, IEEE Int. Conf. Acoust. Speech Signal Process. - Proc.*, Taipei, Taiwan. no. 1, pp. 361–364, 2009. doi: 10.1109/icassp.2009.4959595.

38. Taran, S., and V. Bajaj, "Clustering variational mode decomposition for identification of focal EEG signals," *IEEE Sensors Lett.*, vol. 2, no. 4, pp. 1–4, 2018. doi: 10.1109/lsens.2018.2872415.

39. Ullal, A., and R. B. Pachori, "EEG Signal Classification using Variational Mode Decomposition," *arXiv Prepr.*, vol. arXiv:2003, 2020.

40. Sharma, R., R. B. Pachori, and P. Sircar, "Seizures classification based on higher order statistics and deep neural network," *Biomed. Signal Process. Control*, vol. 59, p. 101921, 2020. doi: 10.1016/j.bspc.2020.101921.

41. Andrzejak, R. G., K. Lehnertz, F. Mormann, C. Rieke, P. David, and C. E. Elger, "Indications of nonlinear deterministic and finite-dimensional structures in time series of brain electrical activity: dependence on recording region and brain state," *Phys. Rev. E. Stat. Nonlin. Soft Matter Phys.*, vol. 64, no. 6 Pt 1, p. 061907, 2001.

42. Cohen, M. E., and D. L. Hudson, "Applying continuous choatic modeling to cardiac signal analysis," *IEEE Eng. Med. Biol.*, vol. 15, no. 5, pp. 97–102, 1996.

43. Cavalheiro, G. L., M. F. S. Almeida, A. A. Pereira, and A. O. Andrade, "Study of age-related changes in postural control during quiet standing through linear discriminant analysis.," *Biomed. Eng. Online*, vol. 8, p. 35, 2009. doi: 10.1186/1475-925X-8-35.

44. Prieto, T. E., J. B. Myklebust, R. G. Hoffmann, E. G. Lovett, and B. M. Myklebust, "Measures of postural steadiness: Differences between healthy young and elderly adults," *IEEE Trans. Biomed. Eng.*, vol. 43, no. 9, pp. 956–966, 1996. doi: 10.1109/10.532130.

45. Pachori, R. B., and S. Patidar, "Epileptic seizure classification in EEG signals using second-order difference plot of intrinsic mode functions," *Comput. Methods Programs Biomed.*, vol. 113, no. 2, pp. 494–502, 2014. doi: 10.1016/j.cmpb.2013.11.014.

46. Pachori, R. B., D. J. Hewson, H. Snoussi, and J. Duchêne, "Analysis of center of pressure signals using empirical mode decomposition and fourier-bessel expansion," *IEEE Reg. 10 Annu. Int. Conf. Proceedings/TENCON*, 2008. doi: 10.1109/TENCON.2008.4766596.

47. Thuraisingham, R. A., Y. Tran, P. Boord, and A. Craig, "Analysis of eyes open, eye closed EEG signals using second-order difference plot," *Med. Biol. Eng. Comput.*, vol. 45, no. 12, pp. 1243–1249, 2007. doi: 10.1007/s11517-007-0268-9.

48. Guo, L., D. Rivero, J. A. Seoane, and A. Pazos, "Classification of EEG signals using relative wavelet energy and artificial neural networks," *2009 World Summit Genet. Evol. Comput. 2009 GEC Summit - Proc. 1st ACM/SIGEVO Summit Genet. Evol. Comput. GEC'09*, Shanghai, China. pp. 177–183, 2009. doi: 10.1145/1543834.1543860.

49. Joshi, V., R. B. Pachori, and A. Vijesh, "Classification of ictal and seizure-free EEG signals using fractional linear prediction," *Biomed. Signal Process. Control*, vol. 9, no. 1, pp. 1–5, 2014. doi: 10.1016/j.bspc.2013.08.006.

7 Soft-computing Approach in Clinical Decision Support Systems

Jyoti Kukreja, Harman Kaur,
and Ahmed Chowdhary
Jagannath International Management School,
New Delhi, India

CONTENTS

7.1 INTRODUCTION

Soft computing is the new regime that has been adopted by the healthcare world and has been proven to be useful, especially in clinical treatment, enabling decision making in patient diagnosis, tracking their treatment, and reducing relapses due to careless monitoring. Domains such as fuzzy logic and neural networks are being deployed to ensure both physical and psychological safety nets for the warriors fighting diseases and the doctors treating them with different medical techniques.

Medical care, starting from the diagnosis of a disease until the recovery, is extremely ambiguous, especially if the particular disease has multiple overlapping symptoms with other diseases and if treatment success of a particular disease shows high inter-patient variability. Sometimes, patients are not able to describe exactly what has happened to them or how they feel. Physicians may not understand or interpret exactly what they hear, and patients may have ambiguous symptoms and signs. Also, patients

DOI: 10.1201/9780367548445-8

may have a large amount of historical data, but physicians have no time to check and investigate all of them. It is very critical to arrive at the most accurate medical diagnosis in a very timely manner because quick and accurate diagnosis and timely initiation of treatment is important to reduce both possible complications and costs. Accurate and fast diagnosis of diseases has been facilitated via technological advancements like the CDSS and experts offering solutions. But unfortunately, diagnostic decisions are still made based upon subjective experiences in many countries around the globe, even though practitioners have become technologically advanced enough to incorporate CDSS in their healthcare system. The irony is that this advent of technology is still treated more as a fashion statement than a means of harnessing the comprehensive features of the applications. In places where it is indeed used efficiently, alert fatigue (a huge number of CDSS patient alerts including a lot of unnecessary alerts) resulting in physician burnout, an increased number of casualties like ADEs (adverse drug events), etc., leads to inefficient patient treatment. Due to these limitations, a new trend of early retirement or profession shift has been observed amongst physicians.

Thus, exploring the applicability and improvement scope of CDSS in disease diagnosis, treatment tracking and prevention of disease relapse is the raison d'être of our review. The contribution of soft computing to the Clinical Decision Support System is now extensively realized. Technology has revolutionized all industries alike, and healthcare has been impacted most profoundly through artificial neural networks and genetic algorithms.

After the development of highly efficient data-analyzing and solution-providing software, the importance of protecting patient information within the system is clear. A hacker may hack into the system and tamper with the outcomes of the CDSS system, hence derailing the treatment of a patient. In the worst case scenario, hacking of the system may lead to injection of highly incorrect doses of medicines, hormones, etc. by the patient-specific hardware connected to the CDSS system. Hence, CDSS security is of utmost importance for the system to be useful to and safe for the patient community.

7.2 LITERATURE REVIEW

Cios and Moore (2000) and Cios et al. (2000) outlined the six steps of problem statement, knowledge discovery for database creation, warehousing, data curation, data mining, synthesizing, and applying knowledge. Data mining had been highlighted as the most significant concept for building an informative database for making sensitive decisions, which can be enabled through devices. Uzoka et al. (2011) mentioned that a number of signs are ignored that build up the issues on one end and push it onto the dead end at the other extreme. The right decisions are delayed, and the diagnosis, which could have helped with or without the availability of specialists, goes haywire. For instance, the deficiency of calcium is ignored in the warning signals of brittle nails and yellow teeth. First, dermatologists and dentists are approached and then the ortho-surgeon, who diagnoses the problem through further tests. Soft computing is said to be a tailor-made customized solution for common ailments. This solution helps to solve the limitation problem of time constraints and precision, and also helps to tackle the problem of the availability of specialists (Çalişir & Dogantekin, 2011). Chang and Liao (2006), reported through their experimental survey that fuzzy logic and neural networks play an important role in figuring out the dimensions of the disease and analyzing those body processes that the patient could not report. It is essential to search through the available medical history about the consequences of the diagnosis and the probability of the patient acquiring any other symptom or disease. This process is enabled through data mining. Advancement in diagnosis through algorithm is unraveled through soft computing. Generally, patients ignore the aggravated indications of the critical stages of diseases and consult doctors only when matters go out of their hands. The problems of geographical distance and overbooking of specialists make it difficult to procure an appointment with specialists, leading to the death of the patients who are at critical stages of a serious disease.

Diagnosis and treatment of Hepatitis B or Hepatitis C depends on the determination of the fibrosis stage from F0 (no damage) to F4 (cirrhosis). A patient needs to undergo liver biopsies in the advanced stages of liver cirrhosis. These are costly and risky as well. Hence, there is a need to have alternative diagnosis methods.

Çalişir and Dogantekin (2011), in their study, found one of the domains of soft computing as highly significant in making the analysis through neuro fuzzy inference system as useful in analyzing by building case comparison on the model of diagnosis and fibrosis disease. Liver fibrosis diagnosis is facilitated through neural networks. Among the different soft computing techniques used in clinical diagnosis are support vector machines or SVMs, fuzzy and artificial neural networks, and decision trees. All of these techniques are further classified in different models. These include expert systems, classification models, and rules-based, case-based, and fuzzy/non-fuzzy models. The methodology for decision-making varies from one model and technique to the other. For instance, in the case of the expert systems, the rules' determination is done based on fuzzy and non-fuzzy methods. 'If-then' structures are utilized in such cases. The classification models rely on the optimized values of the parameters, so that the performance is enhanced and the disease diagnosis can be performed in an effective manner (Gambhir et al., 2016).Tuberculosis, being among the most severe health concerns, has shown to be diagnosed by soft computing techniques (Djam & Kimbi, 2011). The researchers used a fuzzy expert system for TB diagnosis. The fuzzy logic technology application provided the mechanism to obtain significant outcomes from vague and unstructured medical information. This fuzzy expert system could be applied to determine TB, using four major sub-components: Knowledgebase, fuzzification, inference engine, and de-fuzzification (Djam & Kimbi, 2011).

Another research study was conducted that combined two soft computing methodologies for the diagnosis of malaria (Uzoka et al., 2011). The comparison between the Analytical Hierarchy Process and Fuzzy logic was done, and promising results were obtained in the decision support system for the diagnosis of malaria. The hybrid diagnosis system showcases no statistical differences between the two technologies as far as the effectiveness of the diagnosis procedure is concerned. The comparison of every parameter shows that fuzzy logic's performance has an edge over AHP in terms of the mapping of the diagnosis with the medical professional/expert diagnosis (Uzoka et al., 2011).

In the 2018 study published by Sappagh et al., the authors used fuzzy analytical hierarchy process (FAHP) and an adaptive neuro-fuzzy inference system (ANFIS) to model CDSSs in fibrosis stage detection domain.

The use of a mathematical model is done with FAHP to identify the condition. The subtractive clustering technique is utilized for ANFIS and the research experiment shows positive outcomes. Their results showed that these two techniques can be employed successfully in designing a diagnostic CDSS system for fibrosis diagnosis. The two techniques achieve a classification accuracy of 93.3%. This level of accuracy is significant, as other tests like serum markers, imaging tests, and genetic studies have not achieved an acceptable accuracy.

The decision-making process for the medical experts could be simplified by applying the two methods mentioned above (El-Sappagh et al., 2018).

Diabetes, another primary health concern, plagues a significant proportion of the population of many countries. If detected early, the complications of diabetes (i.e., nephropathy, retinopathy, peripheral neuropathy, and even stroke) can be minimized. The data mining techniques, such as SVM, C4.5, Naïve Bayes, etc., can be combined with soft computing techniques, such as GA, and PSO, to have a positive early detection rate (Kaur & Sharma, 2018). The predictive rate of accuracy can be improved by applying the hybrid technique and strategy (Kaur & Sharma, 2018).

7.3   INCORPORATION OF DATABASES INTO THE CDSS SYSTEM TO MAKE IT MORE USEFUL

Text mining systems like the SNP curator (Tawfik & Spruit, 2018) have proven useful in SNP-disease associations. If such a system is integrated within the CDSS, it becomes easier to predict disease outcomes based on the available SNP – disease progression correlation and would enable a medical practitioner to customize a patient's treatment accordingly. Epigenetics and the immunological competency of an individual also play a crucial role in disease emergence, progression, and reaction to standard treatments. Metabolomics has proven very useful in early diagnoses of some cancers, coronary heart disease (Gowda et al., 2008), rheumatoid arthritis (Sasaki et al., 2019), etc. Most importantly, it has been very useful in determining the etiopathogenesis of complex disorders like Parkinson's disease (Shao & Le, 2019). Proteomics analysis for determining disease diagnosis and progression has been used for a while (Lippolis & DeAngelis, 2016). With the help of genomics, proteomics, and metabolomics analyses, personalized treatment regimens can be made for patients, based on their individual profiles. Environmental factors play a huge role in determining the time of emergence, intensity, and the treatment response to a particular genetically inherited disease. One of the most significant examples of late onset diseases that sometimes are triggered by environmental factors are trinucleotide repeat disorders (Van Dellen & Hannan, 2004). Such individuals go on to have children even before the disease manifests itself. Thus, the mutated gene keeps being transferred from one generation to the another. However, an efficient integration of the database of biomarkers associated with these disorders in the CDSS system and their 24X7 monitoring can help such patients know which kind of an environment can trigger their disease.

Use of soft computing to make the CDSS system competent enough to incorporate the SNP, proteomics, metabolomics, and other biomarker databases in order to diagnose and personalize treatment will be a milestone in CDSS technology.

A CDSS system with the complete incorporation of all this emerging data will prove useful in early detection of late onset diseases, diseases that show latency, and metastasizing cancers.

7.4 CDSS ALERTS, ADES, AND PHYSICIAN BURNOUT

Many physicians get CDSS alerts about patient's health such as changes in health parameters, missed drug doses, etc. (Carroll et al., 2012). This system generates many unimportant alerts as well. Hence, physicians tend to ignore these alerts, and in doing so, many important alerts get ignored. This results in miscalculation of drug doses to be given to the patient and leads to adverse drug errors (ADEs) that can severely affect a patient's health. The other facet of the problem is that this high number of alerts leads to physician burnout and early retirement or a change in professions.

Hence, there is an urgent need to improve this integrated software in order to improve the treatment process and prevent the above-mentioned events.

7.5 CDSS SECURITY

The repository of data for a CDSS system needs to be well protected. Any hacking or tinkering with the data might lead to erroneous diagnoses and wrong treatment initiation. A great deal of damage could occur before a particular doctor realizes the errors. A great deal of effort is required to safeguard the security of the system.

7.6 CONCLUSION

Chen et al. (2016) found that comprehension of a fuzzy logic concept was a life saver and well suited for self-analysis to maintain personal health. The collection and analysis of the many diseases one has endured could aid in discovering the probable diseases that an individual could suffer. A checklist for prevention and precaution can be outlined and practiced. Cios and Moore (2000) reported that data mining and knowledge discovery are useful for an individual to discover their set of strengths and weaknesses pertaining to their personal health data so that they are self-guided in making medically correct decisions for themselves. Cios et al. (2000), reported that if minute details are recorded, they can be the best guides to following rules for avoiding the disease. At times, self-analysis is ignored as trust is placed on medical diagnosis. If early warning signs are heeded, then best precautions and medical treatment could be facilitated. Genetic algorithm makes it simple to predict not only physical but also mental health. Fuzzy logic and neural networks connect those invisible dots of the changes in mental conditions that lead to physical pain.

Many cancers show very few symptoms and hence are detected at the advanced stages, with little hope of a successful treatment (Pendharkar et al., 1999). Eom et al. (2008) provided that a mindset also determines listening skills to procure a skill set. Fuzzy logic requires knowledge to deduce the data which is in both pictorial and graphical form.. Data warehousing and data mining can reap benefits, not only for the medical professionals, but for the probable patient as well. If pharmacies retain a copy of the prescription for each patient, then the pharmacist could both caution the patient and build trust in their relationship (Fernandez et al., 2009). It is important to read health as a book so that problem areas can be identified. Gambhir et al. (2016) researched the regular analysis of a rules-based system, which is a segment of CDSS (clinical decision support system). Jung et al. (2015) based their findings on the revolution in the treatment of chronic diseases. Associative methodology (correlation of symptoms and causes) is explained through fuzzy logic. Neural networks help to decipher the causes of the evolution of disease, so that the next generation can take precautions. Kannathal et al. (2015) reported diagnosis of cardiac diseases through neuro fuzzy logic for patients and medical practitioners. Kar et al. (2014) spoke on the review of and future outline for a healthy life of an individual. Khatibi and Montazer (2010) went further and explained fuzzy logic based evidence of a hybrid engine for the coronary effect, while examining symptoms of heart disease for the survival and heart care processes. Kaur and Sharma (2018) carried out their study on one of the most prevalent diseases, diabetes. Both analysis of the data and mining for the results will involve soft computing techniques based on food diagnosis and treatment of the probable side effects that may arise post consumption.

Apart from developing the CDSS system to enhance its efficiency, care has to be taken while integrating the software with technology for patient treatment tracking to strengthen CDSS security to prevent any breach or tinkering.

7.6.1 RESEARCH METHODOLOGY

A set of doctors were interviewed for the awareness of the concept of soft computing in healthcare. A list of private and public hospitals was prepared after scrutinizing the diseases that each hospital is handling. The study of cancer is the most urgent.

7.6.2 FINDINGS

Mental conditions promote physical care.

BIBLIOGRAPHY

Çalişir, D., and Dogantekin, E. 2011.'A new intelligent hepatitis diagnosis system: PCA–LSSVM'. *Expert Syst. Appl.* 38(8): 10705–10708.
Carroll, A. E., Anand, V., and Downs, S. M. 2012. 'Understanding why clinicians answer or ignore clinical decision support prompts', *Appl. Clin. Inform.* 3(3): 309–317. Published 2012 Aug 1. doi: 10.4338/ACI-2012-04-RA-0013
Chang, P. C., and Liao, T. W. (2006).'Combing SOM and fuzzy rule base for flow time prediction in semiconductor manufacturing factory', *Appl. Soft Comput.* 6(2): 198–206.
Chen, L., Li, X., Yang, Y. Kurniawati, H., Sheng, Q. Z. Hu, H. Y., and Huang, N. 2016. 'Personal health indexing based on medical examinations: A data mining approach', *Decis. Support. Syst.* 81: 54–65.
Cios, K. J., and Moore, G.W. 2000. *Medical Data Mining and Knowledge Discovery: An Overview*, Heidelberg: Springer, pp. 1–16.
Cios, K. J., Teresinska, A., Konieczna, S., Potocka, J., and Sharma, S. 2000. 'Diagnosing myocardial perfusion SPECT bull's-eye maps-a knowledge discovery approach'. *IEEE Eng. Med. Biol.* 19(4): 17–25.

Deshpande, A. D., Harris-Hayes, M., and Schootman, M. 2008. 'Epidemiology of diabetes and diabetes-related complications', *Phys. Ther.* 88(11): 1254–1264. doi: 10.2522/ptj.20080020

Djam, X. Y., and Kimbi, Y. H. 2011. 'A decision support system for tuberculosis diagnosis'. *Asia-Pac. J. Sci. Technol.*, 12: 410–4252.

El-Sappagh, S., Ali, F., Ali, A., Hendawi, A., Badria, F. A., and Suh, D. Y. 2018. 'Clinical decision support system for liver fibrosis prediction in hepatitis patients: A case comparison of two soft computing techniques'. *IEEE Access.* 6: 52911–52929. doi: 10.1109/access.2018.2868802

Eom, J. H., Kim, S. C. and Zhang, B. T. (Apta 2008). 'CDSS-E: a classifier ensemble-based clinical decision support system for cardio-vascular disease level prediction', *Expert Syst. Appl.* 34(4): 2465–2479.

Fernandez, A., Jesus, M. J. and Herrera, F. 2009. 'On the influence of an adaptive inference system in fuzzy rule based classification systems for imbalanced data-sets', *Expert Syst. Appl.* 36: 9805–9812.

Gambhir, S., Malik, S. K. and Kumar, Y. 2016. 'Role of soft computing approaches in health care domain: A mini review', *J. Med. Syst.* 40(12). doi: 10.1007/s10916-016-0651-x

Gowda, G.A., Zhang, S., Gu, H., Asiago, V., Shanaiah, N., and Raftery, D. 2008. 'Metabolomics-based methods for early disease diagnostics', *Expert Rev. Mol. Diagn.* 8(5): 617–633. doi: 10.1586/14737159.8.5.617

Jung, H., Yang, J., Woo, J. I., Lee, B.M., Ouyang, J., Chung, K. and Lee, Y. 2015. 'Evolutionary rule decision using similarity based associative chronic disease patients', *Clust. Comput.* 18(1): 279–291.

Kannathal, N., Lim, C. M., Acharya, U. R. and Sadasivan, P. K. 2015. 'Cardiac state diagnosis using adaptive neuro-fuzzy technique', *Med. Eng. Phys.* 28: 809–815.

Kar, S., Das, S., and Ghosh, P. K. 2014. 'Applications of neuro fuzzy systems: a brief review and future outline', *Appl. Soft Comput.* 15: 243–259.

Kaur, P., and Sharma, M. 2018. 'Analysis of data mining and soft computing techniques in prospecting diabetes disorder in human beings: A review', *Int. J. Pharm. Sci. Res.* 9(7). doi. 10.13040/ijpsr.0975-8232.9(7),2700-19

Khatibi, V., and Montazer, G. A. 2010. 'A fuzzy-evidential hybrid inference engine for coronary heart disease risk assessment', *Expert Syst. Appl.* 37(12): 8536–8542.

Kumar, S. U., and Inbarani, H. H. 2016. 'Neighborhood rough set based ECG signal classification for diagnosis of cardiac diseases', *Soft. Comput.* 21: 1–13.

Lai, R.K., Fan, C. Y., Huang, W. H., and Chang, P. C. 2009. 'Evolving and clustering fuzzy decision tree for financial time series data forecasting', *Expert Syst. Appl.* 3: 3761–3773.

Lippolis, R., and DeAngelis, M. 2016. 'Proteomics and human diseases', *J. Proteom. Bioinform.* 9: 063–074. doi: 10.4172/jpb.1000391

Muthukaruppan, S., and Er, M. J. 2012. 'A hybrid particle swarm optimization based fuzzy expert system for the diagnosis of coronary artery disease', *Expert Syst. Appl.* 39: 11657–11665.

Olden, K., Freudenberg, N., Dowd, J., and Shields, A. E. 2011. 'Discovering how environmental exposures alter genes could lead to new treatments for chronic illnesses', *Health Aff* (Millwood) 30(5): 833–841. doi. 10.1377/hlthaff.2011.0078

Papageorgiou, E. I. 2011. 'A new methodology for decisions in medical informatics using fuzzy cognitive maps based on fuzzy rule-extraction techniques', *Appl. Soft Comput.* 11: 500–513.

Patil, B. M., Joshi, R. C., and Toshniwal, D. 2010. 'Hybrid prediction model for type-2 diabetic patients', *Expert Syst. Appl.* 37(12): 8102–8108.

Pendharkar, P. C., Rodger, J. A., Yaverbaum, G. J., Herman, N., and Benner, M. 1999. 'Association, statistical, mathematical and neural approaches for mining breast cancer patterns', *Expert Syst. Appl.* 17: 223–232.

Prasad, V., Rao, T. S., and Babu, M. S. P. 2016. 'Thyroid disease diagnosis via hybrid architecture composing rough data sets theory and machine learning algorithms', *Soft. Comput.* 20(3): 1179–1189.

Sanz, J. A., Galar, M., Jurio, A., Brugos, A., Pagola, M. and Bustince, H. 2014. 'Medical diagnosis of cardiovascular diseases using an interval-valued fuzzy rule-based classification system', *Appl. Soft Comput.* 20: 103–111.

Sasaki, C., Hiraishi, T., Oku, T., Okuma, K., Suzumura, K., Hashimoto, M. et al. 2019. 'Metabolomic approach to the exploration of biomarkers associated with disease activity in rheumatoid arthritis', *PLoS ONE* 14(7): e0219400. doi: 10.1371/journal.pone.0219400

Seera, M., and Lim, C. P. 2014. 'Hybrid intelligent system for medical data classification', *Expert Syst. Appl.* 41: 2239–2249.

Shao, Y., and Le, W. 2019. 'Recent advances and perspectives of metabolomics-based investigations in Parkinson's disease', *Mol. Neurodegener.* 14: 3. doi: 10.1186/s13024-018-0304-2

Shastry, B. S. 2007. 'SNPs in disease gene mapping, medicinal drug development and evolution', *J. Hum. Genet.* 52: 871–880. doi: 10.1007/s10038-007-0200-z

Sweilam, N. H., Tharwat, A. A., and Moniem, N. A. 2010. 'Support vector machine for diagnosis cancer disease: a comparative study', *Egypt. Inform. J.* 11(2): 81–92.

Tawfik, N. S., and Spruit, M. R. 2018. 'The SNPcurator: literature mining of enriched SNP-disease associations', *Database* (Oxford) 2018: bay020. doi:10.1093/database/bay020

Ubeyli, E. D. 2009, 'Adaptive neuro-fuzzy inference Systems for Automatic Detection of breast cancer', *J. Med. Syst.* 33: 353–358.

Übeyli, E. D., and Doğdu, E. 2010. 'Automatic detection of erythemato-squamous diseases using k-means clustering', *J. Med. Syst.* 34(2): 179–184.

Uguz, H. 2012. 'Adaptive neuro-fuzzy inference system for diagnosis of the heart valve diseases using wavelet transform with entropy', *Neural Comput. Applic.* 21: 1617–1628.

Uzoka, F. M. E., Osuji, J., and Obot, O. 2011. 'Clinical decision support system (DSS) in the diagnosis of malaria: a case comparison of two soft computing methodologies', *Expert Syst. Appl.* 38(3): 1537–1553.

Van Dellen, Anton and Hannan, J. Anthony. 2004.' Genetic and environmental factors in Huntington's disease', *Neurogenetics* 5: 9–17. doi: 10.1007/s10048-003-0169-5

Van den Broeck, T., Joniau, S., Clinckemalie, L. et al. 2014. 'The role of single nucleotide polymorphisms in predicting prostate cancer risk and therapeutic decision making', *Biomed. Res. Int.* 2014: 627510. doi: 10.1155/2014/627510

Wang, Y.A., Kammenga, J. E. and Harvey, S. C. 2017. 'Genetic variation in neurodegenerative diseases and its accessibility in the model organism Caenorhabditis elegans', *Hum. Genomics* 11(1): 12. Published 2017 May 25. doi: 10.1186/s40246-017-0108-4

Yeh, D. Y., Cheng, C. H., and Chen, Y. W. 2011. 'A predictive model for cerebrovascular disease using data mining'. *Expert Syst. Appl.* 38(7): 8970–8977.

Zheng, B., Yoon, S. W., and Lam, S. S. 2014. 'Breast cancer diagnosis based on feature extraction using a hybrid of K-means and support vector machine algorithms'. *Expert Syst. Appl.* 41(4): 1476–1482.

8 A Comparative Performance Assessment of a Set of Adaptive Median Filters for Eliminating Noise from Medical Images

Sudhansu Kumar Mishra, Prajna Parimita Dash, and Sitanshu Sekhar Sahu
Birla Institute of Technology – Mesra, Ranchi, India

Ashutosh Rath
ICFAI Business School, Mumbai, India

CONTENTS

8.1 INTRODUCTION

Suppression of various spurious noises from medical images, which is one of the most challenging tasks, is considered to be one preprocessing task in most of the medical image processing operations. The medical imaging technique is used as a non-invasive approach to observe the internal body parts for medical diagnosis. If imaging of one removed part of body is taken for analysis then it does not come under the medical imaging umbrella, and is considered to be a part of pathology. Different modalities of medical imaging include X-ray, CT, MRI, US, etc. It is found

DOI: 10.1201/9780367548445-9

that different unwanted noise signals are added to medical images while acquiring, transmitting, and processing.

Noise is considered to be an unwanted entity that destroys the inherent information of a medical image. Some of the examples of probable noises that mostly affect the medical images are Gaussian, Poisson, Rician, Salt and Pepper noise, etc. In reference [1], it is shown that a 1-dimensional signal like ECG is corrupted due to electromagnetic interference, power line interference, baseline wandering, etc., during acquisition. Similarly, a visual blocking artifacts affect is seen because of the compression of digital images at a very high compression ratio [2]. In reference [3], different approaches for the improvement of old manuscripts having extremely fuzzy backgrounds have been reviewed. The noise present in an MRI image can be attributed to imperfections in the radio-frequency coils or problems related to the acquisition sequences [4, 5]. The presence of impulse noise in an image may be either relatively high or low depending upon the procedure of acquisition and its processing techniques [6]. The salt and pepper noise occurs impulsively at the time of switching, so that it affects different pixel positions randomly [7]. In this article, an adaptive approach to suppress salt and pepper noise from MRI is implemented. This noise may also sometimes be added on when MRI images are acquired, processed, and transmitted [8].

Detection of noise and filtering it from a medical image, and restoration of a noisy pixel to its original one are essential tasks to obtain the proper information. Median Filter (MF) is found to be one of the best linear fixed filters for eliminating salt and pepper noise. In a standard MF, the median value of a mask is calculated, then it replaces the central value of a mask without considering the local features [9]. This filter provides satisfactory results in terms of the signal to noise ratio. However, the noise free pixels are also affected along with the noisy pixel that blurred the image. Therefore, the development of an adaptive digital image filter that eliminates these bottlenecks is required. Adaptive filters are dynamic systems with adaptive parameters and structure. They have the property to change the values of their parameters according to changing situations during processing [10–12]. The goal of the adaptation is to adjust the parameters of the filter according to an interaction with the environment, in order to get the desirable values [13, 14]. In the proposed Circular Adaptive Median Filter (CAMF) technique, only the noisy pixel is replaced, and the size of the mask is changed as per the requirement [15–19]. Thus, an enhanced adaptive and recursive median filter had been developed by exploiting the advantages of the previously developed filters, such as the Cross Adaptive Median Filter (CrAMF), Plus Adaptive Median Filter (PAMF), Conic Adaptive Median Filter (CoAMF), Modified Switching Median Filter (MSWM), Decision Based Filter (DBF), and Recursive Adaptive Modified Filter (RAMF),

In the DBF, proper selection of the adjoined pixel is required, so that their median value will replace the central pixel of the mask. This filter is suitable for images corrupted by high density noise [20], and requires proper selection of neighboring pixels. The MSWM has considered the rank order concept for improving the capability of the median filter [21]. In this filter, if the rank order of the noisy pixel is more than a certain threshold value, then only it will be replaced. Here, the difficulty arises in the optimum selection of the threshold value. The Recursive adaptive median filter, i.e., RAMF, adapts the working window size according to the noise condition/density, i.e., window size increases if all the pixels of any mask are corrupted [22].

This method encouraged us to propose an effective algorithm which minimizes the main demerits of square kernel, i.e., artifacts. Here, we have introduced MCAMF, where the mask used is a circular and dynamic one, and is a modified version of CAMF as proposed in [23]. The proposed MCAMF adapts the working window size according to the noise density, and in the situation where all the pixels are noisy in a mask, its size increases. We have also implemented the Cross Adaptive Median Filter (CrAMF), Plus Adaptive Median Filter (PAMF) and Conic Adaptive Median Filter (CoAMF) that use cross-shaped dynamic kernels, plus and conic respectively. Finally, the performance is investigated by evaluating the output of these filters with that of the conventional rectangular shaped dynamic kernel filters. Here, we have focused on implementation of the filters in suppressing impulse noise from MRI.

8.2 PROPOSED MODIFIED CIRCULAR ADAPTIVE MEDIAN FILTER

In the proposed work, the salt & pepper noise that corrupts some of the pixels of images with either 0 or 255 (i.e., either black or white, respectively) with equal probability, has been considered for suppression. If salt and pepper noise density is considered to be 0.1, then 10 percent of pixels are corrupted. Medanta Hospital, Ranchi, provides us the required medical images for doing research work. We have intentionally introduced salt and pepper noise that has a density of 0.01 for simulation purposes. It classifies the pixels into two categories, defined according to their intensity values, may be noisy or noise-free. In the proposed study, new dynamic kernel blocks like cone, cross, plus, and modified circular have been introduced and applied, having different shapes. Moreover, the pixels present inside the mask are different. The shape of different dynamic kernels like cone, cross, plus, and modified plus are shown in Figures 8.1 and 8.2.

Steps of proposed Modified CAMF algorithm.

Step 1: Noises are artificially included in collected medical images. The mathematical model is expressed as:

$$x = n + y \qquad (8.1)$$

Here, x is the noisy image, n is the noise and original image is y. The mask considered is of circular one denoted by T. The center of the

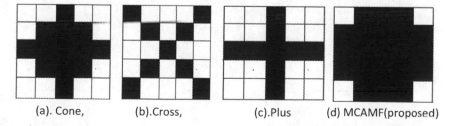

    (a). Cone,   (b).Cross,   (c).Plus   (d) MCAMF(proposed)

FIGURE 8.1   (a) Cone, (b) Cross, (c) Plus, (d) MCAMF (proposed).

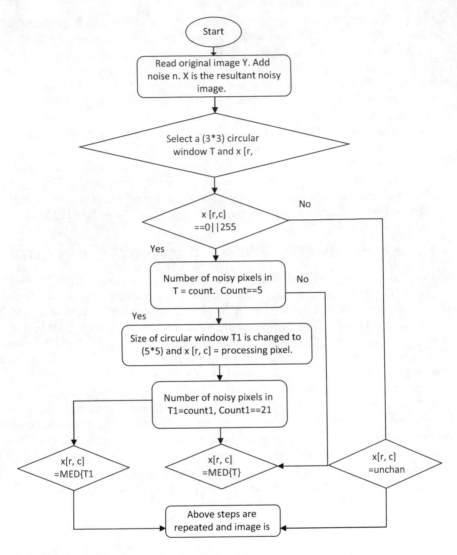

FIGURE 8.2 Proposed Modified CAMF algorithm.

mask is noted as x[r, c] which may or may not have noise. If the central pixel is noisy, then it gets replaced, otherwise, not. Hence, T is updated as:

$$T = \left\{ x\left[r+i, c+j \right], \left(i, j \right) \in w \right\} \qquad (8.2)$$

If it is a (3*3) mask, then it has five numbers in the neighborhood, with the center at x [r, c].
and w ∈ {(−1,0), (0, −1,1), (1,0)}

Step 2: If a circular mask has a noisy pixel at the center, then: x[r, c] = MED(T)

Step 3: If mask size is (3*3), and all the pixels are noisy, then mask size increases to (5*5):

$$T1 = \left\{ x\left[r+i, c+j \right], \left(i, j \right) \in w1 \right\} \tag{8.3}$$

Here, w1 has 21 adjacent pixels in a circle, with the center at x[r, c].
w1 ∈ { (–2, –1:1), (–1:1, –2:2), (2, –1,1)}

Step 4: If the size of the circular mask is (5*5), and has no pixel with noise:,

$$x\left[r, c \right] = MED\left\{ T1 \right\} \tag{8.4}$$

Step 5: If noise is present in every pixels of a mask of size (5*5), then:

$$x\left[r, c \right] = MED\left\{ T \right\} \tag{8.5}$$

Step 6: Repeat these processes from steps 1-5 until the whole image is processed.

8.3   SIMULATION STUDY

The advantage of the Modified Circular Adaptive Median Filter (MCAMF) over several tested methods was observed and demonstrated its image denoising capabilities on thirty-five MRI test images of (512*512) sizes corrupted with noises of varying densities. The detailed results and explanation of the second MRI image MRI2 is only mentioned in this paper due to lack of space. By applying impulse noise of varying densities, the performances of filters are evaluated for different images. The noise densities ranging from 10% to 80% are chosen for investigation. The performance of MCAMF is compared with that of six other competitive filtering methods. The NRDB and PSNR have been considered for objective evaluation. The subjective evaluation is also carried out on the basis of visual quality, and verified by medical experts.

8.3.1   Performance Indices

To obtain the PSNR, which depends on signal power and noise power, the original medical image from Medanta Hospital, Ranchi, India, is considered. The noise power is decided by the density of impulse noise present in the medical image. The present noise is either the noise already inherent inside the collected image or it may be the noise that has been intentionally added. Let *I* be the original image having size of *M*N, K be noisy pixel values and MAX_I be the pixel value, then MSE and PSNR are mathematically expressed as:

$$MSE = \frac{1}{M * N} \sum_{i=0}^{M-1} \sum_{j=0}^{N-1} \left[I\left(i,j \right) - K\left(i,j \right) \right]^{2} \tag{8.6}$$

$$PSNR = 10\log_{10}\left(\frac{MAX_i^2}{MSE}\right) \qquad (8.7)$$

Similarly, the NRDB in decibel is mathematically expressed as:

$$NRDB = 10\log_{10}\left(\frac{MSE_{IN}}{MSE_{OUT}}\right) \qquad (8.8)$$

Where, MSE_{IN} is calculated between original and nosy image, and MSE_{OUT} is found out between filtered and original image.

8.3.2 Simulation Results

We have also given the qualitative results of all the methods for visual analysis of an MRI2 image with 60% noise density in Figure 8.3.

The PSNR and NRDB values obtained from different approaches by considering 60% noise density are dipicted in Tables 8.1 and 8.2. In all the images, the results obtained by the proposed MCAMF filter is superior as compared to others taken for analysis. The proposed method has an improved filtering capability, and has demonstrated superior performance in comparison to others in restoring the important details, i.e., Figure 8.4.

From Table 8.3 it is clear that by varying densities of noise from 10% to 40% both RAMF and MCAMF demonstrated a similar level of performance by considering the PSNR values, but at density of noise with high value, i.e., 50% to 80%, MCAMF has performed better than all the other competetive filters. The supremacy of MCAMF over others has also been justified with the NRDB values, shown in Table 8.4. The DBF approach shows an average level of performance at a high level of noise density, and thus lacks usage procedures for these pixels. The MSWM uses a threshold value which depends on the types of image and noise density. Thus, it is very difficult to optimally predict that threshold. Because of the symmetrical properties of the circular block, in every direction, it is found to be efficient in denoising noise from the edges. The proposed filter has also validated the claim of its superiority in various noise densities over other filters, and minimized the artifacts caused by a square kernel. Figure 8.4 shows satisfactory output subjectively by the MCAMF filter having low to high density of noise. Different values of PSNR obtained by various filters have a noise density between 10% and 80% as shown in Figure 8.5. Figure 8.6 also illustrates an improved restoration capability of the MCAMF, looking at NRDB with the 'MRI2' image corrupted by 10% to 80% fixed-valued impulsive noises (Figure 8.7).

As the computational time of all the methods is calculated and given in Table 8.5, it is observed that MCAMF has a low computational time as compared to RAMF, and it justifies its claim to be a more efficient and effective method for noise filtering.

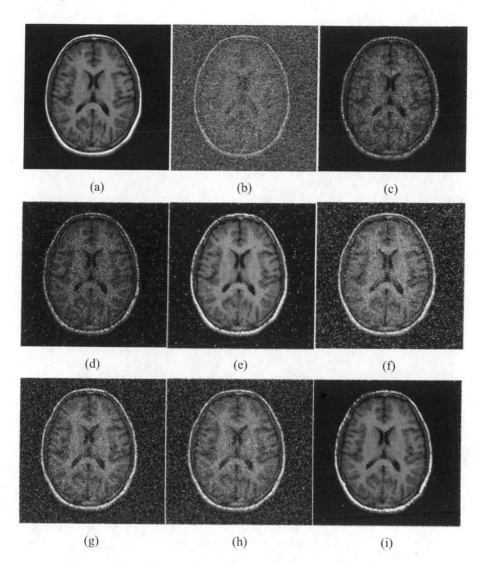

FIGURE 8.3 (a) Original Image (b) Noisy Image (60%), restored images using (c) DBF (d) MSWM (e) RAMF (f) CoAMF (g) PAMF (h) CrAMF and (i) MCAMF (proposed)

The Friedman test, which is used to verify the repeatability of obtained results, has also been performed. The results of the Friedman test is depicted in Table 8.6. By setting displayout = 'On' 0.0112 is obtained to be the critical value, which is less than $\alpha = 0.05$ as shown in Table 8.7.

Two other statistical investigations were run, Sign and Wilcoxon, for obtaining the dominance of one over the other as reported in [24]. By comparing the obtained values in Tables 8.8 through 8.10, it illustrates the dominance of MCAMF over

TABLE 8.1
PSNR Values of Various Filtering Techniques for Fifteen Different Images at 60% Density of Noise

PSNR

IMAGES	FILTERS						
	DBA	SWM	MEDIAN1	CONE	PLUS	CROSS	CIRCULAR
IMG001	26.0045	26.7138	27.5727	26.6099	29.1940	28.6670	29.7718
IMG002	28.9548	28.2219	29.2701	27.5562	29.2364	29.1814	29.3534
IMG003	28.4888	27.1042	29.3290	27.1778	29.1969	29.6952	29.3806
IMG004	29.6800	28.9330	29.2935	26.9911	29.3961	28.9930	29.5770
IMG008	29.1663	28.8653	29.4702	27.8872	29.6217	28.4386	29.9011
IMG015	28.3145	28.6511	29.0872	27.6046	29.2460	28.4765	29.5042
IMG020	28.9886	28.1250	29.0328	27.1395	29.2905	28.3143	29.8386
IMG026	30.3183	29.2953	30.0682	28.3950	29.5599	29.4612	30.7026
HIGHIMAGE	22.5307	22.0825	22.2895	22.0322	23.3622	22.3264	23.6337
MRI1	23.1843	22.8479	23.6590	22.6334	24.2042	23.0127	24.5609
MRI2	**30.1001**	**28.3273**	**32.0155**	**31.3178**	**32.7019**	**31.7281**	**32.8505**
Brain3	22.5359	22.8519	22.9463	22.6343	24.0094	22.9179	24.1016
IMG012	29.2353	28.6409	29.3518	27.2187	29.6779	28.6381	29.5917
IMG014	29.3673	28.1915	29.4108	27.5445	29.2973	28.8468	29.7328
IMG018	29.2610	28.0547	29.2082	27.2938	29.3613	28.6605	29.7366

TABLE 8.2
NRDB Values of Various Filtering Methods for Fifteen Different Images at 60% Density of Noise

NRDB IMAGES	FILTERS						
	DBA	SWM	MEDIAN1	CONE	PLUS	CROSS	CIRCULAR
IMG001	3.1380	3.1139	4.5743	1.5054	4.7880	2.9592	4.8555
IMG002	2.2722	1.7684	2.4801	1.5303	4.6911	2.7176	4.7716
IMG003	1.7812	1.4997	2.1176	1.0210	4.1876	2.3256	4.3815
IMG004	2.4665	2.3672	2.6826	1.4047	4.8493	2.8190	4.6344
IMG008	2.7948	2.8747	3.7571	2.5453	5.6212	3.6439	5.6778
IMG015	3.5360	3.4040	4.5198	3.5648	6.0424	4.4865	6.1023
IMG020	3.7073	3.4298	4.0518	2.8202	5.6575	3.9028	5.8745
IMG026	4.0795	4.4055	5.4388	4.2677	6.1481	5.1816	6.9050
HIGHIMAGE	7.5762	7.0933	7.7445	6.0769	9.5713	7.8090	9.6386
MRI1	2.8977	2.6520	3.5158	2.5427	5.3502	3.4834	5.1942
MRI2	7.3253	7.4846	8.8012	6.2519	7.9696	6.8712	9.8775
Brain3	3.5868	4.3241	4.3555	3.5207	5.8959	4.4238	6.0351
IMG012	3.5187	3.1567	4.2534	3.0512	5.6300	4.1226	5.6766
IMG014	4.4582	4.8968	4.9623	3.881	6.3698	4.6095	6.3717
IMG018	4.6539	4.0467	5.1458	4.1946	6.6661	4.8722	6.7932

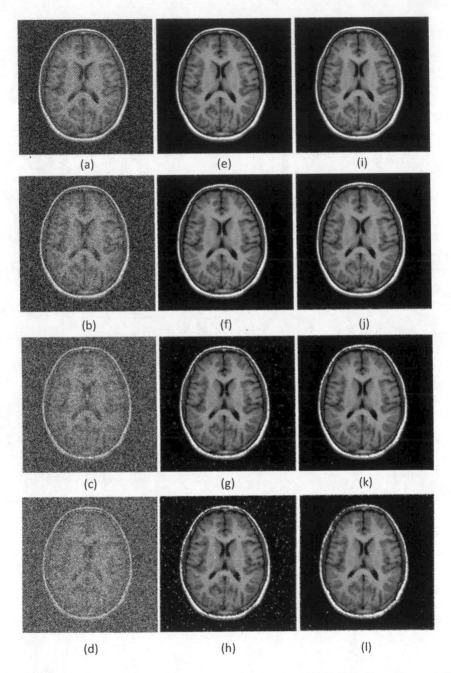

FIGURE 8.4 MRI2 image with noise (a) 20% (d) 40% (g) 50% (j) 60%, filtered image (e-h) RAMF and (i-l) MCAMF

TABLE 8.3

PSNR Values of Various Filtering Techniques for MRI2 Image at Different Density of Noises

PSNR

MRI2				FILTERS				
Noise Density in percentage	DBA	SWM	RAMF	CONE	PLUS	CROSS	CIRCULAR
10	29.5508	27.5718	31.0155	30.2547	31.3219	30.4726	31.8505
20	24.0213	22.6800	27.7820	25.7418	25.9603	25.5047	28.2669
30	20.6662	19.6639	25.7142	21.4805	21.6027	21.2900	26.0652
40	18.1970	17.7458	23.3471	17.7603	17.8768	17.5537	24.3929
50	15.9680	15.9689	21.5754	14.5062	14.4026	14.3877	21.7076
60	14.2507	14.2484	18.4195	11.6898	11.7481	11.6987	19.7717
70	13.4097	12.4019	15.2831	9.2444	9.2434	9.1298	17.5891
80	10.9023	10.7450	11.4222	7.1052	8.0096	7.5401	15.1864

TABLE 8.4

NRDB Values of Various Filtering Techniques for the MRI2 Image for Different Densities of Noise

NRDB

MRI2	FILTERS						
Noise Density in Percentage	DBA	SWM	RAMF	CONE	PLUS	CROSS	CIRCULAR
10	8.2390	7.8548	**8.8012**	6.2614	8.9189	6.8263	**9.8775**
20	7.4510	7.1905	**7.7042**	6.7274	8.4622	6.8366	**8.3761**
30	6.9541	6.8145	**7.5964**	6.5261	7.6333	6.4164	**7.6209**
40	6.0096	5.5198	**6.6365**	6.3292	6.7122	5.7165	**7.2661**
50	5.1749	4.0383	**5.3658**	5.9108	5.4084	4.7713	**6.8066**
60	5.0068	4.0576	**5.1274**	5.5309	4.2433	3.8578	**6.0510**
70	4.1856	3.6441	**4.2455**	5.1856	3.0231	2.7469	**5.3576**
80	3.7620	2.9167	**3.0429**	4.8097	2.9540	2.1056	**4.0973**

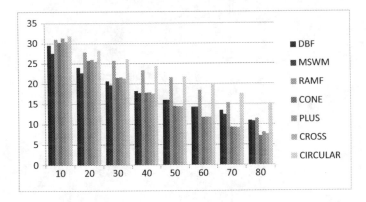

FIGURE 8.5 PSNR values obtained from various filters for the MRI2 Image corrupted with salt and pepper noise at a density varying from 10% to 80%.

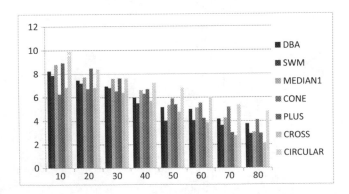

FIGURE 8.6 NRDB values obtained from various filters for the MRI2 Image corrupted with salt and pepper noise at a density varying from 10% to 80%.

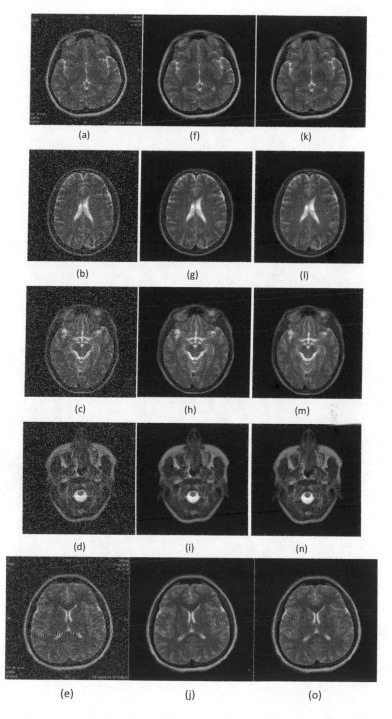

FIGURE 8.7 (a–e) Noisy Images with 20% noise density and their respective. Filtered image (f–j) RAMF and (k–o) MCAMF.

TABLE 8.5

Comparison of Computational Time

CPU TIME (in sec)	DBF	MSWM	RAMF	CONE	PLUS	CROSS	CIRCULAR
	39.759 s	37.533 s	71.543 s	51.680 s	40.495 s	31.899 s	35.990 s

TABLE 8.6

Friedman Test Mean Ranks

Methods	DBF	MSWM	RAMF	CoAMF	PAMF	CrAMF	MCAMF
Mean Ranks	23.7143	16.1429	28.8571	12.8571	32.9286	22.1429	38.3571

TABLE 8.7

Friedman Test Values

Source	Sum of square (SS)	Degree of freedom (dof)	Mean square (MS)	Chi-square	Critical value (p)
Columns	18.5	3	6.16667	11.1	0.0112
Error	1.5	9	0.16667		
Total	20	15			

TABLE 8.8

Results of Sign Tests at $\alpha = 0.05$ and $\alpha = 0.01$ Using PSNR Metric as a Winning Parameter

CIRCULAR	DBA	SWM	MEDIAN1	CONE	PLUS	CROSS
Wins (+)	22	23	23	23	22	23
Loss (−)	1	0	0	0	1	0
Detected differences	$\alpha=0.05$	$\alpha=0.05$	$\alpha=0.05$	$\alpha=0.05$	$\alpha=0.05$	$\alpha=0.05$

TABLE 8.9

Results of Sign Tests at $\alpha = 0.05$ and $\alpha = 0.01$ Using the NRDB Metric as a Winning Parameter

CIRCULAR	DBA	SWM	MEDIAN1	CONE	PLUS	CROSS
Wins (+)	23	23	23	22	19	23
Loss (−)	0	0	0	1	4	0
Detected differences	$\alpha=0.05$	$\alpha=0.05$	$\alpha=0.05$	$\alpha=0.05$	$\alpha=0.05$	$\alpha=0.05$

TABLE 8.10
PSNR Metric As a Winning Parameter in Performing the Wilcoxon Signed Test

Comparison	p Value	h Value
CIRCULAR Vs DBA	2.99E-05	1
CIRCULAR Vs SWM	2.35E-05	1
CIRCULAR Vs MEDIAN1	2.20E-05	1
CIRCULAR Vs CONE	2.25E-05	1
CIRCULAR Vs PLUS	2.89E-05	1
CIRCULAR Vs CROSS	2.19E-05	1

TABLE 8.11
NRDB Metric As a Winning Parameter in Performing Wilcoxon Signed Test Using

Comparison	p Value	h Value
CIRCULAR Vs DBA	2.70E-05	1
CIRCULAR Vs SWM	2.35E-05	1
CIRCULAR Vs MEDIAN1	2.29E-05	1
CIRCULAR Vs CONE	4.01E-05	1
CIRCULAR Vs PLUS	0.002127	1
CIRCULAR Vs CROSS	2.70E-05	1

others. The h-value found in Wilcoxon signed test shows the significant improvement in performance of the proposed approach over others as shown in Tables 8.10 and 8.11.

8.4 CONCLUSION

This chapter presented Modified Circular Adaptive Median Filter (MCAMF), which is a recursive and adaptive median filter having a circular mask for noise suppression from medical images. In this research work, different MRI images collected from the Medanta hospital were tested by our proposed algorithm. Specifically, impulse noises having different densities of noise were considered for testing the denoising capability of the proposed algorithm with others. It is concluded that the MCAMF approach minimizes the artifacts caused by the square kernel, and preserves the edges information with minute details. Here, the PSNR and NRDB are considered to be two performance indices. The MCAMF has low computational complexity, with less time and high accuracy. From different statistical testing, it found that MCAMF dominates all other approaches. The proposed technique can also be tested on various other medical images corrupted with different noise density.

ACKNOWLEDGMENT

We would like to thank Medanta Hospital for providing medical images for research purpose.

REFERENCES

1. Karthikeyan, P., M. Murugappan, and S. Yaacob, "ECG Signal Denoising Using Wavelet Thresholding Techniques in Human Stress Assessment", *International Journal on Electrical Engineering and Informatics*, vol. 4, no. 2, 2012, pp. 306–319.
2. Chacko, Shanty and J. Jayakumar, "Deblocking of Gray Scale Images and Videos Using 3D Spatial Filtering", *International Journal on Electrical Engineering and Informatics*, vol. 6, no. 2, 2014, pp. 421–434.
3. Yahya, Sitti Rachmawati, S. N. H. Sheikh Abdullah, K. Omar, M. S. Zakaria, and C. Y. Liong, "Review on Image Enhancement Methods of Old Manuscript with Damaged Background", *International Journal on Electrical Engineering and Informatics*, vol. 2, no. 1, 2010, pp. 1–14.
4. Ali, H. M. "A new method to remove salt & pepper noise in Magnetic Resonance Images," *2016. 11th International Conference on Computer Engineering & Systems (ICCES)*, Cairo, 2016, pp. 155–160.
5. Deepa, B., and M. G. Sumithra, "Comparative analysis of noise removal techniques in MRI brain images," *2015 IEEE International Conference on Computational Intelligence and Computing Research (ICCIC)*, Madurai, 2015, pp. 1–4.
6. Wei, Y., S. Yan, L. Yang and Y. Fu. "An improved median filter for removing extensive salt and pepper noise," *2014 International Conference on Mechatronics and Control (ICMC)*, Jinzhou, 2014, pp. 897–901.
7. Srisaiprai, P., W. Lee and V. Patanavijit, "An alternative technique using median filter for image reconstruction based on partition weighted sum filter," *2016 13th International Conference on Electrical Engineering/Electronics, Computer, Telecommunications and Information Technology (ECTI-CON)*, Chiang Mai, 2016, pp. 1–6.
8. Vishaga, S. and S. L. Das. "A survey on switching median filters for impulse noise removal," *2015 International Conference on Circuits, Power and Computing Technologies [ICCPCT-2015]*, Nagercoil, 2015, pp. 1–6.
9. Kunsoth, R., and M. Biswas, "Modified decision based median filter for impulse noise removal," *2016 International Conference on Wireless Communications, Signal Processing and Networking (WiSPNET)*, Chennai, 2016, pp. 1316–1319.
10. Chen, Tao, Kai-Kuang Ma and Li-Hui Chen, "Tri-state median filter for image denoising," *IEEE Transactions on Image Processing*, vol. 8, no. 12, 1999, pp. 1834–1838.
11. Sudheesh, K. V., and L. Basavaraj, "Selective weights based median filtering approach for impulse noise removal of brain MRI images," *2016 International Conference on Electrical, Electronics, Communication, Computer and Optimization Techniques (ICEECCOT)*, Mysuru, 2016, pp. 60–65.
12. Ashok, K., and V.R. Vijaykumar. "Adaptive Window Based Multi Stage Impulse Noise Detection for Removal of Random Valued Impulse Noise in Digital Images". *Asian Journal of Information Technology*, vol. 15, 2016, pp. 689–693.
13. Loupas, T., W. N. McDicken, and P. L. Allan, "An adaptive weighted median filter for speckle suppression in medical ultrasonic images," *IEEE Transactions on Circuits and Systems*, vol. 36, no. 1, 1989, pp. 129–135.

14. Chen, Chao-Yu, Chin-Hsing Chen, Chao-Ho Chen, and Kuo-Ping Lin. "An automatic filtering convergence method for iterative impulse noise filters based on PSNR checking and filtered pixels detection", *Expert Systems with Applications*, vol. 63, 2016, pp. 198–207.

15. Kumar, A., and A. Datta., "Adaptive edge discriminated median filter to remove impulse noise," *2016 International Conference on Computing, Communication and Automation (ICCCA)*, Noida, 2016, pp. 1409–1413.

16. Jezebel Priestley, J., V. Nandhini and V. Elamaran, "A decision based switching median filter for restoration of images corrupted by high density impulse noise," *2015 International Conference on Robotics, Automation, Control and Embedded Systems(RACE)*, Chennai, 2015, pp. 1–5.

17. Roy, A., J. Singha, L. Manam and R. H. Laskar. "Combination of adaptive vector median filter and weighted mean filter for removal of high-density impulse noise from colour images," *IET Image Processing*, vol. 11, no. 6, 2017, pp. 352–361, 6.

18. Faragallah, Osama S., and Hani M. Ibrahem. "Adaptive switching weighted median filter framework for suppressing salt-and-pepper noise", *AEU – International Journal of Electronics and Communications*, vol. 70, no. 8, 2016, pp. 1034–1040.

19. Wang, C., L. Li, F. Yang and H. Gong. "A new kind of adaptive weighted median filter algorithm," *2010 International Conference on Computer Application and System Modeling (ICCASM 2010)*, Taiyuan, 2010, pp. V11-667–V11-671.

20. Srinivasan, K. S., and D. Ebenezer, "A new fast and efficient decision-based algorithm for removal of high-density impulse noises," *IEEE Signal Processing Letters*, vol. 14, no. 3, 2007, pp. 189–192.

21. Kang, Chung-Chia, and Wen-June Wang., "Modified switching median filter with one more noise detector for impulse noise removal", *AEU – International Journal of Electronics and Communications*, vol. 63, no. 11, 2009, pp. 998–1004.

22. Meher, Saroj K., and Brijraj Singhawat. "An improved recursive and adaptive median filter for high density impulse noise", *AEU - International Journal of Electronics and Communications*, vol. 68, no. 12, 2014, pp. 1173–1179.

23. Meher, S. "Color Image Denoising with Multi-channel Spatial Color Filtering," *2010 12th International Conference on Computer Modelling and Simulation*, Cambridge, 2010, pp. 284–288.

24. Sagar, Priya, Ashruti Upadhyaya, Sudhansu Kumar Mishra, Rudra Narayan Pandey, Sitanshu Sekhar Sahu, and Ganapati Panda. "A Circular Adaptive Median Filter for Salt and Pepper Noise Suppression from MRI Images", *Journal of Scientific and Industrial Research, National Institute of Science Communication and Information Resources (NISCAIR)*, vol. 79, 2020, pp. 941–944.

9 Early Prediction of Parkinson's Disease Using Motor, Non-Motor Features and Machine Learning Techniques

Babita Majhi and Aarti Kashyap
Guru Ghasidas Vishwavidyalaya, Central University,
Bilaspur, India

CONTENTS

9.1   INTRODUCTION

Parkinson's disease is a chronic, neurological, progressive motor disorder that affects the neurons of the human brain in the substantia nigra (a small area in the brain). Neurons are made up of a number of nerve cells [1]. These nerve cells produce dopamine. Dopamine is a chemical and neurotransmitter substance present in the brain. The signals are transmitted through neurotransmitters to other parts in the brain when all is working properly. Due to the loss of dopamine in the neuron, this lack causes body movement problems that lead to Parkinson's disease. Generally, Parkinson's disease occurs at the age of sixty or above and it is found more often in

DOI: 10.1201/9780367548445-10

men than in women [2]. It is a brain disorder disease that brings shaking, difficulty with walking, stiffness, and coordination in the patient's body. As the symptoms increase, patients face other problems, such as mental changes, sleeping problems, behavior changes, fatigue, depression, etc., which can happen in the last stages of the disease [2]. PD symptoms increase very slowly over time and needs long term treatment. There is currently no cure for Parkinson's disease. More than 50,000 new cases are reported in the United States each year. But there may be even more since Parkinson's is often misdiagnosed. It's reported that Parkinson's complications are the fourteenth largest cause of death as per this Trusted Source in the United States [3].

PD occurs in any patient when there is an imbalance between the inhibitory dopamine and the excitatory acetyl chlorine [4]. There are two types of symptoms in PD – motor symptoms [5] and non-motor symptoms [6]. Motor symptoms such as tremor, bradykinesia, postural instability, and rigidity and non-motor symptoms such as depression, olfaction problems, cognition disorders, and sleep behaviors are the symptoms that affect the PD patient's life. These people, who are suffering from PD, need early detection so that they can receive the necessary treatment.

The objective of this chapter is to develop different Machine Learning based models for early prediction of Parkinson's disease, using motor and non-motor features. The dataset has been collected from the Parkinson's Progression Markers Initiative (PPMI) website [7] which consists of a motor and non-motor features dataset from which some features of motor and non-motor symptoms are used in the development of an ML based early prediction of PD model. The prediction model is a binary class classification problem.

In this chapter, different motor features (captured by MDS-UPDRS Total, MDS-UPDRS Part I, MDS-UPDRS Part I – Patient questionnaire, MDS-UPDRS Part II – Patient questionnaire, MDS-UPDRS Part III [8], Hoehn and Yahr [9], and Modified Schwab & England ADL[10]) and non-motor features (captured by University of Pennsylvania Smell Identification Test (UPSIT) [11], Montreal Cognitive Assessment (MoCA) Score [12], Geriatric Depression Scale Score (GDS) [12], Scales for Outcomes in Parkinson's Disease – Autonomic Dysfunction (SCOPA-AUT) [14] and Neuroimaging markers Single-photon Emission Computed Tomography (SPECT) Striatum binding ratios (SBR) – Left, Right caudate and Left, Right putamen [8] are used. Further models will be developed for the early prediction of PD using different machine learning techniques such as Support Vector Machine [11], Random Forest [11], K-NN [13], Bayesian Network [11] and Decision Tree [8]. Finally, results are compared between different models.

The rest of the chapter is organized as follows: Section 9.2 highlights the recent year's published review of related literature. Section 9.3 deals with the materials and methods where pre-processing of data, normalization, step by step procedure of early prediction of PD, feature description, and classification is explained. In Section 9.4, model descriptions are discussed. Results and discussions are given in Section 9.5. Finally, in Sections 9.6 and 9.7, the conclusion and future scopes are discussed respectively.

9.2 REVIEW OF RELATED LITERATURE

Recently, many researchers applied different Machine Learning techniques for early prediction of Parkinson's disease. Also, they have used different classifiers to classify healthy control (HC) subjects and PD subjects. The summary of their research work is described by year in Table 9.1.

9.3 MATERIALS AND METHODS

9.3.1 DATASET COLLECTION

The dataset is downloaded from the Parkinson's Progression Markers Initiatives website (www.ppmi-info.org.data) [7] which is doing a longitudinal, observational, multi-centre, multi-focus international study with an aim to assess the different features (clinical, neuroimaging, biospecimen, motor and non-motor) to identify whether the subjects are Parkinson Disease (PD) patients or healthy control (HC) (Figures 9.1 and 9.2).

The baseline summary dataset is downloaded from the PPMI website from which sixteen different features out of twenty-one are used for the simulation. The details of the features are given in Tables 9.2 and 9.3.

A total of 541 subjects are considered for simulation purpose where 408 PD (Female – 140, Male – 268) and 133 HC (Female – 42 and Male – 91) subjects are there.

9.3.2 FEATURE DESCRIPTIONS

1. **Movement Disorder Society-Unified Parkinson's Disease Rating Scale (MDS-UPDRS)**

   The MDS-UPDRS is a rating scale test which has four parts: Part I (non-motor experiences of daily living), Part II (motor experiences of daily living), Part III (motor examination) and Part IV (motor complications). Part I has two components: (IA) concerns various practices that are surveyed by the examiner, with all relevant data from patients and guardians, and (IB) is finished by the patient with or without the guide of the guardian, however autonomously from the specialist. These segments can, not withstanding, be investigated by the rater to guarantee that all inquiries are addressed unmistakably and the rater can help clarify any apparent ambiguities. Part II is a self-administered questionnaire like Part IB, but can be looked into by the specialist to guarantee fulfillment and clearness. The official versions of Part IA, Part IB, and Part II do not have separate "ON" or "OFF" ratings. However, for individual programs, the same questions can be used separately for "ON" and "OFF". Part III has guidelines for the rater to give or show to the patient; it is finished by the rater. Part IV has directions for the rater and guidelines to be read to the patient. This part incorporates inferred data, with the rater's clinical perceptions and decisions, and is finished by the rater. [25]

TABLE 9.1
Related Literature Review by Year

Authors	Year	Dataset Used	Data Extraction	Features Used	Feature Selection Algorithms	Models Used	Result Obtained
Mohammad R. Salmanpour et al. [8]	2020	204 PD subjects (149 M, 55 F)	From PPMI website	18 clinical features (MDS-UPDRS, part-I,II and III, (1-6), demographics(7-8), DAT SPECT images, Left, Right Putamen, Left, Right Caudate (9-16) and disease duration (17-18)	Genetic Algorithm (GA), Ant Colony Optimization (ACO), Particle Swarm Optimization (PSO), Simulated Annealing (SA), Deferentail Evolution (DE) and Non-dominated Sorting Genetic Algorithm NSGAII	Predictor Algorithms – LOLIMOT(Local Linear Model Trees), Radial Basis function(RBF), Multilayer perceptron-backpropagation (MLP-BP), Least Absolute Shrinkage & Selection Operator-Least angle Regression (LASSOLAR), Random Forest (RFA), Recurrent Neural Network(RNN), Bayesian Ridge Regression(BRR), Decision Tree Classification (DTC), Passive Aggresive Regression (PAR), Thiel-Sen Regression and Adaptive Neuro Fuzzy Inference System (ANFIS)	LOLIMOT predictor –lowest Absolute Error (4.32±0.19), in all features (4.15±0.46)
Zehra Karapinar Senturk [15]	2020	Total 195 voice dataset, speech signal of 31 people (23 PD and 8 healthy Control Group)	UCI Machine Learning Repository	MDVP:Fo (Hz) – Average vocal fundamental frequency, MDVP:Fhi(Hz)- Maximum vocal fundamental frequency, MDVP:Jitter (Abs) fundamental frequency etc	Univariate Selection(US), Recursive Feature Elimination (RFE) and Feature Importance (FI)	Classification and Regression Trees (CART), Support Vector Machine (SVM) and Artificial Neural Network (ANN)	Highest accuracy - SVM (93.84%), CART(90.76%) and ANN(91.54%)

Author	Year	Subjects	Data source	Features	Technique	Classifier/Method	Results
Yingchuan Chen et al. [16]	2020	Total 200 subjects (69 HC and 131 PD)	Beijing Tiantan Hospital, Capital Medical University, China	MRI features	Thalamus segment algorithm	Support Vector Machine (SVM)	Accuracy (95%) Sensitivity (97.44%) specificity (90.48%)
Filippo Cavallo et al. [17]	2019	30 HC(25 M, 5 F) 30 IH(Idiopathic Hyposmia (21 M, 9 F) 30 PD (25 M, 5F)	Data collected from SensHand V1 device	48 Upper limbs features	-	SVM, RF and NB	RF 0.97 F-measure, 0.97 SVM 79% accuracy
Mohammad R. Salmanpour et al. [18]	2019	184 total PD subjects (123 M, 61 F) average age (67.9±9.60)	From PPMI website	93 features used with MoCA score	Genetic Algorithm (GA), Ant Colony Optimization (ACO), Particle Swarm Optimization (PSO), Simulated Annealing (SA), Deferential Evolution (DE) and Non-dominated Sorting Genetic Algorithm NSGAII	Predictor Algorithms: LOLIMOT, RBF, MLP-BP, LASSOLAR, RFA, RNN, BRR, DTC, PAR, Thiel-Sen Regression and ANFIS	LOLIMOT (1.68±0.12) MLP-BP (2.12±0.15) RFA (2.36±0.17)
M. Wenzel et al. [19]	2019	645 total subjects (207 HC and 438 PD)	PPMI Website	FP-CIT SPECT (DatScan) images features	-	Deep Convolutional Neural Network Automated Anatomic Labelling(AAL) Hottest Voxels (HV)	PPMI-SBR (mean±SD) (0.970±0.017) AAL-SBR (0.957±0.017) HV-SBR (0.966±0.011) CNN (0.972±0.014)
T. I. Pedrosa et al. [13]	2018	16 total records High Amplitude Tremor(8 individuals)	Physionet Bank database website	Tremor level features	Fast Fourier Transform (FFT) Discrete Fourier Transform (DFT)	KNN and SVM	Both algorithms achieves Accuracy (92.85%)

(Continued)

TABLE 9.1
(Continued)

Authors	Year	Dataset Used	Data Extraction	Features Used	Feature Selection Algorithms	Models Used	Result Obtained
		Low Amplitude Tremor(8 individuals)					
Indira Rustempasic et al. [1]	2018	total voice -31 people, 23 (PD) 195 instances (48 normal, 147 PD)	Max Little of the University of Oxford, National Centre for Voice and Speech, Denver, Colorado	Biomedical voice measurement features	-	Fuzzy C-Means Clustering (FCM)	Accuracy (68.04%) Sensitivity (75.34%) Specificity (45.83%)
Haijun Lei et al. [12]	2018	Total 208 subjects (56 Normal Control, 123 PD, 29 SWEDD)	PPMI Website	Fractional Anisotropy(FA), Coefficient of diffusion weighted tensor imaging (DTI), Cerebrospinal fluid (CSF) and Gray matter (GM) of Magnetic Resonance Imaging, MoCA features	Least Absolute Shrinkage and selection operator method (LASSO), Elastic Net regularized regression method and Multi-modal Multi-task	Support Vector Regression (SVR) and Support Vector Classification (SVC)	MoCA score achieves best performance with multi-model feature GCD (0.661 corelation coefficient and 1.707 root mean squared error)
Bo Peng et al. [20]	2017	69 PD patients and 103 Normal Control patients	PPMI Website	T1-weighted Brain Magnetic Resonance images features (MRI)	Filter and Wrapper based feature selection method	Multilevel-Regions of Interest(ROI) feature based Machine learning algorithm, Multi-Kernal Support Vector Machine	Accuracy (85.78%) Specificity (87.79%) Sensitivity (87.64%)

Author	Year	Subjects	Source	Features	Statistical tools	Methods	Results
R. Prashanth et al. [21]	2016	Normal Subjects (183) and PD subjects (401)	PPMI Website	Non-motor features – Rapid Eye Movement(REM) Sleep Behaviour Disorder (RBD) and Olfactory loss, Cerebrospinal Fluid (CSF) measurements, UPSIT. SBR	Statistical tools – Mean and Standard Deviation	Naïve Bayes, Logistic Regression, Boosted Trees, Random Forests, SVM	SVM achieves highest result Accuracy 96.40% Sensitivity 95.01% Specificity 95.01%
K.N. Reddy Challa et al. [11]	2016	Total 586 subjects (184 Normal patients and 402 early PD subjects)	PPMI Website	Non-motor features – UPSIT, RBDSQ, CSF, Neuroimaging markers – Single-photon emission computed tomography (SPECT)	–	Multilayer Perceptron, Bayesian Network, Random Forest and Boosted Logistic Regression	Boosted Logistic Regression achieves high accuracy 97.159% and area under the ROC curve 98.9%
Gurpreet Singh et al. [22]	2015	831 T1-weighted MRIs data of total 831 subjects	PPMI Website 2011	MRIs features	Kohonen self organizing map feature extraction method (KSOM)	Least Squares Support Vector Machine (LS-SVM)	Accuracy (99%) confidence level (99.9%)
R. Prashanth et al. [23]	2014	Total 674 subjects (Normal 181 and Early PD 493)	PPMI Website 2013	Non-motor Striatal Binding Ration(SBR) features	–	SVM using RBF and Liner Kernel	Accuracy RBF (96.14%) Linear (92.28%)
Laura Silveira Moriyama et al. [24]	2008	106 PD and 118 Controls data for SS-16(Sniffin Sticks Identification Test) and 95 PD and 109 controls for UPSIT-40	General Hospital of the University of Sao Paulo Medical School, Brazil	Non-motor features University of Pennisylvaniz Smell Identification Test (UPSIT)	–	Multiple Linear Regression and Logistic Regression	SS-16 Specificity – 89.0% Sensitivity 81.1% UPSIT-40 Specificity – 83.5% Sensitivity 82.1%

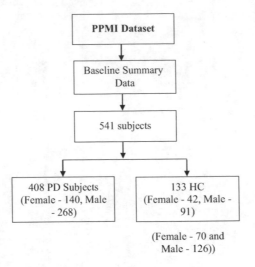

FIGURE 9.1   PPMI dataset.

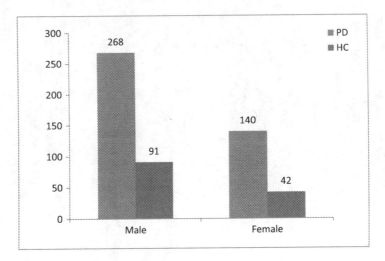

FIGURE 9.2   Graphical representation of dataset used in simulation.

2. **University of Pennsylvania Smell Identification Test (UPSIT – Total Score)**

The University of Pennsylvania Smell Identification Test is a test used to identify olfactory dysfunction in PD subjects. Olfactory dysfunction is a loss of the ability to detect odors. By smelling scratch-off blocks in a set of booklets, a subject can identify each smell. Most of the people suffering from olfactory loss have Parkinson's disease. The test does not mean that every person who is suffering from olfactory loss has PD [26]. UPSIT is one of the most reliable tests to identify PD. UPSIT test consist of four different booklets, each with ten pages: each page has a different odor. The subject

TABLE 9.2

Baseline Summary Features

S.No.	Feature Names	Description
1.	Subject Number	–
2.	Age(Years)	–
3.	Number of Relatives with PD	–
4.	Subject Diagnosis	–
5.	Years of Education	–
6.	Duration of Disease (Months)	–
7.	MDS-UPDRS Total	Movement Disorder Society-Unified Parkinson's Disease Rating Scale Total
8.	MDS-UPDRS Part I	Movement Disorder Society-Unified Parkinson's Disease Rating Scale Part I
9.	MDS-UPDRS Part I – Patient questionnaire	Movement Disorder Society-Unified Parkinson's Disease Rating Scale Part I – Patient Questionnaire
10.	MDS-UPDRS Part II - Patient questionnaire	Movement Disorder Society-Unified Parkinson's Disease Rating Scale Part II – Patient Questionnaire
11.	MDS-UPDRS Part III	Movement Disorder Society-Unified Parkinson's Disease Rating Scale Part II – Patient Questionnaire
12.	Hoehn & Yahr	–
13.	Modified Schwab & England ADL	Modified Schwab & England Activities of Daily Living
14.	UPSIT – Total Score	University of Pennsylvania Smell Identification Test Total Score
15.	MoCA Score*	Montreal Cognitive Assessment Score
16.	GDS Score	Geriatric Depression Scale Test Score
17.	SCOPA-AUT	Scales for Outcomes in Parkinson's Disease – Autonomic Dysfunction
18.	SBR – Left Caudate	Striatum binding ratios – Left Caudate
19.	SBR – Right Caudate	Striatum binding ratios – Right Caudate
20.	SBR – Left Putamen	Striatum binding ratios – Left Putamen
21.	SBR – Right Putamen	Striatum binding ratios – Right Putamen

scratches a particular block on a page and smells it. Each page consists of one question and four answer options from which to select one option, according to what they smell. The same process is repeated for each of the pages of the UPSIT booklet. After completion of the test, the UPSIT score is calculated by adding the scores. If all the odors are correctly identified, the maximum score is forty. The test takes only a few minutes, and an observer can predict whether a subject has PD. In this chapter, the UPSIT score is used for the prediction of PD collected from a baseline summary dataset, which is taken from PPMI [25].

3. **Montreal Cognitive Assessment Score (MoCA Score)**

The Montreal Cognitive Assessment Score looks for non-motor Parkinson disease symptoms. It is a quick screening instrument designed for the prediction of cognitive dysfunction that evaluates diverse psychological spheres, such as consideration and focus, leadership capacities, memory, language, visuoconstructional abilities, theoretical reasoning, computations, and

TABLE 9.3

Features (Motor, Non-Motor, and Neuro-imaging Markers) Used in Simulation Study

Sl.No.	Feature Names
	Motor Features
1.	MDS-UPDRS Total
2.	MDS-UPDRS Part I
3.	MDS-UPDRS Part I – Patient questionnaire
4.	MDS-UPDRS Part II – Patient questionnaire
5.	MDS-UPDRS Part III
6.	Hoehn & Yahr
7.	Modified Schwab & England ADL
	Non-Motor Features
8.	UPSIT – Total Score
9.	MoCA Score
10.	GDS Score
11.	SCOPA-AUT
	Neuroimaging Features/Biomarkers
12.	SBR – Left Caudate
13.	SBR – Right Caudate
14.	SBR – Left Putamen
15.	SBR – Right Putamen
	Target
16.	Subject Diagnosis (PD and HC)

PD – Parkinson's disease and HC – Healthy Control

direction [27]. Time to obtain the MoCA score is around ten minutes. The maximum score is thirty points. The MoCA test consists of eleven questions of various patterns, which a subject has to answer by using their IQ. These questions are – 1. Alternating Trail Making, 2. Visuoconstructional Skills (Cube), 3. Visuoconstructional Skills (Clock), 4. Naming, 5. Memory, 6. Attention, 7. Sentence repetition, 8. Verbal fluency, 9. Abstraction, 10. Delayed recall, and 11. Orientation. After completion of these questions, a total score is the sum of the answers for all the questions. The MoCA score is included in the battery of tests for the early prediction of PD and can be downloaded from PPMI [25].

4. **Geriatric Depression Scale Score (GDS Score)**

The Geriatric Depression Scale Score is used to test the depression level of subjects. It is a self-rating instrument test. This test consists of fifteen questions where subjects have to choose the best answer for how they felt about the past week. The answer is in the form of 0 and 1, where 0 represent "No," indicating depression for questions 1, 5, 7, 11, and 13, and 1 represents "Yes," indicating depression for questions 2, 3, 4, 6, 8, 9, 10, 12, 14, and 15. The range of depression is represented as: 0–4(Normal), 5–7(Slight Depression), 8–11 (Medium Depression) and 12–15 (Serious Depression) [12, 25].

5. **Scales for Outcome in Parkinson's Disease-Autonomic Dysfunction (SCOPA-AUT)**

Scales for Outcome in Parkinson's disease is a self-report questionnaire with twenty-three items of autonomic dysfunction. There are different questions related to upper and lower gastro-intestinal function, cardiocirculatory function, urinary function, sexuality, and other autonomic problems like sweating and light sensitivity, etc. All items of SCOPA-AUT are grouped into different domains, such as gastrointestinal functioning (7), thermoregulatory functioning (4), urinary functioning (6), pupillomotor functioning (1), cardiovascular functioning (3) and sexual function (2 for men, 2 for women). The maximum score for this questionnaire is sixty-nine [25].

6. **Single-photon Emission Tomography (SPECT Images)**

Single-photon emission tomography (SPECT) is generally used for the diagnosis of Parkinson disease. SPECT imaging uses I-Ioflupane, which provides information based on local binding of presynaptic dopamine transporters (DaTs) with I-Ioflupane. The specific binding ratio (SBR) is the ratio between the concentrations of the specific binding radioactivity of a target brain region with the non-specific binding radioactivity of the reference region. The specific binding ratio (SBR) is calculated from SPECT. This feature is generally used in the images. Some features are extracted and calculated using Left, Right Caudate and Left, Right Putamen. These features are considered in this research work.[8, 25]

9.3.3 DATA PRE-PROCESSING

The filtration, removal, and handling of noisy data are known as data pre-processing. In this process, missing values in the data are filled by zeros. A total of twenty-one variables are available in the dataset, from which sixteen variables are used as input for development of prediction models. The attribute "subject diagnosis" is taken as the target value both for PD and HC. The implementation of models is completed in MATLAB 16a. After preprocessing, the data undergo a normalization process.

Data Normalization

For the purpose of classification, there is a need to normalize the data. Without scaling, it is difficult to analyze the data. If the data are scaled properly, the classifier will perform better. So, to achieve this, the dataset is normalized by using the Z-Score normalization method (Figure 9.3).

Steps of Prediction of PD

Classification

Classification is the process of separating the data into different groups on the basis of their features. It is a supervised Machine Learning technique as the class label is available. The prediction of PD patients is a binary classification problem. The class labels "1" represent the PD subjects, and "0" represents the HC subjects. The classification has two phases: 1. Training phase and 2. Testing phase. During the training phase the classifiers are trained using the past data, until the error square is minimized. Once the

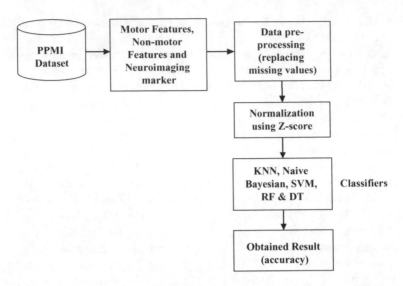

FIGURE 9.3　Step-by-step process of development of the early prediction of PD models.

error value is minimized, then testing of the classifier is done with the testing dataset, and performance of the classifier is obtained. In this study, ten-fold cross validation is performed.

Cross Validation method

Cross validation is a statistical method used to evaluate the performance of Machine Learning models. Details of the method as given in Figure 9.4:

1. **Hold-out Method:** Hold-out is the point at which the dataset is divided into 'train' and 'test' sets. The train set is used in model development, and the test set is utilized to perceive how well that model performs on concealed information. A typical split when utilizes the hold-out strategy, 70% of information is used for training and 30% of the information for testing [28].

2. **'K-fold Cross-validation:** 'K-fold cross-validation is the point at which the dataset is split into 'k' groups randomly. One of the group is utilized as the test set, and the rest are utilized as the training set. The model is prepared on the training set and scored on the test set. At that point the process continues until every group has been utilized as the test set. Ten-fold cross validation

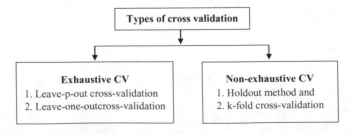

FIGURE 9.4　Types of cross validation.

is done in this research work. The dataset is split into ten groups containing an equal number of values, and the model is trained and tested ten times separately. Thus, each group has a chance for the test set [28].

9.4 MODEL DESCRIPTION

Different types of Machine Learning models are employed to estimate the performance of Parkinson's disease, and comparative analysis between them is also conducted. The models such as KNN, Naive Bayesian, Decision Tree, Support Vector Machine, and Random Forest are applied to develop different classifiers. Following are the descriptions of different models used in the research work.

K-Nearest Neighbor (KNN) is a supervised Machine Learning algorithm used for both classification and regression. It is a simple and easy to implement algorithm. KNN finds the nearest neighbors by calculating the Euclidean distance between the data points. KNN has two properties – (1) Lazy learning algorithm and (2) Non-parametric learning algorithm [29].

Naive Bayes (NB) is a probabilistic AI model that is utilized for classification tasks [29]. Utilizing Bayes hypothesis, this model discovers the likelihood of an occurrence, given that B has happened. Here, B represents evidence, and A represents hypothesis. The supposition made here is that the indicators/highlights are free. That is, the nearness of one specific element doesn't influence the other. Consequently, it is called Naïve. The Bayes equation is given as

$$P(A|B) = \frac{P(B|A)P(A)}{P(B)}$$

Support Vector Machine (SVM) is a supervised machine learning algorithm which aims to find a hyperplane in the N-dimensional space – a plane which has the maximum margin is to be chosen. Vectors are information focuses that are nearer to the hyperplane and impact the position and direction of the hyperplane. Utilizing these help vectors, the edge of the classifier is expanded. Erasing the help vectors will change the situation of the hyperplane. These are the focuses that assist in building the SVM [29].

Decision Tree (DT) is a tree-like structure used for classification and regression. It is a supervised Machine Learning algorithm used in decision making. The objective of utilizing a DT is to make a preparation model that can be used to foresee the class or estimation of the objective variable by taking in basic choice principles gathered from earlier data (training information). In DT, for anticipating a class name for a record, one has to start from the foundation of the tree. We look at the estimations of the root property with the record's characteristic. Based on correlation, one follows the branch and jumps to the next node [29].

Random Forest (RF) Random Forest is one of the supervised Machine Learning techniques. It is an ensemble of decision trees where it builds a "forest" and is trained using different methods, such as "bagging",

"AdaBoostM1", "LogitBoost," etc. LogitBoost is used in the implementation. The main advantage of Random Forest is that it is used for both classification and regression problems [29].

9.5 RESULT AND DISCUSSION

The percentage of accuracy obtained by the different Machine Learning based models is shown in Table 9.4. Figure 9.5 represents the graphical view of the comparison among the five models. It is shown that SVM and Random Forest have obtained the high accuracy result of 99.45% and 99.63% respectively, compared to the KNN and Naive Bayes models. The Decision Tree based model also has the accuracy of 99.02%. Out of these five models, Random Forest is the best with the highest accuracy and ranked number one, followed by SVM and DT, with ranks two and three, respectively. The Naïve Bayes and KNN based models are fourth and fifth.

TABLE 9.4

Comparative Analysis of PD Prediction Models Using Ten-fold Cross Validation

		Error During Testing			
S. No.	Model Name	Value	(%)	Accuracy	Rank
1.	KNN	0.0298	2.98%	97.02%	5
2.	Naive Bayes	0.0133	1.33%	98.67%	4
3.	SVM(with polynomial kernel)	0.0055	0.55%	**99.45%**	2
4.	Decision Tree	0.0098	0.98%	99.02%	3
5.	Random Forest	0.0037	0.37%	**99.63%**	1

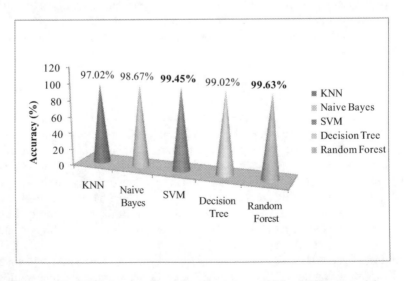

FIGURE 9.5 Graphical representation of comparative analysis of different models

9.6 CONCLUSION

In this chapter, for the early prediction of Parkinson's disease, different Machine Learning algorithms are used. The baseline summary dataset collected from the PPMI website is mainly used for simulation. A total of 541 subjects are taken from which 408 are PD subjects and 133 are HC subjects. Sixteen variables are selected out of twenty-one available features for model development. Missing values are handled by imputing zeros. Five different Machine Learning algorithm based models are developed for early prediction of Parkinson's disease and comparative analysis is done among them. It is observed from comparative analysis that SVM and Random Forest are the best algorithms, obtaining the most accurate results as compared to the other three models used for the simulation study.

9.7 FUTURE WORK

Early prediction of Parkinson's disease is still a challenging task. Day to day, different researchers have been doing new research in this field. Researchers try to develop different new models, using Machine Learning techniques to obtain more accurate results. Now, there is a need of applying new techniques, such as swarm intelligences, deep learning, semi supervised learning, quantum computing, etc., for the early prediction of Parkinson disease. Also, there is a need to find out new features for the correct diagnosis of the PD patients.

REFERENCES

1. Indira Rastempasic and Mahmet Can. Diagnosis of Parkinson's Disease using Fuzzy C-Means Clustering and Pattern Recognition. *Southeast Europe Journal of Soft Computing*. 2. 2013. doi:10.21533/scjournal.v2i1.44.
2. N. Sietsk Heyn, *Parkinsons Disease*, Medically reviewed on January 24 2020. http://www.medicinenet,com/parkinsons_disease/article.htm
3. https://www.healthline.com/health/parkinsons
4. Dorsey, E.R., and B.R. Bloem. The Parkinson pandemic – a call to action, *JAMA Neurology* 75 (2018) 9–10, doi:10.1001/jamaneurol.2017.3299
5. Fahn, S. Clinical aspects of Parkinson disease, in: R. Nass, S. Przedborski (Eds.), *Park. Dis. Mol. Ther. Insights from Model Syst*, 1st ed., Elsevier Inc., (2008), pp. 3–48.
6. Schapira, A.H., R.K. Chauduri, and P. Jenner, Non-motor features of Parkinson disease, *Nature Reviews Neuroscience* 18 (2017) 435–450.
7. www.ppmi-info.org
8. Salmanpour, Mh. R., M. Shamsaei, A. Saberi, I. S. Klyuzhin, J Tang, V. Sossi and A. Rahmim, Machine learning methods for optimal prediction of motor outcome in Parkinson's disease, *Physica Medica*, Elsevier, 69 (2020) 233–240, doi:10.1016/j.ejmp.201912.022.
9. Martinez-Martin, Pablo, Luis Prieto, and Maria Joao Forjaz, Longitudinal metric properties of disability rating scales for Parkinson's disease, International Society for Pharmacoeconomics and Outcomes Research (ISPOR), Value in health 9, 6 (2006) 386–393, doi: 10.1111/j.1524-4733.2006.00131.x
10. Siderowf, A. Schwab and England activities of daily living scale, Encyclopedia of Movement Disorders. (2010), doi:10.1016/B978-0-12-374105-9.00070-8.

11. Challa, Kamal Nayan Reddy, Venkata Sasank, Pagolu, Ganapati Panda and Babita Majhi, *An improved approach for prediction of Parkinson's Disease using machine learning techniques*, International conference on Signal Processing, Communication, Power and Embedded System (SCOPES), Paralakhemundi, India. 2016.

12. Lei, H., Z. Huang, T. Han, Q. Luo, Y. Cai, G. Liu and B. Lei, Joint regression and classification via relational regularization for Parkinson's disease diagnosis, *Technology and Health Care*, 26 (2018) S19–S30, doi:10.3233/THC-174540.

13. Pedrosa, T.I., F.F. Vasconcelos, L. Medeiros and L.D. Silva, Machine learning application to quantify the tremor level for Parkinson's disease patients, *Procedia Computer Science, Science Direct*, Elsevier, 138 (2018) 215–220, doi:10.1016/j.procs.2018.10.031.

14. Bostantjopoulou, S., Z. Katsarou, I. Danglis, H. Karakasis, D. Milioni, and C. Falup-Pecurariu. Self-reported autonomic symptoms in Parkinson's disease: properties of the SCOPA-AUT scale, *Hippokrata*, 20(2) (2016) 115–120.

15. Senturk, Zehra Karapinar. Early diagnosis of parkinson's disease using machine learning algorithms, *Medical Hypotheses*, Elsevier, 138 (2020) 109603, doi: 10.1016/j.mehy.2020.109603

16. Chen, Y., G. Zhu, D. Liu, Y. Liu, T. Yuan, X. Zhang, Y. Jiang, T. Du and J. Zhang, The morphology of thalamic subnuclei in parkinson's disease and the effects of machine learning on disease diagnosis and clinical evaluation, *Journal of the Neurological Sciences*, 411 (2020) 116721. doi:10.1016/j.jns.2020.116721.

17. Cavallo, F., A. Moschetti, D. Esposito, C. Maremmani and E. Rovini, Upper limb motor pre-clinical assessment in Parkinson's disease using machine learning, *Parkinsonism and Related Disorders*, Elsevier, 63 (2019) 111–116, doi:10.1016/j.parkreldis.2019.02.028.

18. Salmanpour, Mh. R., M. Shamsaei, A. Saberi, I. S. Klyuzhin, J. Tang, V. Sossi and A. Rahmim, Optimized machine learning methods for prediction of cognitive outcome in Parkinson's disease, *Computers in Biology and Medicine*, Elsevier, 111 (2019) 103347, doi:10.1016/j.compbiomed.2019.103347.

19. Wenzel, M., F. Milletari, J. Kruger, C. Lange, M. Schenk, I. Apostolova, S. Klutmann, M. Ehrenburg and R. Buchert, Automatic classification of dopamine transporter SPECT: deep convolutional neural networks can be trained to be robust with respect to variable image characteristics, *European Journal of Nuclear Medicine and Molecular Imaging*, Springer, (2019) doi:10.1007/s00259-019-04502-5.

20. Peng, Bo, Suhong Wang, Zhiyong Zhou, Yan Liu, Baotong Tong, Tao Zhang, and Yakang Dai, A multilevel-ROI-features-based machine learning method for detection of morphometric biomarkers in Parkinson's disease, *Neuroscience Letters*, 651 (2017) doi:10.1016/j.neulet.2017.04.034.

21. Prashanth, R., S. D. Roy, P.K. Mandal and S. Ghosh, High-accuracy detection of early parkinson's disease through multimodal features and machine learning, *International Journal of Medical Informatics*, Elsevier, 90 (2016) 13–21, doi:10.1016/j.ijmedinf.2016.03.001.

22. Singh, Gurpreet, and Lakshminarayan Samavedham. Unsupervised learning based feature extraction for differential diagnosis of neurodegenrative diseases: A case study on early-stage diagnosis of Parkinson disease, *Journal of Neuroscience Methods*, Elsevier, 30 (2015). doi:10.1016/j.jneumrth.2015.08.011.

23. Prashanth, R., S. D. Roy, P.K. Mandal and S. Ghosh. Automatic classification and prediction models for early parkinson's disease diagnosis from SPECT imaging, *Expert Systems with Applications, Elsevier*, 41 (2014) 3333–3342, doi:10.1016/j.eswa.2013.11.031.

24. Silveira-Moriyama, Laura, Margarete de Jesus Carvalho, Regina Katzenschlager, Aviva Petrie, Ronald Ranvaud, Egberto Reis Barbosa, and Andrew J. Lees. The use of smell identification tests in the diagnosis of Parkinson's disease in Brazil, *Movement Disorders*, Wiley InterScience, 23, 16 (2008) 2328–2334, doi:10.1002/mds.22241

25. http://www.ppmi-info.org/wp-content/uploads/2017/05/PPMI-General-Operations-Manual.pdf

26. Posen, M.M., D. Stoffers, J. Booij, B.L. Van Eck-Smit, E.C. Wolters, and H.W. Berendse. Idiopathic hyposmia as a preclinical sign of parkinson's disease, *Annals of Neurology*, 56 (2004) 173–181.

27. Camargo, C.H. Ferreria, Ed. Souza Tolentino, Comparison of the use of screening tools evaluating cognitive impairment in patients with Parkinson's disease, *Dement Neuropsychol*, 10(4), (2016) 344–350.

28. https://medium.com/@eijaz/holdout-vs-cross-validation-in-machine-learning-7637112d3f8f

29. Han, Jiawei, Micheline Kamber and Jian Pei. *Data Mining, Concepts and Technology*, Waltham, MA: Elsevier, 3rd edition.

Part II

Deep Learning Techniques in Biomedical and Health Informatics

10 Deep Neural Network for Parkinson Disease Prediction Using SPECT Image

Biswajit Karan, Animesh Sharma,
and Sitanshu Sekhar Sahu
Birla Institute of Technology, Mesra, Ranchi, India

Sudhansu Kumar Mishra
Birla Institute of Technology, Mesra, Ranchi, India

CONTENTS

10.1 INTRODUCTION

Parkinson's disease (PD) is a progressive neuro related disorder [1]. It is the most common neuron related disease after Alzheimer's [2]. It affects about ten million people worldwide with 1–2 % of the total population in the age group of fifty to sixty years [3]. The neurological progress of PD patients is evaluated by the unified Parkinson disease rating scale (UPDRS) and the Hehn & Yahr staging scale, which includes motor and non-motor symptoms [4, 5]. PD patients show many motor and non-motor types of symptoms, such as tremor, speech disorder, sleeping disorders, and difficulty in muscular movement. Currently, the concrete clinical diagnosis is based on bradykinesia, rigidity, and tremor at rest. The clinical diagnosis gives

DOI: 10.1201/9780367548445-12

159

accurate results in the mature stage of the disease. However, in the mild stage, an accurate assessment is difficult [6]. Many signal processing and Machine Learning techniques are used for PD detection. From different literature surveys it was found that speech signal, handwriting task (spiral drawing), and SPECT images can be used for the PD analysis and detection shown in Figure 10.1.

Using speech signal, PD detection has been performed by many authors [3–10] from the last two decades. Most of the authors used the signal processing algorithm along with the Machine Learning method. The details of studies are summarized in Table 10.1.

Some studies related to PD detection have been performed using hand-writing tasks. These studies have utilized handwriting images for analysis and detection of PD. The summary and important findings are presented in Table 10.2.

Some related studies of SPECT image for PD detection are presented in the following section. Klyuzhin, Ivan, et al. [17] used deep convolutional neural networks based features of DaTscan SPECT images to predict PD progression. R. Prashanth et al. [18] has developed a model for automatic classification and prediction for PD. Authors have utilized Striatal Binding Ratio (SBR) values of SPECT scans and support vector machine (SVM) and logistic regression in the model building process and obtained an accuracy of 96 % in classifying early PD and healthy subjects. Adeli, Ehsan, et al. [19] proposed robust diagnosis method, joint feature-sample selection

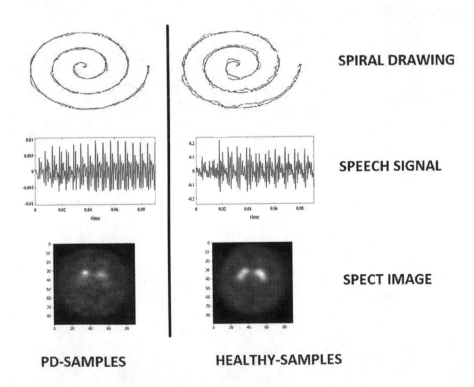

FIGURE 10.1 Biomedical Signal Used for PD Detection.

TABLE 10.1
Summary of Studies for PD Detection and Analysis Using Speech Database

References	Database	Signal Processing Techniques Used	Classifier Used	Findings
Little et al. [3]	Speech database	Acoustic features	SVM	91.4% accuracy is obtained with sustained vowels.
A.Tsanas et al. [4]	Speech database	Dysphonia measures	SVM	99% accuracy is obtained with sustained vowels.
Sakar et al. [5]	Speech database	Acoustic features	SVM	82.4% accuracy is obtained with sustained vowels.
Orozco-Arroyave et al. [7]	Speech database	Spectral and cepstral measures	SVM	76% accuracy with sustained vowel and 89% with word.
C. O Sakar et al. [8]	Speech database	Tunable Q-factor wavelet based cepstral coefficients	SVM	83% accuracy with sustained vowel.
Vásquez-Correa et al. [9]	Speech database	Intrinsic mode based features	SVM	76% classification accuracy using Diadochokinetic Task
Karan et al. [10]	Speech database	Hilbert spectrum based features	SVM	90% accuracy with sustained vowel and 91% with word

TABLE 10.2
Summary of Studies Related to Handwriting Image for PD Analysis and Detection

References	Database	Signal Processing Techniques Used	Classifier Used	Findings
P.Drotar et al. [11]	Hand-writing image	pressure exerted by person on writing surface acts as feature	SVM	81.3% accuracy is obtained
M.Diaz et al. [12]	Hand-writing image	CNN based features	SVM	89.7% accuracy is obtained.
Souza et al. [13]	Hand-writing image	Structural co-occurrence matrix based features	SVM	85.54% accuracy is obtained.
Taleb et al. [14]	Hand-writing image	CNN-BLSTM based features	Softmax	97.6% accuracy.
P.Sharma et al. [15]	Hand-writing image	Modified grey wolf optimization technique	----	94.83% accuracy.
Vásquez-Correa et al. [16]	Hand-writing image	Deep learning-based features	Softmax	97.6% accuracy

(JFSS) for PD detection using MRI data. JFSS does optimal selection of subset of samples and features. These optimal features have given a reliable diagnosis model. Jesús, et al. [20] proposed a new computer aided system based on feature extraction and classification of the DaTSCAN imaging. They reported accuracy up to 94.7% and 91.3% with two different databases. Rojas et al. [21] used empirical mode decomposition (EMD) based features obtained from SPECT images utilized to classification of PD. Accuracy of 95 % is reported by the authors. Thomas J et al. [22] have proposed a model based on enhanced probabilistic neural network (EPNN) for diagnosis of PD based on three types of features related to motor, non-motor, and neuro-imaging The other studies in [23–28] used the SPECT image for PD detection and found remarkable result. It was observed in different study in [29–34] that the unconventional method using different biosignal like speech, handwriting and SPECT image add useful results in the diagonosois of PD.

In this chapter, Convolutional Neural Networks (CNN) is used for binary classification as well as multiclass classification. SPECT images along with CNN proves to be a faster approach for classification of PD and healthy subjects. This method works directly on raw image data and CNN extract features from the image. It functions as an automatic and intelligent feature extractor using filters, convolution operation, and pooling operation.

10.2 METHODOLOGY

10.2.1 DATABASE

Researchers currently believe that dopamine loss can be detected by single-photon emission computed tomography (SPECT). The data used in our studies has been collected from the Parkinson's Progression Markers Initiative (PPMI) database at https://www.ppmi-info.org. We downloaded data of ninety-eight PD, ninety-six SWEDD and 102 healthy patients' SPECT images (Figure 10.2).

The SPECT imagining of PD, healthy, and SWEDD subjects are shown in Figure 10.3. It is noticed that it is easy to differentiate between PD and healthy subjects but difficult to distinguish between healthy and SWEED.

10.2.2 IMAGE PROCESSING

Since we are working directly on images and wanted to show the power of CNN for working on images, we kept the image processing task as minimal as possible.

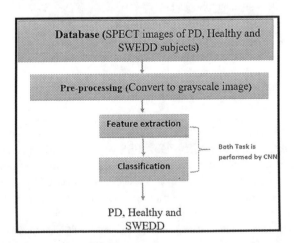

FIGURE 10.2 Workflow of SPECT image classification using CNN.

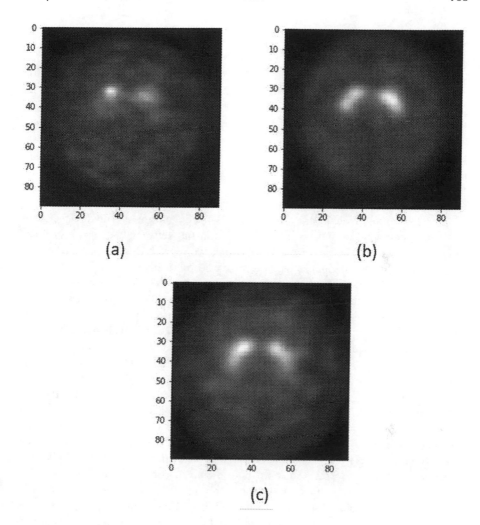

FIGURE 10.3 (a), (b) and (c) are the SPECT images of PD, Healthy and SWEDD subjects respectively.

SPECT scan consists of ninety-one transaxial slices, out of which we had to select one slice for a patient, as many slices were irrelevant for our study since no clear dopaminergic effect was visible. We found out that around thirty-five to forty-eight slices show relevant dopaminergic effect, Thus, we chose to select the forty-second slice for further analysis. This slice was further converted to a grayscale image as retaining the color channels of the image was meaningless since it had a black background and a white region depicting the dopaminergic effect. Thus, we got rid of the color channels to make our CNN classifier computationally less expensive. Since the original image is already similar to grayscale, converting it to grayscale did not remove any important information that we could exploit.

10.2.3 Methodology

10.2.3.1 Convolutional Neural Networks (CNN)

Convolutional Neural Networks: The human brain basically looks for features while trying to categorize an image. Depending upon the types and number of features it recognizes, the brain switches among one or more options. A classic example of this is a picture depicting a creature that appears to be a rabbit to some people and a duck to others. The different areas of the image have different features. Based on those, the brain can classify the image as one of the two possibilities [35]. The aim of Convolutional Neural Networks is to act like a human brain while categorizing images. The basic function is taking an image as input to the CNN which gives the output as an image class. The first step is the convolution operation.

The CNN architecture is shown in Figure 10.4. Convolution is a fundamental tool in signal processing for understanding the relationship between two signals or functions and how one affects the other [36]. First, we represent our input image as a matrix consisting of ones and zeros. This matrix is applied in a convolution operation with a feature detector, also known as a kernel or filter. A feature detector is also a square matrix. The choice of dimensions is arbitrary. The filter is applied to the input image and we obtain a value that represents how many matching elements are present in the two matrices as a sum of the products of the corresponding elements. The step size is called the stride and is the distance between consecutive matching operations, measured in pixels. Conventionally, people select two pixels as the stride. The result is a convolved feature, or activation map or feature map. One of the important functions of convolution is to reduce the size of the image. The reduction is directly proportional to the size. This makes the processing of an image faster. There is a risk to lose some information; however, we get an idea of where a particular feature is dominant in the image [37]. Our classification is driven by features. Hence, the importance of the convolution step is paramount. Also, the spatial relationship between pixels is preserved. We have several feature maps to create the first convolution layer. The network decides on some important features through its training, which results in a number of feature maps. To get each feature map, a different filter was applied to the input image. We are filtering the input image through the numerous filters, but our main goal is to detect the features represented by the feature detectors. The convolution step is followed by the rectifier function. The purpose is to increase non-linearity in the network. Images are highly non-linear, because of the presence of different elements, colors, and backgrounds. The transitions between pixels might be

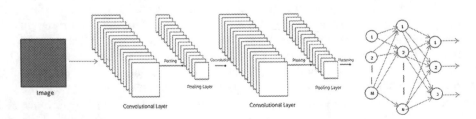

FIGURE 10.4 CNN architecture.

non-linear. Convolution introduces linearity in the image; therefore, a rectifier function is used afterwards [38].

Neural networks will have problems classifying the images if the features it is looking for are tilted or oriented in a different direction compared to the features the network determined from other images. Therefore, the network must have spatial invariance. In other words, the CNN does not attach importance to the differences in the feature position in terms of sense of direction, orientation, or texture if it finds the feature in the image. If the detected feature is rotated or squashed as compared to the feature being detected, the network should find it nonetheless. Hence, we apply pooling. A square is taken and applied to the feature detector. The size of the box and the stride in moving the box are chosen as needed. We are removing some information that does not have importance to the feature, since we only take into account the maximum value in the box and disregard the other pixels. However, we are preserving the information about the features since the maximum values represent the position where we found the entity at its closest point to a feature. Also, because we are taking the maximum values, we are accounting for distortions in the feature in different images. As a result, we are able to preserve the features, implement spatial invariance and most importantly, reduce the size. We are reducing the number of parameters that go in our final layers. Reducing the information is beneficial since it decreases the possibility of overfitting. There are several types of pooling: max pooling (described above), sum pooling, average pooling. Pooling is also known as down sampling. Flattening involves taking a pooled feature map from the pooling layer and converting it into a column, row by row. The column is the input for an Artificial Neural Network (ANN) for the purpose of optimizing the process further. We add a whole ANN to our CNN. Fully connected layers are a special type of hidden layer. For classification problems, we need to have an output, or neuron, per class. After the prediction is made, the error is calculated with regards to the true value, and then it is backpropagated through the network. The feature detectors and the weights are adjusted accordingly, and the process is repeated.

Experimental paradigm: The experimental paradigm of the proposed method is shown in Figure 10.5. The total samples are divided into 80% training and 20% testing samples. The model is developed from 80% of the samples, while the remaining 20% is used for testing the performance of the model.

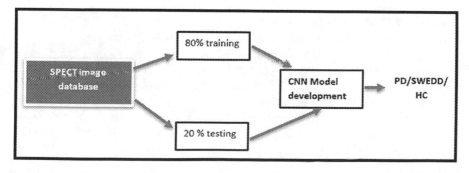

FIGURE 10.5   Experimental paradigm.

Classification Metrics: The testing performance is represented by following parameters.

Let us define some terminologies before describing the various metrics.

- TP (True Positive): Sample with true value 1 and predicted as 1
- TN (True Negative): Sample with true value 0 and predicted as 0
- FP (False Positive): Sample with true value 0 and predicted as 1
- FN (False Negative): Sample with true value 1 and predicted as 0

1. Accuracy: It is the ratio of the number of correctly predicted samples and total number of samples [39].

$$\text{Accuracy} = (\text{TP} + \text{TN}) / (\text{TP} + \text{TN} + \text{FP} + \text{FN}) \qquad (10.1)$$

2. Precision: It is the ability of a classifier to not predict a sample as positive if it is actually negative.

$$\text{Precision} = \text{TP} / (\text{TP} + \text{FP}) \qquad (10.2)$$

3. Recall: It is the ability of a classifier to correctly predict all positive samples.

$$\text{Recall} = \text{TP} / (\text{TP} + \text{FN}) \qquad (10.3)$$

4. AUC: It is the area under the ROC curve. ROC stands for Receiving Operator Characteristic and is the curve of True Positive Rate (Recall) vs. False Positive Rate (FP / (FP + TN)). It is the probability that the classifier ranks a random positive sample higher than a random negative sample. It provides an aggregate measure of how well a classifier performs for all possible classification thresholds.

10.2.3.2   Results and Discussion

For classifying these images, we split our data into training and testing.

We used the training data to train our CNN classifier by passing the SPECT images after converting them into grayscale. Since the images were originally in a grayscale fashion, it was intuitive to get rid of the RGB color channels and use the grayscale format. We used Tensorflow and Keras in Python to train our model. We passed the images into the CNN architecture discussed earlier, which was run for 50 epochs with a batch size of 32. LeakyReLU was used as an activation function in the convolution and hidden layers. Softmax activation was used in the hidden layer to predict the probability of an image belonging to one of the 3 classes, i.e., PD, SWEDD, and Healthy. We also used a constant learning rate reduction with a factor of 0.5 if the training accuracy did not improve after every 3 epochs to help our classifier converge faster. Finally, we saved our model using the H5 file format of Keras, through which we can easily make predictions without having to train the model again. The saved weights were then loaded and convolution was performed on some sample images to analyze the features that our CNN was able to extract and how it differentiated between the three classes. The results of this experiment have been shown in Figure 10.6. We can clearly observe from the convolution outputs that the CNN classifier was able to extract the Region of Interest in the image without us having to do it for the classifier. Thus, it works as an automatic feature extractor and classifier, eliminating the time consuming task of feature extraction from the image. We have also compared our CNN classifier with a SVM classifier by flattening the images into a large vector of one dimension and analyzed the results without applying any feature extraction. We wanted to compare how well these models perform on raw image data for prediction of Parkinson's disease. The results of applying the SVM and CNN classifier on the gray-scale SPECT images have been summarized in Table 10.3.

ROC Curves: To check the classifier performance, the region of convergence (ROC) is an alternative representation, plotted in terms of true positive rate (TPR) and false-positive rate (FPR).

From Figure 10.7 it is observed that CNN is superior to classifier SVM. The AUC value for PD is same for both CNN and SVM classifier. But for SWEDD and the healthy image, the performance of CNN is better than SVM.

10.2.3.3 Comparison with Related Work

We also attempted binary classification using CNN, i.e., classifying PD and SWEDD or Healthy patients, as we wanted to see how well CNN performs in comparison to other works that have been done in this area, most of which have attempted binary classification of patients using feature based methods. The results are presented in Table 10.4 and Figure 10.8.

The related works in this area have been tabulated in Table 10.5 with the size of data they have used, the methods that they have applied on their data, and their reported performance.

The prosed model performance is comparable with similar studies performed with SVM. In all the aforementioned papers, the authors had to extract some features from images through various techniques, for instance, SVD, GLCM, striatal binding ratio, shape analysis, surface fitting, etc. and then use those features as input to their classifier for prediction. In our prosed work, Convolutional Neural Networks performed both as a feature extractor and classifier. The high accuracy reported by our CNN model indicates that Deep Learning based classification can go a long way in helping to classify and arrive at an early diagnosis of Parkinson's disease using SPECT images.

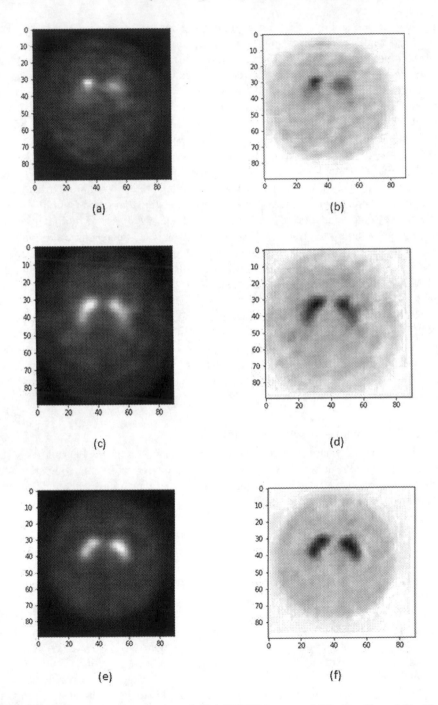

FIGURE 10.6 (a), (c), (e) are the original SPECT images of PD, Swedd, and Healthy patients respectively. (b), (d) and (f) are the output after applying the convolution operator by using the kernel learned by the CNN classifier while training.

TABLE 10.3

Various Classification Metrics for the Training and Test Data

Method	Accuracy (%)	Precision			Recall		
		PD (%)	Swedd (%)	Healthy (%)	PD (%)	Swedd (%)	Healthy (%)
SVM	76.67	100	14.28	92.31	98.72	50	66.67
CNN	88.33	100	92.86	76.92	100	68.42	95.24

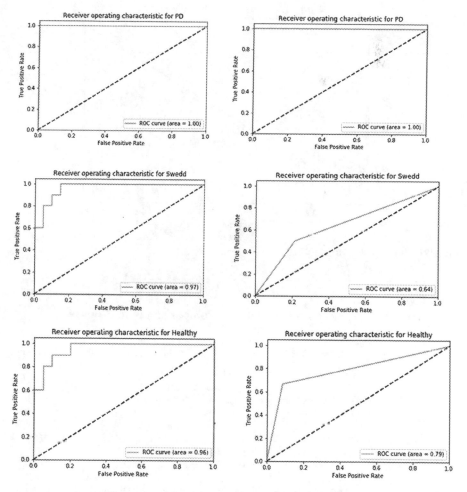

FIGURE 10.7 The first columns represents the ROC curve of CNN and second columns represents the ROC curve of SVM of three subjects: PD, SWEDD, and healthy.

TABLE 10.4
Performance of the Proposed Method in Binary Classes (PD and Control/SWEDD)

Parameters

Accuracy	98.33%
Precision	95.24%
Recall	100%
AUC	98.75%

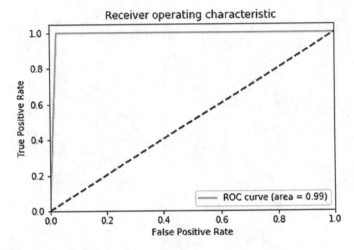

FIGURE 10.8 ROC Curve of Binary Classifier.

TABLE 10.5
Details of Previously Reported Result and Comparison with Proposed Approach

Past Study	Sample Size	Method Used	Accuracy
Segovia et al.	95 PD, 94 Control	voxels based features followed by classification using SVM.	94.7%
Illan et al.	100 PD, 108 Control	Voxels based features and SVM	96.81%
Prashanth et al.	369 PD, 179 Control	striatal binding ratio based features, and classification using SVM	96.14%
Oliveira et al.	445 PD, 209 Control	voxels based features followed by classification using SVM.	97.86%
Martinez-Murcia et al.	158 PD, 111 Control	Haralick texture based features SVM classifier	97.4%
Prashanth et al.	427 PD, 208 Healthy, 80 SWEDD	Shape Analysis and Surface Fitting based features and RBF-SVM as classifier.	97.29%
Proposed approach	98 PD, 96 Swedd and 102 Healthy patient's SPECT images	Convolution neural network based features	98.33%

10.3 DISCUSSION

The overall accuracy of the SVM model on test data is 76.67% and that of the CNN model is 88.33%. It can be observed from Table 10.1 that both the SVM and CNN models are able to classify the PD patients quite accurately, which is clearly seen by the precision and recall score for PD class. On the other hand, the precision and recall scores are not up to the mark for Healthy and SWEDD classes for SVM model. It is quite clear that the CNN model outperforms the SVM model in classifying the Healthy and SWEDD class. The precision score of Swedd for the SVM model is 14.28% and recall score is 50%, and the precision score of Healthy for SVM model is 92.31% and recall score is 66.67%, which makes it quite evident that it is not able to classify these two classes accurately. The precision score of SWEDD for the CNN model is 76.92% and the recall score is 68.42%, and the precision score of Healthy for CNN model is 95.24% and recall score is 66.67%. Since the gap between these two classes is minimal, SVM applied on raw image data without any feature extraction is not able to distinguish one from the other. CNN performs better than SVM but is still not able to classify SWEDD and Healthy patients with the same accuracy as it predicts PD. The Dopamine secretion for PD is quite low with respect to SWEDD and healthy. Thus, it is easily classified by both the models with 100% precision score for both CNN and SVM and recall score of 98.72% for SVM and 100% for CNN.

10.4 CONCLUSION

A convolution neural network (CNN) based classifier is proposed for Parkinson's disease identification. SPECT images can be directly fed into the CNN classifier for prediction. An accuracy of 88.33% is achieved when classifying patients into the classes of PD, Healthy, and SWEDD and an accuracy of 98.33% is produced when distinguishing between PD and Healthy or SWEDD. The CNN does not require any feature extraction or feature engineering and is faster in diagnosis of Parkinson's disease. The important inference of this study is that the SPECT image with CNN may be the good method for PD analysis and detection. Compared to speech and handwriting image, SPECT image gives an accurate identification of PD.

REFERENCES

1. Booij J, Tissingh G, Boer GJ, et al. "[123I]FP-CIT SPECT shows a pronounced decline of striatal dopamine transporter labelling in early and advanced Parkinson's disease." *Journal of Neurology, Neurosurgery, and Psychiatry.* 62(2): 133–140, (1997).
2. Dashtipour K, Tafreshi A, Lee J, and Crawley R. Speech disorders in Parkinson's disease: Pathophysiology, medical management and surgical approaches. *Neurodegenerative Disease Management.* 5(5): 337–348, (2018).
3. Little MA, McSharry PE, Hunter EJ, Spielman J, and Ramig LO. "Suitability of dysphonia measurements for telemonitoring of Parkinson's disease." *IEEE Transactions on Biomedical Engineering.* 56(4): 1015–1022, (2009).
4. Tsanas A, Little MA, McSharry PE, Spielman J, and Ramig LO. "Novel speech signal processing algorithms for high accuracy classification of Parkinsons disease." *IEEE Transactions on Biomedical Engineering.* 59: 1264–1271, (2012).

5. Sakar BE, Isenkul ME, Sakar CO, Sertbas A, Gurgen F, Delil S, et al. "Collection and analysis of a Parkinson speech dataset with multiple types of sound recordings." *IEEE J Biomed Heal Informatics.* 17: 828–834, (2013).
6. Booij J, and Knol RJJ. "SPECT imaging of the dopaminergic system in (premotor) Parkinson's disease." *Parkinsonism & Related Disorders.* 13: S425–S428, (2007).
7. Orozco-Arroyave JR, Hönig F, Arias-Londoño JD, Vargas-Bonilla JF, and Nöth E. "Spectral and cepstral analyses for Parkinson's disease detection in Spanish vowels and words." *Expert System.* 32: 688–697, 2015.
8. Sakar CO, Serbes G, Gunduz A, et al. "A comparative analysis of speech signal processing algorithms for Parkinson's disease classification and the use of the tunable Q-factor wavelet transform." *Applied Soft Computing.* 74: 255–263, (2019).
9. Vásquez-Correa, J. C., Rios-Urrego, C. D., Rueda, A., Orozco-Arroyave, J. R., Krishnan, S., and Nöth, E. Articulation and empirical mode decomposition features in diadochokinetic exercises for the speech assessment of Parkinson's disease patients. In *Iberoamerican Congress on Pattern Recognition* (pp. 688–696). Springer, Cham. (2019, October).
10. Karan, Biswajit, et al. "Hilbert spectrum analysis for automatic detection and evaluation of Parkinson's speech." *Biomedical Signal Processing and Control.* 61: 102050, (2020).
11. Drotár, Peter, et al. "Evaluation of handwriting kinematics and pressure for differential diagnosis of Parkinson's disease." *Artificial intelligence in Medicine* 67: 39–46, (2016).
12. Diaz, Moises, et al. "Dynamically enhanced static handwriting representation for Parkinson's disease detection." *Pattern Recognition Letters* 128: 204–210, (2019).
13. de Souza, João WM, et al. "A new approach to diagnose parkinson's disease using a structural cooccurrence matrix for a similarity analysis." *Computational Intelligence Neuroscience.* (2018) Article ID 7613282. doi:10.1155/2018/7613282
14. Taleb, Catherine, et al. "Detection of Parkinson's disease from handwriting using deep learning: a comparative study." *Evolutionary Intelligence* (2020): 1–12, (n.d.). doi:10.1007/s12065-020-00470-0
15. Sharma, Prerna, et al. "Diagnosis of Parkinson's disease using modified grey wolf optimization." *Cognitive Systems Research.* 54: 100–115, (2019).
16. Vásquez-Correa, Juan Camilo, et al. "Multimodal assessment of Parkinson's disease: A deep learning approach." *IEEE journal of biomedical and health informatics.* 23.4: 1618–1630, (2018).
17. Klyuzhin, Ivan, et al. "Use of deep convolutional neural networks to predict Parkinson's disease progression from DaTscan SPECT images." *Journal of Nuclear Medicine.* 59.Supplement 1: 29, (2018).
18. Prashanth, R., et al. "Automatic classification and prediction models for early Parkinson's disease diagnosis from SPECT imaging." *Expert Systems with Applications.* 41.7: 3333–3342, (2014).
19. Adeli, Ehsan, et al. "Joint feature-sample selection and robust diagnosis of Parkinson's disease from MRI data." *Neuro Image.* 141: 206–219, (2016).
20. Martínez-Murcia, Francisco Jesús, et al. "Automatic detection of Parkinsonism using significance measures and component analysis in DaTSCAN imaging." *Neurocomputing.* 126: 58–70, (2014).
21. Rojas, A., et al. "Application of empirical mode decomposition (EMD) on DaTSCAN SPECT images to explore Parkinson disease." *Expert Systems with Applications.* 40.7: 2756–2766, (2013).
22. Hirschauer, Thomas J., Hojjat Adeli, and John A. Buford. "Computer-aided diagnosis of Parkinson's disease using enhanced probabilistic neural network." *Journal of medical systems.* 39.11: 179, (2015).

23. F. Segovia, J. M. Gorriz, J. Ramirez, I. Alvarez, J. M. Jimenez-Hoyuela, and S. J. Ortega, "Improved parkinsonism diagnosis using a partial least squares based approach." *Medical Physics*. 39: 4395–4403, (2012).

24. I. A. Illan, J. M. Gorrz, J. Ramirez, F. Segovia, J. M. Jimenez-Hoyuela, and S. J. Ortega Lozano, "Automatic assistance to Parkinson's disease diagnosis in DaTSCAN SPECT imaging." *Medical Physics*. 39: 5971–5980, (2012).

25. R. Prashanth, S. Dutta Roy, P. K. Mandal, and S. Ghosh, "Automatic classification and prediction models for early Parkinson's disease diagnosis from SPECT imaging." *Expert System with Appllications*. 41: 3333–3342, (2014).

26. F. P. Oliveira, and M. Castelo-Branco, "Computer-aided diagnosis of Parkinson's disease based on [123I] FP-CIT SPECT binding potential images, using the voxels-as-features approach and support vector machines." *Journal of Neural Engineering*. 12: 026008, (2015).

27. F. Martinez-Murcia, J. Górriz, J. Ramírez, M. Moreno-Caballero, M. Gómez-Río, and P. S. P. M. Initiative, "Parametrization of textural patterns in 123I-ioflupane imaging for the automatic detection of Parkinsonism." *Medical Physics*. 41: 012502, (2014).

28. R. Prashanth, S. Dutta Roy, P. K. Mandal, and S. Ghosh, "High accuracy classification of parkinson's disease through shape analysis and surface fitting in 123I-Ioflupane SPECT imaging" *IEEE Journal of Biomedical and Health Informatics*. 21. 3: 794–802, (2017).

29. C. G. Goetz, W. Poewe, O. Rascol, C. Sampaio, G. T. Stebbins, C. Counsell, N. Giladi, R. G. Holloway, C. G. Moore, G. K. Wenning, M. D. Yahr, and L. Seidl, "Movement disorder society task force report on the Hoehn and Yahr staging scale: Status and recommendations." *Movement Disorders*. 19. 9: 1020–1028, (2004).

30. Movement Disorder Society, State of the art reviews the Unified Parkinson's Disease Rating Scale (UPDRS), "Status and recommendations". In: *Movement Disorders*. 18. 7: 738–750, (2003).

31. M. C. de Rijk, "Prevalence of Parkinson's disease in Europe: A collaborative study of population-based cohorts". *Neurology*. 54: 21–23, (2000).

32. De Lau, L. M. and Breteler, M. M. "Epidemiology of Parkinson's disease". *Journal of Lancet Neurology*.5. 6: 525–535, (2006).

33. Madsen, Mark T. "Recent advances in SPECT imaging." *Journal of Nuclear Medicine*. 48.4: 661, (2007).

34. Noyce, Alastair John, Andrew John Lees, and Anette-Elconore Schrag. "The prediagnostic phase of Parkinson's disease." *Journal of Neurology, Neurosurgery, and Psychiatry*. 87.8: 871–878, (2016).

35. Guo, Xiaojie, Liang Chen, and Changqing Shen. "Hierarchical adaptive deep convolution neural network and its application to bearing fault diagnosis." *Measurement*. 93: 490–502, (2016).

36. Lo, Shih-Chung B., et al. "Artificial convolution neural network for medical image pattern recognition." *Neural Networks* 8.7-8 (1995): 1201–1214.

37. Lo, S.-C.B., et al. "Artificial convolution neural network techniques and applications for lung nodule detection." *IEEE Transactions on Medical Imaging* 14.4: 711–718, (1995).

38. Wang, Jianfeng, and Xiaolin Hu. "Gated recurrent convolution neural network for OCR." *31st International Conference on Neural Information Processing Systems*: pp. 334–343, (2017).

39. Karan, Biswajit, Sitanshu Sekhar Sahu, and Kartik Mahto. "Parkinson disease prediction using intrinsic mode function based features from speech signal." *Biocybernetics and Biomedical Engineering* 40.1: 249–264, (2020).

11 An Insight into Applications of Deep Learning in Bioinformatics

M. A. Jabbar
Vardhaman College of Engineering, Hyderabad, India

CONTENTS

11.1 INTRODUCTION

Hesper and Hogeweg first introduced the term Bioinformatics in 1970 [1]. Bioinformatics have emerged recently to study of biological sequences. Mathematization of biology led to the development of Bioinformatics.

Bioinformatics involves the development of new methods and software tools to understand the biological phenomena of biological data. Essential issues in bioinformatics are 1) Protein sequence prediction, 2) multiple sequence alignment, and 3) Phylogenic inferences [1]. Bioinformatics is also called a subfield of computational biology, which focuses on biological aspects.

DOI: 10.1201/9780367548445-13

Bioinformatics is useful for:

1. Expression and regulation of proteins and genes.
2. Understanding the basic fundamentals of evolution in molecular biology.
3. Studying the bio-pathways and networks that are important to systems biology.
4. Aiding in the modeling of DNA, RNA, protein structures, and the interactions of molecules [1].

Methods from bioinformatics are widely used to leverage observation-based biology. To solve the problems faced in the Bioinformatics field, Deep Learning methods offer powerful approaches for use. The Bioinformatics field [2] is dependent on artificial intelligence (AI) to handle the complex data [3]. Without the field of Bioinformatics, the modern developments in biology and medicine could not have been possible [4].

Deep Learning has emerged since the early 2000s [5]. Deep Learning is a powerful Machine Learning approach with roots in artificial neural networks (ANNs) and has been increasingly used in Bioinformatics to mine large and complex relationships hidden in biomedical and biological data. Unlike Machine Learning methods, Deep Learning can work well with imbalanced, heterogeneous, high dimensional data, and can produce better results than ML. Due to multiple hidden layers in Deep Learning, small changes in input data are not affected, making the Deep Learning model robust. Deep Learning has the power to capture multiple levels of data abstraction and processing power within the cells [6].

Multiple hidden layers in Deep Learning are responsible for mapping the relationship between the complex data. Deep Learning architectures (like Recurrent Neural Network, Convolution Neural Network, Deep, Deep Boltzmann Machine, Autoencoder, Generative adversarial networks, and Deep Belief Network) are widely used in different areas of Bioinformatics, such as gene expression, prediction of protein structure, and disease prediction, drug prediction, drug discovery, medical image recognition, gene annotation, and health care management.

This chapter will discuss Deep Learning models that can be applied in Bioinformatics and the challenges faced by them.

11.2   MODELS IN DEEP LEARNING

In this section we will provide the background of Deep Learning models, along with their applications.

11.2.1   CONVOLUTIONAL NEURAL NETWORKS (CNN)

The name of CNN is derived from its layers: convolutional and fully connected layers and pooling layers [7, 8]. CNN is the most widely used model in the field of image processing. The main key benefit of this algorithm is that it identifies the significant features without the guidance of humans. The architecture comprises a stack of convolutional layers, pooling, and fully connected layers in the sequence followed by each other. Features can be extracted by employing convolution layer and pooling

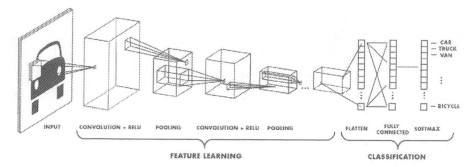

FIGURE 11.1 CNN architecture [9].

layer. Then these extracted features are mapped to the output by a fully connected layer. A CNN, due to the filters contained in it, can conquer both temporal and spatial dependencies of the image. CNN is the most widely used and accepted algorithm for the image dataset because of its dimensionality reduction feature. Figure 11.1 shows the architecture of CNN.

Convolution layer: Mathematically, it is defined as a function used to merge the data of two different kinds of groups. This layer in CNN will play an essential role; it is made up of a stack of operation functions such as convolutions and linear functions. It swaps the window across the image subsequently it calculates the input and dot product of pixels and a specific feature is detected then feature maps are produced [10]. This enables convolutions to highlight important features. Working of convolution is highlighted in Figure 11.2.

Pooling layer: This layer is applied to minimize the dimension of the image, i.e., dimensionality reduction is performed by this layer. Pooling can be done in two ways, depending on the requirement: one is max pooling and the other is avg pooling. Among these two functions, max-pooling is used the most. The outcome of max-pooling is the maximum value of an image wrapped by the kernel, whereas avg pooling yields the average value of an image wrapped by the kernel. A max pooling layer in addition to dimensionality reduction removes the noise from the image i.e., noise suppressant. How the pooling layer works is shown in Figure 11.3.

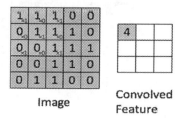

Image

Convolved Feature

FIGURE 11.2 How the pooling layer works [9].

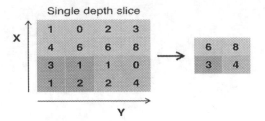

FIGURE 11.3   How the pooling layer works [9].

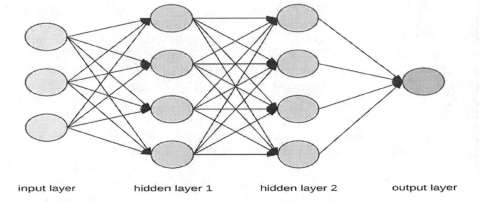

FIGURE 11.4   The architecture of the fully connected layer [9].

Fully connected layer: This layer is known as the feed-forward layer. These
layers are the final layers of the network. The output of the pooling layer is
fed as the input to this layer. The main function of this layer is to flatten the
output of the pooling layer and convert it into a single value. Then, weights
are applied to determine the appropriate label [11]. Figure 11.4 describes
the Fully connected layer.

11.2.2   RECURRENT NEURAL NETWORK (RNN)

RNN comprises hidden layers that are proficient in analyzing data flow [12]. This
technique can be implemented in various applications where output is computed
based on preceding computations, for example: text analysis, speech analysis, music
composition, and so on. RNN network is often supplied with training data that pos-
sesses firm interdependencies, in order to preserve the previous data. The drawback
of this network is it experiences a vanishing gradient problem. To eradicate this problem, an RNN variant was developed known as Long Short Term Memory (LSTM)
[13, 14]. How RNN works is shown in Figure 11.5.

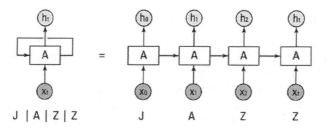

FIGURE 11.5 How RNN Works [15].

11.2.3 AUTOENCODER

An autoencoder is an unsupervised learning algorithm that employs a back-propagation technique equalizing target values to input values. This network is implemented to minimize the input size; if original data is needed, then it can be restored from compressed data [16–18]. The main aim of this network is to extract the coding and features through a data-driven technique. In the case of multi-dimension data, it is very difficult to load the entire data into the network, and it is a time-consuming process. Therefore, data compression and dimensionality reduction are necessary to compress the data so that it can be loaded without any loss of data [19, 20]. An autoencoder model compresses the data and can even encode the input information to small code. After certain processing steps, the small code is decoded to the input. The following figure demonstrates the architecture and working of autoencoders. Architecture of an Autoencoder is shown in Figure 11.6.

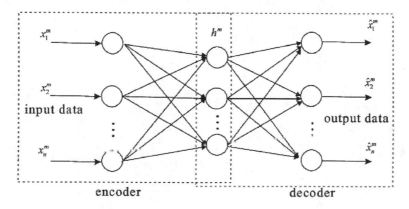

FIGURE 11.6 Architecture of an Autoencoder [21].

11.2.4 Deep Belief Network (DBN)

DBN is a type of DL algorithm developed by a stack composed of Restricted Boltzmann Machines (RBMs). In this network, every hidden layer is connected to the RBM layer. It is achieved, along with the greedy algorithm, by being the layer on top of unsupervised learning. Later, the algorithm is refined, relying on output [22]. Deep Belief Networks are utilized to identify the clusters, to create video sequences, and to capture motion data. A DBN network is shown in Figure 11.7.

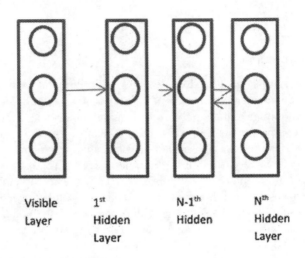

FIGURE 11.7   DBN network [23].

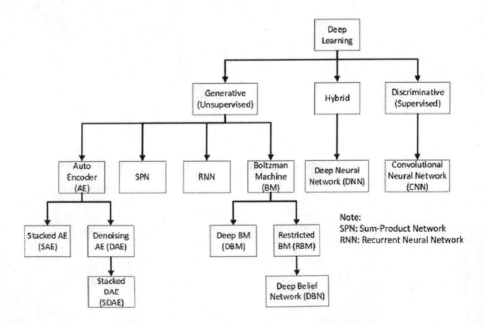

FIGURE 11.8   Classification of deep learning [25].

As seen in the above figure, DBNs' topmost two layers have indirect connections and the last layers have direct connections. Training is performed in a greedy method, along with fine-tuning of weight, to extract the features from the input image [24]. Classification of Deep Learning is shown pictorially in Figure 11.8.

11.3 DEEP LEARNING IN BIOINFORMATICS

Deep Neural Network (DNN) models include Convolution Neural Network (CNN), Recurrent Neural Network (RNN), Deep Autoencoders (DA), Deep Belief Networks (DBN's), Restricted Boltzmann Machine (RBM), LSTM, GAN, etc. Selecting the appropriate DNN for the task at hand is a challenging one.

Although various models of DNN exist, most DNNs are classified as:

1. DNN for supervised learning.
2. DNN for unsupervised learning.
3. Deep Neural Network for semi-supervised learning [26].

State-of-the-art applications of Deep Learning for biological data are discussed as follows:

11.3.1 DEEP LEARNING FOR OMICS

Deep Learning models are extensively used in Omics such as metabolomics, proteomics, and genomics.

Fakoor et al. [27] proposed Deep autoencoder (DA) based methods for gene expression data.

DA methods have been used for feature extraction in diagnosing cancer. The proposed method is used for the classification and detection of various cancer types, based on gene expression data.

In [28] authors applied a Stacked Denoising Autoencoder (SDAE) model for gene expression data. That model is used to transform high-dimensional and noisy data to lower-dimensional data. The proposed method adopted SDAE to improve the accuracy of the cancer gene expression data. In [29] authors developed a deeply stacked denoising autoencoder (DSDA) for protein structure reconstruction. A stacked denoising autoencoder was trained for the target protein.

Lee and Yoon [30] proposed a DBN based unsupervised method to perform the auto prediction. The authors used a restricted Boltzmann machine (RBM) to address the conventional contrastive divergence problem for class-imbalanced prediction.

In [31] authors proposed a Deep Learning and active learning-based feature selection method to select genes/miRNAs based on expression profiles. Deep Belief Network is used to generate a high-level representation of the genes. Two hidden layers were used in the proposed model. The authors experimented on three data sets, namely prostate cancer, colon cancer, and ovarian cancer data sets.

L. Chen, et.al developed a model for bioinformatics [32]. This model is based on the responses received from rat cells treated with the same stimuli. The proposed method performed well in simulating cellular signaling systems. A Deep Learning model for bioinformatics is proposed by [33]. The proposed model uses a deep belief network and a hybrid convolution neural network. The authors used thirty-one large-scale CLIP-seq datasets, for experiments.

In [34] authors presented a Deep Learning-based model to infer the expression of target genes. The proposed approach uses the Omnibus dataset (microarray-based Gene Expression data set). Their model recorded achieved a lower error rate than LR.

A Deep Learning-based framework for predicting binding sites of RBPs and modeling structural binding preferences was introduced by [35]. DL-based methods outperformed the non-DL based method in predicting RBP. Single-cell DNA methylation states prediction was done by [36]. A computational deep neural network (DNN) was proposed in the paper. The proposed model was used to estimate the effect of single nucleotide changes in DNA.

Quang et al. [37] proposed a deep neural network (DNN) model to annotate and identify pathogenicity in GV. DNNs are used to capture the non-linear relationships among features. Their model achieved 19% reduction in the error rate and 14% relative increased AUC.

Prediction of structures in proteins was proposed by Rhys Heffernan et al. [38]. The proposed model uses an iterative Deep Learning neural network with three iterations for prediction. The proposed model achieved an accuracy of 82%. In [39] authors proposed a compound-Protein interactions prediction using Deep Learning. Their proposed model achieved remarkable accuracy on both balanced and imbalanced datasets.

In [40] a Deep Learning and feature embedding based combined model was designed for predicting compound-protein interactions. Their model learns the low-dimensional features from unlabeled data. The RF-based method outperforms the CNN-based method when applied to the DUD-E data set.

Cancer type classification based on Deep Learning was proposed by Yuchen Yuan et al. [41]. The proposed model uses three steps for classification, namely (1) clustered gene filtering (CGF), (2) indexed sparsity reduction (ISR), and (3) DNN classifier to extract high-level features. Experiments were conducted on a TCGA dataset and outperformed 24% in testing accuracy.

Chemical genomics-based virtual screening (CGBVS) based on DNN was proposed by Masatoshi Hamanaka et al. [42]. Accuracy of the model was recorded as 98.2%

In [43] authors proposed modeling of gene expressions using Deep Learning. The model was built using CNN on the activity of TF binding on 20Kb gene neighborhoods of target gene expression. The proposed model recorded an accuracy of 77%.

Prediction of DNA- RNA-binding proteins (DeepBind) was experimented with by Babak Alipanahi et al. [44]. The authors performed training on in vitro data, and the results show that the Deep Learning model outperformed the other state-of-the-art methods.

In [45] authors introduced a package named Basset to apply CNNs. The package is used to learn the activity of DNA sequences from genomics data. A CNN Model for predicting the DNA sequence binding was proposed by Haoyang Zeng et al. [46].

The proposed model is based on a Cloud-based framework to solve the problems in computational biology.

A Deep Learning model for noncoding variants named DeepSEA (http://deepsea. princeton.edu/) was introduced by Jian Zhou et al. [47]. This model is used to improve the prioritization of functional variants and also disease-associated variants. In [48] authors proposed a model (LINSIGHT) to predict the deleterious noncoding variants from genomic data. LINSIGHT improved the prediction of noncoding nucleotide sites.

Prediction of a protein secondary structure was proposed by Sheng Wang et al. [49]. Authors developed a Deep Convolutional Neural Fields model for protein prediction. This model is an extension of CNN and is the combination of shallow NN and Conditional Random Fields (CRF). The proposed model obtained 84% Q3 accuracy.

In [50] authors proposed a model (deepMiRGene) to predict Precursor microRNA. The proposed model was designed based on recurrent neural networks (RNN), and long short-term memory networks (LSTM). Experiments were carried out on three recent benchmark datasets taken from [51]. Prediction of Micro RNA using Deep Learning was introduced in [52]. The authors designed the model using RNN based on sequence-sequence interaction learning and auto-encoding.

A Reinforcement Learning and Particle Swarm Optimization based Operon Prediction model was introduced in [53]. Three genomes, Bacillus subtilis, Staphylococcus aureus, and Pseudomonas aeruginosa PA01, were used to test the prediction performance of the model. The proposed model recorded the highest accuracy compared with existing approaches.

The reinforcement-based model for the Bio Agents system was introduced in [54]. Experiments were carried out on Paullinia cupana - Guaraná plant and Paracoccidioides brasiliensis – Pb fungus. The proposed model overcomes the drawbacks of the lack of samples for training purposes and data size.

The Fragment Assembly Problem using a Q-learning agent-based approach was introduced in [55]. A new reinforcement learning approach was used in the proposed method. The proposed model outperforms other models.

In [56] authors analyzed a protein interaction network in cancer data. In the network, Edges denote interactions and Nodes arc used to represent proteins. Analysis results show that the network node degree distribution is scale-free.

11.3.2 DEEP LEARNING FOR BIOMEDICAL IMAGING

DNN has been used in several areas of biomedical imaging. Deep Learning has been used for analyzing medical images and shows remarkable performance in various applications.

In [57] authors demonstrated the applications of Deep Learning to brain imaging data. The proposed method uses a novel constraint-based approach to visualize the neuroimaging data. Classification of lung nodules using Deep Learning was experimented with in [58]. A convolution neural network was used to classify computed tomography images. The proposed method outperforms the conventional feature computing CAD frameworks.

In [59] authors designed a model that was Deep Learning-based for feature representation with a stacked auto-encoder. Experiments were carried out using the ADNI dataset. The proposed model achieved good diagnostic accuracy for AD/MCI classification.

Image segmentation is an important area in computer vision and image processing. Various Deep Learning algorithms have been developed for image segmentation. Automatic phenotyping of developing embryos was proposed in [60]. The proposed system will detect segments, and locates cells and nuclei in microscopic images. A convolutional network was trained to classify the pixels.

In [61] authors trained the convolutional network for image segmentation. The proposed method was applied to 3D segmentation. The proposed method learns affinity graphs from the raw EM images. The proposed method improved segmentation accuracy. In [62] authors discussed partitioning specific structures in biomedical imaging. The proposed method is based on crowd-sourced manual annotation and Deep Learning. In [63] authors addressed the problem of neuroanatomy. The proposed method uses a DNN as a pixel classifier. The DNN is trained on a $512 \times 512 \times 30$ segmentation using plain gradient descent.

Voxel classification integrating 3-2D CNNs were presented in [64]. The proposed method was applied to low field knee MRI scans. The proposed method outperformed 3D multi-scale features. In [65] authors proposed a fully automatic brain tumor segmentation method using Deep Learning.

Pancreas segmentation using deep convolutional networks (ConvNets) was proposed by Holger R. Roth et al. [66]. Experiments were carried out using CT scans. The proposed approach is based on a probabilistic bottom-up approach.

Parallel Multi-Dimensional LSTM was introduced in [67]. The proposed method is used for Biomedical Volumetric Image Segmentation and is easy to parallelize. Convolutional network based computer-aided detection was experimented with in [68]. Experiments were carried out on 3D medical images. Detection of Sclerotic Spine Metastases using Deep Learning was formulated in [69]. Experiments were carried out using CT images of fifty-nine patients. Authors designed a two-tiered framework to operate a successful candidate generation system. The proposed method recorded AUC of 0.834.

Medical image classification using customized CNN was proposed in [70]. Experiments were carried out on lung images. The proposed method automatically learns intrinsic image features from the data set. In [71] authors proposed detection of Mitosis in Breast Cancer using Deep Neural networks. The proposed model is based on deep max-pooling. Predicting the neoadjuvant chemotherapy response using Deep Learning was proposed by Petros-Pavlos Ypsilantis et al. [72]. Experiments were carried out using 107 esophageal cancer patients' data. Proposed model recorded sensitivity of 80.7% with 3S-CNN.

In [73] authors proposed a deep learning approach to annotate gene expression data. Experiments were carried out to extract features. Features are extracted from ISH images of developing mouse brain. The proposed method recorded AUC of 0.894 ± 0.014. Detection of carcinoma cancer using Deep Learning was proposed by Angel Alfonso Cruz-Roa et al. [74]. Experiments were carried out using 1,417 images of skin. Haar-based wavelet transforms (Haar) and discrete cosine transform

(DCT) image representation techniques were used. The proposed method recorded 91.4% in balanced accuracy.

Chest pathology identification using Deep Learning was proposed by Yaniv Bar, et al. [75]. A convolutional neural network is trained to identify pathologies in chest x-ray images. Experiments were carried out on ninety-three image data set. Simulation results recorded (AUC) of 0.93.

A max-pooling CNN model was developed in [76] to detect the mitosis in breast histology images. Convolutional neural networks are trained to classify the pixels in the images.

In [77] authors addressed the automatic segmentation of neuronal structures using Deep Learning. A deep artificial neural network was used as a pixel classifier and was trained by plain gradient descent on a $512 \times 512 \times 30$ stack.

Protein subcellular localization using Deep Learning was performed in [78]. The proposed system uses an eleven-layer neural network to map yeast proteins, and the system achieved an accuracy of 91%. In [79] authors undertook counting bacterial colonies using Deep Learning. A 28.5k Microbiology Imaging data set was used for training and validation.

Deep multiple instance learning was applied on microscopy images in [80]. The proposed method is the combination of multiple instance learning (MIL) and Convolution neural networks (CNNs). Experiments were carried out on mammalian and yeast datasets.

In [81] authors proposed imaging flow cytometry using Convolution Neural networks and on linear dimension reduction (DeepFlow). Deep Flow is applied on a large dataset of cell-cycling Jurkat cells. DeepFlow's predictions are fast compared with existing measures.

Classification of neural progenitor cells (NPCs) using Convolution neural networks was proposed in [82]. Convolution neural networks are trained to classify NPCs and non-NPCs.

11.3.3 Deep Learning for Biomedical Signal Processing

Biomedical signal processing is a domain area recorded activity from the human body that is used to solve problems. Deep Learning has been an effective way to decode human brain activity from neurological data. Various deep learning approaches have been developed for brain decoding.

Unsupervised approach to learn Neural Correlates in ECoG Electrophysiology was proposed in [83]. Incremental learning of a Deep Belief Network was applied on real patient data.

A DL model for Classification of EEG was proposed in [84] and the method is based on the Motor Imagery task. Deep Belief Network (DBN) was trained during the process. Restricted Boltzmann Machines are stacked in the model. The proposed method outperforms with eight hidden layers.

Authors in [85] proposed a Deep Belief Networks (DBN) based model from EEG signals. A Restricted Boltzmann Machine (RBM) is tuned to optimal EEG channels. Experiments were carried out on Deep Dataset. A semi-supervised framework using DL was implemented in [86]. The proposed model was based on the Restricted

Boltzmann Machine (RBM). The experiments were conducted on a real EEG dataset for classification.

An EEG-based emotion recognition task was performed in [87]. The p method is based on a deep belief network (DBN). DBN was trained for recognizing three emotions: negative, positive, and neutral. The proposed model was compared with KNN, LR, and SVM classifiers, and the proposed model recorded an average accuracy of 86.08%.

Automatic emotion recognition using a deep learning network (DLN) was proposed in [88]. A model was developed with a stacked autoencoder (SAE). Power spectral densities of EEE signals were given as input to the model. The proposed model achieved an accuracy of 46.03%.

Rhythm perceptions in Electroencephalography (EEG) recording were experimented with in [89]. A convolution neural network with stacked denoising autoencoders was tested on data of thirteen adults, with a mean age of twenty-one. The model is trained to classify African and Western rhythms.

Rhythm perception in Electroencephalography (EEG) recordings was discussed in [90]. A CNN model is used to analyze EEG data. The data set comprises twelve East African and twelve Western rhythmic stimuli. An individual rhythm from the EEG was recorded with an accuracy of 50%.

Classification of electroencephalographics (EEG) using Convolutional Neural Network was discussed in [91]. Six electrodes on two subjects were used to obtain the results. In [92] authors proposed a model to detect P300 waves in the human brain. The proposed model is based on CNN, with seven classifiers. The model is tested on Data set II from the third BCI competition. The model obtained a recognition rate of 95.5 %. In [93] authors proposed a model for emotion detection, using EEG signals. Power spectral features from an EEG signal were used to detect valence levels. The model is used to detect subtle affective responses.

The Gaussian Bernoulli restricted Boltzmann machine (RBM) based Deep Learning model is used to decode the ECoG based Brain Computer Interface (BCI) system [94]. A deep feature learning method is used on unlabeled and mixed labeled data sets.

In [95] authors proposed a model using Deep Feature Learning for EEG Recordings. The proposed model addressed the challenges for feature learning. Experiments were carried out on the OpenMIIR dataset of EEG recordings. Auto-encoders were used to focus on features, to take the advantage of individual feature learning.

A deep learning model to decode human brain activity was proposed in [96]. The model is based on LSTM-CNN to extract the representations of electroencephalograms (EEG). The model is evaluated in terms of the quality of the generated images. A generative adversarial network (SNGAN) is used for the EEG images.

11.3.4 TRANSFER LEARNING FOR BIO INFORMATICS (TL FOR BIOINFORMATICS)

Transfer learning is considered a ML technique where a trained model is re-purposed on a second task. Transfer learning has the ability to utilize existing knowledge in the target task.

There are three kinds of transfer learning strategies, namely, (1) inductive TL, (2) Unsupervised TL, and (3) Transductive TL. Figure 11.9 shows the various techniques

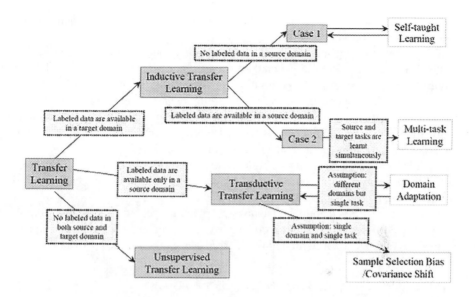

FIGURE 11.9 Transfer learning strategies [97].

of transfer learning. Transfer learning has gained notice through DL techniques [97, 98]. Transfer Learning strategies have been highlighted in Figure 11.9.

Lung Pattern Classification using an ensemble of transfer learning and Deep Learning was executed in [98]. Experiments were carried out on a dataset of 14,696 image patches from 120 CT Scans. The proposed model recorded an accuracy of 85.5%.

The LSTM model for a biomedical named entity was proposed in [99, 100]. The proposed method is based on gold-standard corpora (GSCs) and reduced the error rate by approximately 11%.

Prediction of phenome–genome association was discussed in [101]. Dual Label Propagation (DLP) was used in the proposed model. Transfer learning framework (tlDLP) uses this Dual Label Propagation (DLP). The proposed model with gene ontology improved the performance.

In [102] authors proposed an algorithm for disease gene discovery using transfer learning. Mutual Interaction-based Network Propagation is designed for disease gene discovery. Protein-protein interactions (PPI) using transfer learning was proposed in [103]. Authors used Collective Matrix Factorization to test on Helicobacter pylori and Human data sets. The proposed method performed well compared with other methods.

In [104] authors proposed a model for molecular recognition features in proteins by transfer learning(s) (MoRFs). The proposed method is useful in identifying the functionally disordered regions in proteins.

11.3.5 DEEP REINFORCEMENT LEARNING FOR BIOINFORMATICS

Reinforcement learning is considered as the training given to Machine Learning models in order to make a sequence of decisions. Reinforcement learning applied with

Deep Learning shows less complexity compared with the supervisee and genetic algorithms.

Deep reinforcement learning for sequence alignment system was proposed in [105]. The proposed (DQNalign) method proved that there is a good possibility of aligning two long sequences by combining conventional algorithms. In [106] authors proposed peptide sequencing using Deep Reinforcement Learning. Experiments were carried out on public data sets and achieved higher precision. In [107] authors trained the cells and control signals using Reinforcement learning. The proposed model is designed to speed up collective cell migration.

In [108] authors proposed Deep Reinforcement Learning for the DNA fragment assembly problem. The proposed approach achieved good performance compared with other approaches.

In [109] agent-based modeling using Deep Reinforcement Learning was proposed. The proposed model was tested on two real developmental processes and performed well for agent-based modeling for large datasets generated by live imaging.

11.3.6 DEEP FEW SHOT LEARNING FOR BIOINFORMATICS

Due to data scarcity in biology and biomedicine, few shot learning combined with deep Learning is successfully applied to bioinformatics. Few-shot learning trains a ML model with less data.

In [110] authors proposed a model for predicting the Enzyme Commission number (DEEPre) using few shot and Deep Learning. Experiments were carried out on a separate low-homology dataset.

In [111] a model for processing phenotypic data was designed. The proposed model is fully automatic in robust phenotypic drug discovery. This model is used for many disease areas for drug discovery. One-shot learning is used to lower the quantity of data required for precise predictions in drug discovery [112].

11.3.7 DEEP LEARNING FOR PUBLIC HEALTH

Public health targets preventing diseases, and promotes healthcare by analyzing social behaviors and the spread of diseases in relation to environmental factors. There is a growing interest in deploying Deep Learning techniques to target public health outcomes.

A deep recurrent neural network (DRNN) based model is designed to predict PM2.5 [113]. This model is designed for time series prediction. The authors developed a model to predict the concentration of PM2.5, which is the main air pollutant substance in Japan. Input to the model is the data collected from sensors from over forty-two cities.

In [114] authors developed a Social Restricted Boltzmann Machine (SRBM) to predict Human Behavior. Date from three events (environmental events, social influences, and self-motivation) are pulled into three layers: visible historical, visible, and hidden layers.

The deep belief networks based model is designed to categorize antibiotic-related Twitter posts [115]. The proposed method is applied to training with 412 labeled and

TABLE 11.1
Deep Learning Libraries

S.No	Deep Learning Libraries	Languages	Advantages
1	Caffe	Python, C++, Matlab	Support for Pre-trained Models, Firm implementation and user friendly.
2	DL ToolBox	Matlab	User friendly.
3	TensorFlow	Python, R, Java, C++,	Heterogeneous distributed computing, reinforced for models, and flexible
4	MXNet	Python, R, Julia, C++,	Distributed support
5	Neon	Python	Speed
6	DL4J	Python, Java, Scala	Distributed support
7	Torch	C++, Lua	Fast implementation and flexible for models, functional extensionality
8	Keras	Python, R	User friendly.
9	Theano	Python	User friendly, along with higher level wrappers, extensible.

150 000 unlabeled examples. Antibiotic-related Twitter posts are categorized into nine classes. There is a growing interest to track human behavior using mobile phone metadata. In [116] authors proposed a model based on convolutional neural network (CNN) to predict the demographic information from mobile phone data. The CNN consists of five layers, followed by two dense layers and a vertical convolution filter.

In [117–120] authors investigated the applications of Deep Learning in health informatics. Few authors developed the Deep Learning models for the social media users' health data set. Authors developed the models using semi-supervised Deep Learning frameworks, which exhibited good performance.

Deep Learning libraries mainly used by the researchers are discussed in Table 11.1.

11.4 DEEP LEARNING IN BIOINFORMATICS: CHALLENGES AND LIMITATIONS

Even though deep neural networks based models are widely applied in bioinformatics, there are few challenges and limitations in applying the deep neural networks in biomedical applications.

1. DL requires a huge amount of data for processing. As in many cases health data is not freely available (public). For some rare diseases, the data base many contain a small set of tuples, which leads to misclassification.
2. Lack of excessive computational resources may take time.
3. Deep Learning architecture is often treated as a black box [121]. Researchers a sometimes unable to solve misclassifications, which in turn leads to low accuracy.
4. Over fitting is also one of the limitations in Deep Learning.

5. Vanishing gradient may occur sometimes. This will reduce the accuracy of the Deep Learning model
6. Processors for high computational processing are required in order to improve the complexity, which will increase the cost.
7. Uncertainty scaling is also a challenging task.
8. Catastrophic forgetting: Deep Learning models are unable to learn the new knowledge without interfering with the new knowledge [122].
9. The biological data is imbalanced, and training the neural network with imbalanced data may result in undesirable results.

11.5　CONCLUSION

Deep Learning in bioinformatics is a new field. Research is actively taking place in informatics to solve the problems faced in this area. This chapter comprehensively discussed the various applications of Deep Learning in bioinformatics. We provided various tools in adopting different methods in Deep Learning. Although Deep Learning produced promising results in bioinformatics, there are still some challenges in applying the Deep Learning in bioinformatics like misclassification, overfitting, imbalanced data, and interpreting the results. Challenges in Deep Learning are also highlighted in this chapter. We believe that this comprehensive review will provide an insight and help the research community in further development toward applying Deep Learning architectures in bioinformatics.

REFERENCES

1. Manisekhar, S. R. et al., *Introduction to bioinformatics, Statistical Modelling and Machine Learning Principles for Bioinformatics Techniques, Tools, and Applications, Algorithms for Intelligent Systems*, Mumbai, India: Springer. (2020) doi:10.1007/978-981-15-2445-5_1
2. Hapudeniya, M. "Artificial neural networks in bioinformatics". *Sri Lanka Journal of Bio-Medical Informatics*, 1: 2. (2010).
3. Can, T. Introduction to bioinformatics. *Methods in Molecular Biology* 1107: 51–71. (2014).
4. Srinivasa, K. G., Siddesh, G. M., & Manisekhar, S. R. (Eds.). *Statistical Modelling and Machine Learning Principles for Bioinformatics Techniques, Tools, and Applications*. Algorithms for Intelligent Systems. Mumbai, India. doi: 10.1007/978-981-15-2445-5. (2020).
5. S. R. Manisekhar, G. M. Siddesh and Sunilkumar S. Manvi, *Introduction to Bioinformatics, Book Statistical Modelling and Machine Learning Principles for Bioinformatics Techniques, Tools, and Applications*. Algorithms for Intelligent Systems. Mumbai, India. doi:10.1007/978-981-15-2445-52020
6. Seonwoo Min, Byunghan Lee and Sungroh Yoon, "Deep learning in bioinformatics" in Briefings in Bioinformatics Advance Access published July 29, 2016.
7. Y. LeCun, L. Bottou, Y. Bengio, and P. Haffner, "Gradient-based learning applied to document recognition." *Proceedings of the IEEE*, 86(11),2278–2324, Nov. 1998.
8. D. H. Hubel and T. N. Wiesel, "Receptive fields, binocular interaction and functional architecture in the cat's visual cortex." *The Journal of Physiology*, 160(1), 106–154, 1962.

9. Mojammil Husain, "Convolutional Neural Network Tutorial (CNN/ConvNets)", 2016
10. Zeiler, M. D., and Fergus, R. "Visualizing and understanding convolutional networks," in *Proc. Eur. Conf. Comput. Vision*, Zurich, Switzerland: pp. 818–833.2014.
11. Szegedy, C. et al., "Going deeper with convolutions," in *Proc. Conf. Comput. Vis. Pattern Recognit.*, pp. 1–9. 2015.
12. Williams, R. J. and Zipser, D. "A learning algorithm for continually running fully recurrent neural networks," *Neural Computing and Applications*, 1(2), 270–280, 1989.
13. Bengio, Y., Simard, P., and Frasconi, P. "Learning long-term dependencies with gradient descent is difficult," *IEEE Transactions on Neural Networks and Learning Systems*. 5(2), 157–166, 1994.
14. Hochreiter, S. and Schmidhuber, J. "Long short-term memory," *Neural Computing and Applications*. 9(8), 1735–1780, 1997.
15. Kevin Vu, "Recurrent neural networks (RNN): Deep learning for sequential data", *KDNuggets tutorial, Sensors, MDPI*, 16(10), 2016.
16. Hinton, G. E. and Salakhutdinov, R. R. "Reducing the dimensionality of data with neural networks," *Science*. 313(5786), 504–507, 2006.
17. Poultneyet, C. et al., "Efficient learning of sparse representations with an energy-based model," in *Proc. Adv. Neural Inf. Process. Syst.*, China, 1137–1144. 2006.
18. Vincent, P., Larochelle, H., Bengio, Y., and Manzagol, P.-A. "Extracting and composing robust features with denoising autoencoders," in *Proc. Int. Conf. Mach. Learn.*, Helsinki, Finland, 1096–1103. 2008.
19. Rifai, S., Vincent, P., Muller, X., Glorot, X. and Bengio, Y. "Contractive auto-encoders: Explicit invariance during feature extraction," in *Proc. Int. Conf. Mach. Learn.*, Bellevue, Washington, 833–840. 2011.
20. Masci, J., Meier, U., Ciresan, D. and Schmidhuber, J. "Stacked convolutional autoencoders for hierarchical feature extraction," in *Proc. Int. Conf. Artif. Neural Netw.*, Finland, 52–59. 2011.
21. Ma, Tao et al., "A hybrid spectral clustering and deep neural network ensemble algorithm for intrusion detection in sensor networks", *KDNuggets tutorial, Sensors, MDPI*, Volume 16(10); 2016, Oct, 5.
22. Yuming, Hua, Junhai, Guo & Hua, Zhao. (2015). "Deep Belief Networks and deep learning". *Proceedings of 2015 International Conference on Intelligent Computing and Internet of Things*, United States.
23. Navamani, T. M., "Efficient deep learning approaches for health informatics", *Deep Learning and Parallel Computing Environment for Bioengineering Systems*, Elsevier, Australia: 5. 2019.
24. LeCun, Y., Bottou, L., Bengio, Y. and Haffner, P. "Gradient-based learning applied to document recognition," *Proceedings of the IEEE*.86(11), 2278–2324, 1998.
25. Aminanto, M. E. and Kim, K. "Deep learning in intrusion detection system: An overview," *International Research Conference on Engineering and Technology*, 2016, 2016.
26. Bengio, Y., and Goodfellow, I, J., *Deep Learning*, Courville A., MIT, Switzerland. 2015.
27. Fakoor, R., Ladhak, F., Nazi, A. and Huber, M., "Using deep learning to enhance cancer diagnosis and classification," in *Proc. ICML*, 2013. IBIS Hotel Kuta-Bali, Indonesia.
28. Danaee, P., Ghaeini, R. and Hendrix, D. A. "A deep learning approach for cancer detection and relevant gene identification," in *Proc. Pacific Symp. Biocomput.*, 22, 219–229, 2016.
29. Li, H., Lyu, Q. and Cheng, J. "A template-based protein structure reconstruction method using deep autoencoder learning," *Journal of Proteomics& Bioinformatics*. 9(12), 306–313, 2016.

30. Lee, T. and Yoon, S., "Boosted categorical restricted Boltzmann machine for computational prediction of splice junctions," in *Proc. ICML*, Atlanta, USA, pp. 2483–2492, 2015.
31. Ibrahim, R., Yousri, N. A., Ismail, M. A. and El-Makky, N. M. "Multilevel gene/MiRNA feature selection using deep belief nets and active learning," in *Proc. IEEE EMBC*, 3957–3960, 2014.
32. Chen, L., Cai, C., Chen, V. and Lu, X. "Trans-species learning of cellular signaling systems with bimodal deep belief networks," *Bioinformatics*. 31(18), 3008–3015, 2015.
33. Pan, X. and Shen, H.-B. "RNA-protein binding motifs mining with a new hybrid deep learning based cross-domain knowledge integration approach," *BMC Bioinformatics*. 18(1), 136, 2017.
34. Chen, Y., Li, Y., Narayan, R., Subramanian, A. and Xie, X. "Gene expression inference with deep learning," *Bioinformatics*. 32(12), 1832–1839, 2016.
35. Zhang, S. et al., "A deep learning framework for modeling structural features of rna-binding protein targets," *Nucleic Acids Research*. 44(4), e32, 2016.
36. Angermueller, C., Lee, H. J., Reik, W. and Stegle, O. "DeepCpG: Accurate prediction of single-cell DNA methylation states using deep learning," *Genome Biology*. 18(1), 67, 2017.
37. Quang, D., Chen, Y. and Xie, X., "DANN: A deep learning approach for annotating the pathogenicity of genetic variants," *Bioinformatics*. 31(5), 761–763, 2015.
38. Heffernan, R. et al., "Improving prediction of secondary structure, local backbone angles, and solvent accessible surface area of proteins by iterative deep learning," *Scientific Reports*. 5, 11476, 2015.
39. Tian, K., Shao, M., Zhou, S. and Guan, J. "Boosting compound-protein interaction prediction by deep learning," *Methods*. 110, 64–72, 2016.
40. Wan, F. and Zeng, J. "Deep learning with feature embedding for compound-protein interaction prediction," bioRxiv, 086033, 2016.
41. Yuan, Y. et al., "DeepGene: An advanced cancer type classifier based on deep learning and somatic point mutations," *BMC Bioinformatics*. 17(17), 476, 2016.
42. Hamanaka, M. et al., "CGBVS-DNN: Prediction of compound-protein interactions based on deep learning," *Molecular Informatics*. 36(1–2),1600045, 2017, doi: 10.1002/minf.201600045.
43. Denas, O., and Taylor, J. "Deep modeling of gene expression regulation in erythropoiesis model," in *Proc. ICMLRL*, 1–5, 2013.
44. Alipanahi, B., Delong, A., Weirauch, M. T. and Frey, B. J., "Predicting the sequence specificities of DNA- and RNA-binding proteins by deep learning," *Nature Biotechnology*. 33(8), 831–838, 2015.
45. Kelley, D. R., Snoek, J. and Rinn, J. L. "Basset: learning the regulatory code of the accessible genome with deep convolutional neural networks," *Genome Research*. 26(7), 990–999, 2016.
46. Zeng, H., Edwards, M. D., Liu, G. and Gifford, D. K. "Convolutional neural network architectures for predicting DNA–protein binding," *Bioinformatics*. 32(12), 121–127, 2016.
47. Zhou, J. and Troyanskaya, O. G. "Predicting effects of noncoding variants with deep learning-based sequence model," *Nature Methods*. 12(10), 931–934, 2015.
48. Huang, Y. F., Gulko, B. and Siepel, A. "Fast, scalable prediction of deleterious noncoding variants from functional and population genomic data," *Nature Genetics*, 49(4), 618–624, 2017.
49. Wang, S., Peng, J., Ma, J. and Xu, J. "Protein secondary structure prediction using deep convolutional neural fields," *Scientific Reports*. 6, 18962, 2016.

50. Park, S., Min, S., Choi, H. and Yoon, S. "Deep MiRGene: Deep neural network based precursor microrna prediction," CoRR.abs/1605.00017, 9. Apr. 2016.

51. Tran Van Du, T., Tempel, S., Zerath, B., Zehraoui, F., Tahi, F. miRBoost: boosting support vector machines for microRNA precursor classification. *RNA*. 21(5), 775–785, 2015. doi:10.1261/rna.043612.113

52. Lee, B., Baek, J., Park, S. and Yoon, S. "DeepTarget: End-to-end learning framework for microRNA target prediction using deep recurrent neural networks," CoRR. abs/1603.09123, 9, Aug. 2016.

53. Chuang, L.-Y., Tsai, J.-H. and Yang, C.-H. "Operon prediction using particle swarm optimization and reinforcement learning," in *Proc. ICTAAI*, France, 366–372, 2010.

54. Ralha, C. G., Schneider, H. W., Walter, M. E. M. T. and Bazzan, A. L. C. "Reinforcement learning method for BioAgents," in *Proc. SBRN*, UK, Singapore, US East and West coasts, 109–114, 2010.

55. Bocicor, M.-I., Czibula, G., and Czibula, I.-G. "A reinforcement learning approach for solving the fragment assembly problem," in *Proc. SYNASC*, London, UK, 191–198, 2011.

56. Zhu, F., Liu, Q., Zhang, X. and Shen, B. "Protein-protein interaction network constructing based on text mining and reinforcement learning with application to prostate cancer," in *Proc. BIBM*, Kolkata, India, 46–51, 2014.

57. Plis, S. M., Hjelm, D. R., Salakhutdinov, R. et al. Deep learning for neuroimaging: a validation study. *Frontiers in Neuroscience* 2014.

58. Hua, K.-L., Hsu, C.-H., Hidayati, S. C. et al. Computer-aided classification of lung nodules on computed tomography images via deep learning technique. *Onco Targets and Ttherapy* 2015.

59. Suk H-I, Shen D. "Deep learning-based feature representation for AD/MCI classification." *Medical Image Computing and Computer-Assisted Intervention–MICCAI*. 2013, 583–590, Springer, 2013.

60. Ning, F., Delhomme, D., LeCun, Y. et al. Toward automatic phenotyping of developing embryos from videos. *Image Processing, IEEE Transactions on*, 14(9), 1360–1371, 2005.

61. Turaga, S. C., Murray, J. F., Jain, V. et al. Convolutional networks can learn to generate affinity graphs for image segmentation. *Neural Computation*. 22(2), 511–538.

62. Helmstaedter, M., Briggman, K. L., Turaga, S. C. et al. "Connectomic reconstruction of the inner plexiform layer in the mouse retina". *Nature*. 500 (7461): 168–74. 2013.

63. Ciresan, D., Giusti, A., Gambardella, L. M. et al. Deep neural networks segment neuronal membranes in electron microscopy images. In: *Advances in neural information processing systems*. F. Pereira, C. J. C. Burges, L. Botto, and K. Q. Weinberger (eds.), Lake Tahoe, NV: Harrahs and Harveys, 2843–2851, 2012.

64. Prasoon, A., Petersen, K., Igel, C. et al. "Deep feature learning for knee cartilage segmentation using a triplanar convolutional neural network". *Medical Image Computing and Computer-Assisted Intervention–MICCAI*. Springer, 2013, 246–253, 2013.

65. Havaei, M., Davy, A., Warde-Farley, D. et al. Brain tumor segmentation with deep neural networks. arXiv preprint arXiv: 1505.03540 2015.

66. Roth, H. R., Lu, L., Farag, A. et al. "Deeporgan: Multi-level deep convolutional networks for automated pancreas segmentation". *Medical Image Computing and Computer-Assisted Intervention–MICCAI Springer*, 2015, 556–64, 2015.

67. Stollenga, M. F., Byeon, W., Liwicki, M. et al. Parallel multi-dimensional LSTM, with application to fast biomedical volumetric image segmentation. arXiv preprint arXiv:1506.07452 2015

68. Roth, H. R., Lu, L., Liu, J.et al. Improving computer-aided detection using convolutional neural networks and random view aggregation. arXiv preprint arXiv:1505.03046 2015.

69. Roth, H. R., Yao, J., Lu, L. et al. Detection of sclerotic spine metastases via random aggregation of deep convolutional neural network classifications. *Recent Advances in Computational Methods and Clinical Applications for Spine Imaging*. Cambridge, USA: Springer, 2015, 3–12.

70. Li, Q., Cai, W., Wang, X. et al. Medical image classification with convolutional neural network. In: *Control Automation Robotics & Vision (ICARCV), 2014 13th International Conference on*. Lake Tahoe Nevada, 2014. 844–848. IEEE.

71. Cireşan, D. C., Giusti, A., Gambardella, L. M. et al. "Mitosis detection in breast cancer histology images with deep neural networks." *Medical Image Computing and Computer-Assisted Intervention–MICCAI*. Springer, 2013, 411–418, 2013.

72. Ypsilantis, P.-P., Siddique, M., Sohn, H.-M. et al. "Predicting response to neoadjuvant chemotherapy with PET imaging using convolutional neural networks". *PloS one*. 10(9), e0137036, 2015.

73. Zeng, T., Li, R., Mukkamala, R. et al. Deep convolutional neural networks for annotating gene expression patterns in the mouse brain. *BMC Bioinformatics*. 16(1),1–10, 2015.

74. Cruz-Roa, A. A., Ovalle, J. E. A., Madabhushi, A. et al. A deep learning architecture for image representation, visual interpretability and automated basal-cell carcinoma cancer detection. *Medical Image Computing and Computer-Assisted Intervention–MICCAI*. Springer, 2013, 403–10, 2013.

75. Bar, Y., Diamant, I., Wolf, L. et al. Deep learning with non-medical training used for chest pathology identification. In: *SPIE Medical Imaging, 94140V-V-7*. Taipei, Taiwan, International Society for Optics and Photonics. 2015.

76. Ciresan, D., Giusti, A., Gambardella, L. M. and Schmidhuber, J. "Mitosis detection in breast cancer histology images with deep neural networks," in *Proc. MICCAI*, Montreal, Quebec, 411–418, 2013.

77. Ciresan, D., Giusti, A., Gambardella, L. M. and Schmidhuber, J. "Deepneural networks segment neuronal membranes in electron microscopy images," in *Proc. NIPS*, Tampa, Florida, 2012, pp. 2843–2851.

78. Pärnamaa, T. and Parts, L. "Accurate classification of protein subcellular localization from high-throughput microscopy images using deep learning," *G3*.7(5), 1385–1392, 2017.

79. Ferrari, A. Lombardi, S. and Signoroni, A. "Bacterial colony counting with convolutional neural networks in digital microbiology imaging," *Pattern Recognition*. 61, 629–640, 2017.

80. Kraus, O. Z., Ba, J. L. and Frey, B. J. "Classifying and segmenting microscopy images with deep multiple instance learning," *Bioinformatics*. 32(12), i52–i59, 2016.

81. Eulenberg, P. et al., "Deep learning for imaging flow cytometry: Cell cycle analysis of Jurkat cells," bioRxiv, 081364, Oct. 2016.

82. Jiang, B., Wang, X., Luo, J., Zhang, X., Xiong, Y. and Pang, H. "Convolutional neural networks in automatic recognition of trans differentiated neural progenitor cells under bright-field microscopy", in *Proc. IMCCC*, 122–126, 2015.

83. Freudenburg, Z. V., Ramsey, N. F., Wronkeiwicz, M. et al. Real-time naive learning of neural correlates in ECoG Electrophysiology. Inernaional. *Journal of Machine Learning and Computing* 2011.

84. An, X., Kuang, D., Guo, X. et al. A Deep Learning Method for Classification of EEG Data Based on Motor Imagery. *Intelligent Computing in Bioinformatics*. Springer, 203–10, 2014.

85. Karim, M. R., Beyan, O., Zappa, A., Costa, I. G., Rebholz-Schuhmann, D., Cochez, M. and Decker, S., 2021. Deep learning-based clustering approaches for bioinformatics. *Briefings in Bioinformatics*, 22(1), pp. 393–415.

86. Jia, X., Li, K., Li, X. et al. A Novel Semi-Supervised Deep Learning Framework for Affective State Recognition on EEG Signals. In: *Bioinformatics and Bioengineering (BIBE), 2014 IEEE International Conference on*, 30–37. IEEE. 2014.

87. Zheng, W.-L., Guo, H.-T., Lu, B.-L. Revealing critical channels and frequency bands for emotion recognition from EEG with deep belief network. In: *Neural Engineering (NER), 2015 7th International IEEE/EMBS Conference on*, 154–157. IEEE. 2015.

88. Jirayucharoensak, S., Pan-Ngum, S., Israsena, P. EEG-based emotion recognition using deep learning network with principal component based covariate shift adaptation. *The Scientific World Journal*: 12. 2014, 2014.

89. Stober, S., Cameron, D. J., Grahn, J. A. Classifying EEG recordings of rhythm perception. In: *15th International Society for Music Information Retrieval Conference (ISMIR'14)*. China, 649–654, 2014.

90. Stober, S., Cameron, D. J., Grahn, J. A. Using Convolutional Neural Networks to Recognize Rhythm. In: *Advances in Neural Information Processing Systems*. China, 1449–1457, 2014.

91. Cecotti, H, Graeser, A. Convolutional neural network with embedded Fourier transform for EEG classification. In: *Pattern Recognition, 2008. ICPR 2008. 19th International Conference on*. China, Nanchang, 1–4. IEEE. 2008.

92. Cecotti, H, Gräser, A. "Convolutional neural networks for P300 detection with application to brain-computer interfaces". *Pattern Analysis and Machine Intelligence, IEEE Transactions on*, 33(3), 433–45. 2011.

93. Soleymani, M., Asghari-Esfeden, S., Pantic, M. et al. Continuous emotion detection using EEG signals and facial expressions. In: Multimedia and Expo (ICME), *2014 IEEE International Conference on*. 1–6. Indore, India, IEEE. 2014.

94. Wang, Z, Lyu, S, Schalk, G. et al. Deep feature learning using target priors with applications in ECoG signal decoding for BCI. In: *Proceedings of the Twenty-Third international joint conference on Artificial Intelligence*. 1785–1791. Atlantic City, USA, AAAI Press. 2013.

95. Stober, S., Sternin, A., Owen, A. M. et al. "Deep feature learning for EEG recordings". arXiv preprint arXiv:1511.04306 2015.

96. Zheng, Xiao, et al. "Decoding human brain activity with deep learning." *Biomedical Signal Processing and Control* 56: 13. 101730, 2020.

97. Pan, Sinno Jialin, and Qiang Yang. "A survey on transfer learning." *IEEE Transactions on Knowledge and Data Engineering*. 22(10), 1345–1359, 2009.

98. O'Shea, J. P., Chou, M. F., Quader, S. A., Ryan, J. K., Church, G. M., and Schwartz, D."pLogo: a probabilistic approach to visualizing sequence motifs", *Nature Methods* 10, 1211–1212, 2013. doi: 10.1038/nmeth.2646

99. Anthimopoulos, M., Christodoulidis, S., Ebner, L., Christe, A., and Mougiakakou, S. "Lung pattern classification for interstitial lung diseases using a deep convolutional neural network," *IEEE Transactionson Medical Imaging* 35, 1207–1216, 2016.doi: 10.1109/TMI.2016.2535865

100. Giorgi, J. M., and Bader, G, D. "Transfer learning for biomedical named entity recognition with neural networks," *Bioinformatics* 34, 4087–4094, 2018. doi: 10.1093/bioinformatics/bty449

101. Raphael, Petegrosso, Sunho, Park, Tae, Hyun Hwang, Rui, Kuang, "Transfer learning across ontologies for phenome–genome association prediction," *Bioinformatics*. 33(4), 529–536, 15 February 2017. doi: 10.1093/bioinformatics/btw649

102. Hwang, T. and Kuang, R. "A heterogeneous label propagation algorithm for disease gene discovery," in *Proc. 10th SIAMInternational Conference on Data Mining*, Columbus, 583–594, Apr. 2010.

103. Xu, Q., Xiang, E.W. and Yang, Q. "Protein-protein interaction prediction via collective matrix factorization," in *Proc. IEEE Interna-tional Conference on Bioinformatics and Biomedicine*, Hong Kong, 62–67, Dec. 2010.

104. Hanson, Jack, et al. "Identifying molecular recognition features in intrinsically disordered regions of proteins by transfer learning," *Bioinformatics*. 36(4), 1107–1113, 2020.

105. Song, Yong Joon, et al. "Pairwise heuristic sequence alignment algorithm based on deep reinforcement learning." arXiv preprint arXiv:2010.13478 2020.

106. Fei, Zhengcong. "Novel Peptide Sequencing With Deep Reinforcement Learning." *2020 IEEE International Conference on Multimedia and Expo (ICME)*. IEEE, 2020.

107. Hou, H., Gan, T., Yang, Y.et al. Using deep reinforcement learning to speed up collective cell migration. *BMC Bioinformatics* 20, 571, 2019. doi:10.1186/s12859-019-3126-5

108. Bocicor, M., Czibula, G. and Czibula, I. "A Reinforcement Learning Approach for Solving the Fragment Assembly Problem," *2011 13th International Symposium on Symbolic and Numeric Algorithms for Scientific Computing*, Timisoara, 191–198, 2011. doi: 10.1109/SYNASC.2011.9.

109. Wang, Z., Wang, D., Li, C., Xu, Y., Li, H., Bao, Z. "Deep reinforcement learning of cell movement in the early stage of C.elegans embryogenesis," *Bioinformatics*. 34(18), 3169–3177, 2018. doi: 10.1093/bioinformatics/bty323. PMID: 29701853; PMCID: PMC6137980.

110. Yu Li, Sheng, Wang, Ramzan, Umarov, Bingqing, Xie, Ming, Fan, Lihua, Li, Xin, Gao, "DEEPre: sequence-based enzyme EC number prediction by deep learning," *Bioinformatics*. 34(5), 760–769, 01 March 2018. doi: 10.1093/bioinformatics/btx680.

111. Joslin, J., Gilligan, J., Anderson, P., et al. "A fully automated high-throughput flow cytometry screening system enabling phenotypic drug discovery. *SLAS DISCOVERY: Advancing the Science of Drug Discovery*. 23(7),697–707, 2018. doi:10.1177/2472555218773086

112. Altae-Tran, Han, et al. "Low data drug discovery with one-shot learning," *ACS Central Science* 3(4), 283–293, 2017.

113. Ong, B. T., Sugiura, K. and Zettsu, K. "Dynamically pre-trained deep recurrent neural networks using environmental monitoring data for predicting pm2. 5," *Neural Computing and Applications*. 27, 1–14, 2015.

114. Phan, N., Dou, D., Piniewski, B. and Kil, D. "Social restricted Boltzmann machine: Human behavior prediction in health social networks," in *Proc. IEEE/ACM Int. Conf. Adv. Social Netw. Anal. Mining*, 424–431, Aug. 2015.

115. Kendra, R. L., Karki, S., Eickholt, J. L. and Gandy, L. "Characterizing the discussion of antibiotics in the Twittersphere: What is the bigger picture?" *Journal of Medical Internet Research*. 17(6), Art. no. e154, 2015.

116. Felbo, B., Sundsøy, P., Pentland, A., Lehmann, S. and de Montjoye, Y.-A. "Using deep learning to predict demographics from mobile phone metadata," Feb. 2016. [Online]. Available:http://arxiv.org/abs/1511.06660

117. Zou, B., Lampos, V., Gorton, R. and Cox, I. J. "On infectious intestinal disease surveillance using social media content," in *Proc. 6th Int. Conf. Digit. Health Conf.*, 157–161, 2016.

118. Garimella, V. R. K., Alfayad, A. and Weber, I. "Social media image analysis for public health," in *Proc. CHIConf. Human Factors Comput. Syst.*, 5543–5547, 2016. [Online]. Available: doi 10.1145/2858036.2858234

119. Zhao, L., Chen, J. Chen, F., Wang, W., Lu, C.-T. and Ramakrishnan, N., "Simnest: Social media nested epidemic simulation via online semisupervised deep learning," in *Proc. IEEE Int. Conf. Data Mining*, 639–648, 2015.

120. Horvitz, E. and Mulligan, D., "Data, privacy, and the greater good," *Science*. 349(6245), 253–255, 2015.

121. Daniele, Ravi, Charence, Wong, Fani, Deligianni, Melissa, Berthelot, Javier, Andreu-Perez, Benny, Lo, and Guang-Zhong, Yang, "Deep Learning for Health Informatics" in *IEEE Journal of Biomedical and Health Informatics*, 21(1), Jan. 2017.
122. Kirkpatrick, J., Pascanu, R., Rabinowitz, N., Veness, J., Desjardins, G., Rusu, A. A., Milan, K., Quan, J., Ramalho, T., Grabska-Barwinska, A., Hassabis, D., Clopath, C., Kumaran, D. and Hadsell, R. "Overcoming catastrophic forgetting in neural networks," *Proceedings of the National Academy of Sciences of the United States of America*, 114(13), 3521–3526, 2017.

12 Classification of Schizophrenia Associated Proteins Using Amino Acid Descriptors and Deep Neural Network

Sushma Rani Martha

Odisha University of Agriculture and Technology, Bhubaneswar, India

Tusar Kanti Dash and Ganapati Panda

CV Raman Global University, Bhubaneswar, India

Snehasis Mallick

Odisha University of Agriculture and Technology, Bhubaneswar, India

Manorama Patri

Ravenshaw University, Cuttack, India

CONTENTS

12.1 INTRODUCTION

The application of Neural networks in healthcare is of great concern in the modern era with the invention of numerous Machine Learning approaches. At the same time, numerous data have been generated from the molecular biology laboratories across the world on every existing and emerging human disorder. Schizophrenia is the most commonly reported neurological disorder where physicians are facing a considerable amount of difficulties in the diagnosis and treatment of the disorder [1]. Identifying the genes and proteins involved in Schizophrenia is also a critical issue even after so many genome-wide expression studies have already been done and reported in the public domain. Combining computational analysis with traditional research may help in resolving this issue to a greater extent.

Through this chapter, it is the goal to familiarize the readers with all the databases available in the public domains from where genome-wide molecular level data can be extracted regarding human diseases. Thereafter, the amino acid features through which the protein sequences can be converted to numerical vectors are discussed. Lastly, the importance and reliability of the Deep Neural Network in the classification of the proteins into three classes based on their degree of relatedness with the disorder is shown.

Although some common biological functional pathways may be shared by similar proteins, every protein deciphers a different function from others. A peptide sequence of length one hundred has $i = 1$ to 100 positions, where every position of i has the probability of accommodating or occupying one of the twenty amino acids. So, the probability of occurrence of each of twenty amino acids at every position is 1/20 or 0.05. Biological systems mimic the (Deoxyribonucleic Acid) DNA with so much precision that after transcription and translation of every gene, a specific protein will be expressed with a unique arrangement of the twenty amino acids throughout the protein sequence. Based on these criteria, the amino acid composition differs from protein to protein. This difference enhances the ability of a protein to attain a difference in its physicochemical properties [2].

12.2 PROTEIN DATASET PREPARATION

The experimental design for the primary level protein classification and dataset preparation is precisely illustrated in Figure 12.1. Protein sequences were downloaded from eight databases providing molecular-level information on human genetic disorders [3]. The proteins found after second- tier screening were assembled into the Positive or Schizophrenic dataset, and the proteins obtained after first- tier screening were assigned the Intermediate dataset. Some proteins that did not fall into any of these categories and have not been reported as human schizophrenia-associated proteins were retrieved from the UniProt protein sequence database to form the Negative

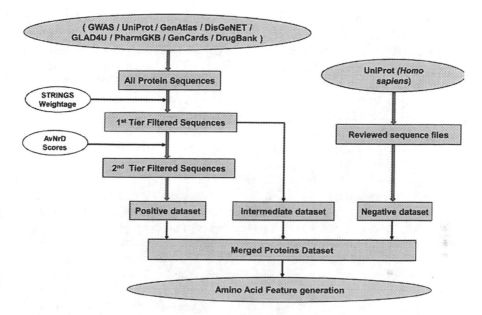

FIGURE 12.1   Experimental Design for preparation of positive, intermediate, and negative protein datasets.

or Non Schizophrenic dataset. All three datasets were merged and subjected to DNN after the generation of the feature table.

12.2.1   PROTEIN SEQUENCE DATABASES

1. GWAS (https://www.ebi.ac.uk/gwas/): This is a database jointly hosted by European Molecular Biology Laboratory (EMBL) and National Human Genome Research Institute (NHGRI). It provides information about genome-wide expression data on specific human traits collected from publications made through different online search engines [4]. Searching for Schizophrenia after setting P-value $\leq 5*10^{-8}$ resulted in 339 proteins in the hit list.

2. GLAD4U (glad4u.zhang-lab.org/): It is a repository of proteins collected mainly from the PUBMED (https://www.ncbi.nlm.nih.gov/pubmed/) literature data base of NCBI [5]. GLAD4U provided a hit list of 417 proteins for Schizophrenia after adjusting the cut off score as >1.

3. PharmGKB (https://www.pharmgkb.org/): This is a well-known database of Pharmacogenomics. It provides the ease of searching for the proteins on a specific human disorder, along with the information on drug and chemical conditions of the disorder [6]. We could collect 366 human proteins for schizophrenia from PharmGKB.

4. DisGeNET (www.disgenet.org/)]: It is a database that has information on human genetic disorders and is prepared by text mining several curated expert databases [7]. It provides a list of 259 proteins after the cut-off score

representing the association between the resulted proteins and schizophrenia is set as >1

5. GenAtlas (genatlas.medecine.univ-paris5.fr/): This database is targeted to display the consequences of gene mutation on the human physiology and resulting disorders. Alongside, it also provides information on the structure, function, and expression of the genes and drug-related information [8]. We found a total of 200 proteins from GeneAtlas.

6. DrugBank (https://www.drugbank.ca/): It is a repository of FDA (Food and Drug Administration) approved drugs that may be administered to humans for their physical and mental well-being. These drugs of DrugBank may have passed any level of the clinical trial process. Some drugs are approved; some are investigational, and some are withdrawn from medical practice. This database furnishes information about the pharmacological properties of drugs, as well as about the target proteins of the drugs [9]. On searching for Schizophrenia, 233 target proteins corresponding to about 130 drugs were found that have been deposited in DrugBank for Schizophrenia.

7. GenCards (https://www.genecards.org/): GenCards is a storehouse of information on human genes and proteins. It hosts disease-related data and corresponding drug-related data through a project named MalaCards [10]. Unlike DrugBank, it includes a much larger number of chemical compounds for every disease. In addition to the Drug, it stores other compounds which have drug-like properties and hence are likely to be used successfully as drugs.

8. UniProt (https://www.uniprot.org/): Protein Information Resource (PIR), Swiss Institute of Bioinformatics (SIB), and European Bioinformatics Institute (EBI) collaboratively formed the UniProt database. It is the most widely used protein sequence database because of its user-friendly data representations and the most precise annotations provided about each protein. It also renders cross-references to other databases to the users who are interested in acquiring deeper knowledge about the protein [11]. 142 Schizophrenia proteins and also the non-schizophrenic proteins were collected from UniProt, all of which were reviewed sequences. Schizophrenic proteins were subject to the screening process, while non-schizophrenic proteins were taken into the negative dataset.

12.2.2 STRING NETWORK ANALYSIS

After merging the results of eight databases discussed in the previous section and eliminating the redundancy of data, a superset of above 1,200 proteins was attained. Thereafter, the protein datasets obtained from each database were separately subjected to STRING (https://string-db.org/) network analysis to identify the most important proteins within that group. STRING output is in the form of interconnected networks joining several nodes, where each node represents one protein [12]. The protein nodes with a higher number of first order connectivity, and thence higher weightage scores, were selected, and the rest were rejected. Figure 12.2 illustrates the STRING networks that resulted from the UNIPROT protein sequence database before and after filtering. Nodes that were present as an outlier in the first part of the

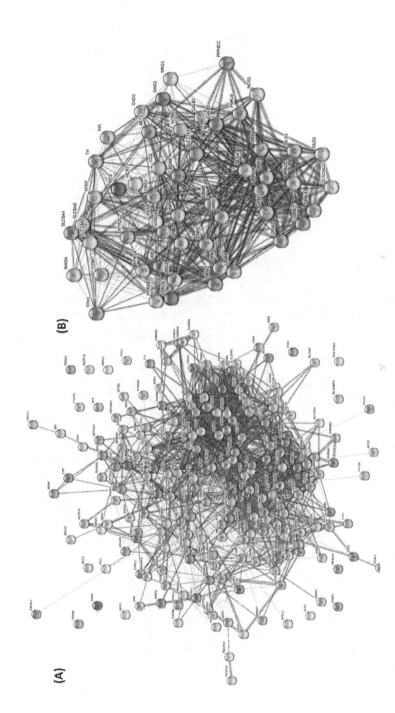

FIGURE 12.2 STRING output used for network analysis and first tier gene set screening: (a) Presence of a number genes forming outliers are clearly observable. Genes with more number of interactions are joined with thicker bands. (b) All the non-interacting genes forming outliers have been eliminated after screening.

figure and not connected with other nodes are missing in the second part of the figure. Similarly, sequences from all the databases are subjected to this screening process. Now the number of proteins reduced from 399, 142,200, 259, 417, 366, 231 and 233 to 40, 31, 32, 32, 55, 54, 53 and 53 for GWAS, UniProt, GenAtlas, DisGeNET, GLAD4U, PharmGKb, GenCards, and DrugBank databases respectively.

The second level of screening was done on the basis of Average Normalized Databases Scores. The weightage scores for the proteins being associated with schizophrenia are collected from the above-mentioned databases. The scores of every database depend on the algorithms used by those databases. These scores were normalized, and then the average score was computed for each protein. The scores of the filtered twenty-six proteins are shown in Table 12.1. Twenty-six genes found after the second level of screening were comprised of five Dopamine receptor proteins, five Serotonin receptor proteins, eleven Glutamate receptor proteins, two Sodium dependant transporters, one Cannabinoid receptor, and two enzymes viz. Glutamate decarboxylase1 and Tyrosine 3-monooxygenase (TH) are also accommodated within this

TABLE 12.1

Scores from Databases Representing the Probability of the Genes Being Involved with Schizophrenia and Their Normalized Average Values

Filtered Proteins	GLAD4U	DisGeNET	GenCards	DrugBank	Avg. Nrml. Score
CNR1	2.50426	7.30576	3.137862	2.744814	3.923176
DRD1	3.97004	4.55613	3.942529	5.323275	4.447994
DRD2	**19.95984**	**8.73668**	**3.965254**	**5.085129**	**9.436725**
DRD3	13.45183	7.93660	5.121752	4.940665	7.862713
DRD4	5.55587	6.78536	4.473642	5.367636	5.545629
DRD5	1.62456	0.59498	3.670659	5.150762	2.76024
GAD1	3.27512	4.89257	3.441716	3.659752	3.81729
GRIA1	0.51849	0.21796	4.906277	0.731950	1.593669
GRIA2	0.64516	0.26455	4.165579	0.714098	1.447348
GRIA4	1.05649	0.48837	5.939044	1.829876	2.328446
GRIN1	2.62803	3.14523	4.043532	2.227675	3.011118
GRIN2A	1.17829	4.25014	3.170689	2.195851	2.698741
GRIN2B	2.69769	4.87501	3.780922	2.195851	3.387369
GRM1	0.33296	0.09805	3.958521	3.659752	2.012321
GRM2	1.68881	3.92270	4.345703	7.319504	4.319178
GRM3	**6.36724**	**9.05112**	**5.748819**	**7.319504**	**7.121672**
GRM4	0.34455	0.22939	5.456749	7.319504	3.337549
GRM7	0.88209	0.44735	0.221367	7.319504	2.217578
HTR1A	2.77371	0.61314	3.069684	3.450623	2.476789
HTR1B	0.91383	0.47208	3.417307	2.439834	1.810764
HTR2A	**12.06196**	**10.2396**	**4.365063**	**4.171330**	**7.709488**
HTR2C	5.75313	1.13481	3.734628	3.903735	3.631577
HTR7	0.78856	3.99702	3.157221	4.722260	3.166264
SLC6A3	3.16583	7.95291	2.969522	2.024543	4.028201
SLC6A4	4.19623	6.85964	2.996456	1.742739	3.948766
TH	1.66540	0.93284	2.799498	2.439834	1.959394

group of 26. DRD2 is found as the most important protein within these 26 proteins, acquiring an AvNrD score of 9.437, which is the highest value.

12.2.3 THREE DIMENSIONAL STRUCTURE OF DRD2

The 3-D structure of DRD2 has been retrieved from PBD (www.rcsb.org/) as well as through GPCRdb (https://gpcrdb.org/). The structure of DRD2 obtained from GPCRdb has been refined, validated using Procheck (https://www.ebi.ac.uk/thorntonsrv/software/PROCHECK/) and visualized through Discovery Studio (https://www.3dsbiovia.com/products/collaborative-science/biovia-discovery-studio/) respectively. PDB is the databank of three-dimensional structures of Biological macromolecules predicted by NRM spectroscopy or X-ray crystallography [13]. GPCRdb contains computationally predicted three-dimensional structures of G Protein-Coupled receptor proteins [14]. Procheck is an online software capable of predicting the Ramachandran Plot for 3-D structures [15]. Discovery Studio is a 3-D visualization tool providing editing in the structure to some extent [16].

The 3D structures of the PDB (6CM4: 2.867Å resolution) have a large segment with missing residues, creating a split in the structure. Hence, the structure predicted through GPCRdb is shown in Figure 12.3. This predicted structure also has a split of smaller size as compared to the experimental PDB structure. As labeled in Figure 12.3, the split is in the third intracellular loop of the protein structure, which may have occurred due to the unavailability of an appropriate experimentally predicted template structure matching with the IC3 loop region of the protein. Like most schizophrenia-associated proteins, DRD2 is a GPCR protein with seven transmembrane regions. Apart from the split at the IC3 region, an extra helix is observed at the C terminal of the protein. This structure has 91.6 % of the residues in the allowed region and 7.6 % of the residues in the additionally allowed region of the Ramachandran plot.

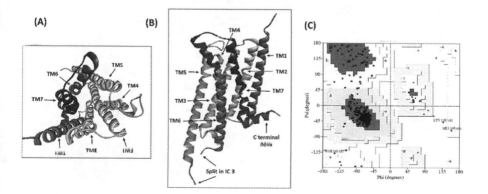

FIGURE 12.3 Structure of DRD2: (a) Top to bottom (Extra cellular to intracellular) view of the protein 3D structure. (b) Longitudinal view of the 3D structure of DRD2 obtained through GPCRdb. (c) Ramachandran Plot obtained using Procheck.

12.3 FEATURE TABLE GENERATION

Protein sequences are the collection of amino acids which are represented through twenty letters. To perform any kind of computational analysis, it is necessary to convert the amino acid sequences into numeric values. These numeric features can be used as the input data for the Neural Network classifiers [17]. Features are considered for the transformation of all of the positive datasets, negative datasets, and test datasets into the numeric feature vectors. Figure 12.4 represents the overview of all the amino acid descriptors used for the feature table generation. All the amino acid properties and their corresponding features are discussed in this section. R programming language is used for generating all the features. 'Proctr' package was used for deriving AAC, PAAC, and CTD descriptors, and 'Peptides' package was used to compute the physicochemical based descriptors like aaComp, Charge, InstaIndex, and molwt.

12.3.1 AMINO ACID COMPOSITION

1. **Monopeptide composition:** The frequency of occurrence of twenty amino acids comprising the polypeptide chain varies from protein to protein. This property is based on the preference for the usage of certain amino acids by specific proteins to attain specific stable structural and functional conformation. The frequency of amino acids when multiplied by one hundred and divided by the length of the polypeptide gives the Amino Acid composition values. Twenty numerical values are obtained for twenty amino acids.

$$AAC(i) = \frac{\sum_{i}^{20} N_i}{L} \times 100 \qquad (12.1)$$

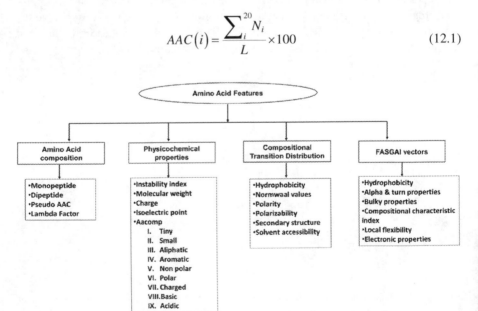

FIGURE 12.4 Overview of all the amino acid descriptors used for preparation of feature table.

Where N_i is the number of times one of the twenty amino acid residues, i occurs within the protein with length L. AAC of a protein is also known as the percentage of occurrence of the twenty amino acid residues in that protein [18].

2. **Dipeptide composition:** This is based on the predominance of occurrence of certain amino acid pairs as compared to other in specific polypeptides.

$$DiAAC(i,j) = \frac{\sum N_{i,j}}{\sum N_i + N_j} \times 100 \tag{12.2}$$

Where, $N_{i,j}$ is the number times residues i and j occur together in paired form as a dipeptide. N_i and N_j are the number of amino acid residue i and the number of amino acid residue j respectively [18]. Here the normalization of the frequency of dipeptide can also be done by changing the denominator from $(N_i + N_j)$ to L, where L is the length of the protein.

Each amino acid can be in pairing with every other amino acid, for which there can be $20 * 20 = 400$ possible combinations. As shown in Figure 12.5, all the possible dipeptides can be represented through a matrix with 400 components. The first row shows the combination of amino acid A with itself and every other amino acid. The second row shows a combination of C with itself and every other amino acid and so on. Similarly, the amino acid composition of K spaced amino acid pairs can be obtained, and these features can be used for protein classification [19].

3. **Pseudo-amino acid composition:** Sometimes the proteins cannot be distinctively defined by using only AAC features, as it may fail to extract proper information from the proteins. So pseudo amino acid composition (pAAC) is considered to make up for this failure. pAAC is the modified form of AAC, which includes twenty AAC values along with λ tier correlation factors (θ_λ) resulting in the formation of twenty $+ \lambda$ components [20]. In our study, we have got $20 + 30 = 50$ numerical vectors. The λ factor depends

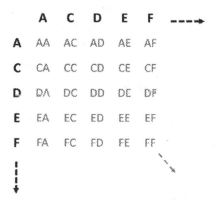

FIGURE 12.5 Possible combinations amino acid residues to form 400 possible dipeptide.

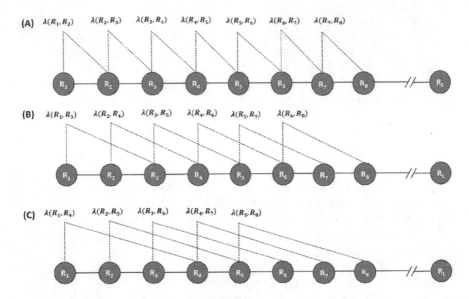

FIGURE 12.6 Correlation between the contiguous amino acids of a protein: (a) Correlation between i & i + 1 positioned amino acid. (b) Correlation between i & i + 2 positioned amino acids. (c) Correlation between i & i + 3 positioned amino acids.

on the hydrophobicity and molecular weight of the amino acids. Factor λ is an integer between 0 to N where N is the length of the protein sequence.

$$X = \left[X_1, X_2, X_3 \ldots X_{20}, X_{20+1}, X_{20+2}, X_{20+3} \ldots X_{20+\lambda} \right] \qquad (12.3)$$

When $\lambda = 0$, then the 20 dimensional aminoacid composition values are the output $(X_1, X_2, X_3 \cdots X_{20})$ and depending upon the number of correlation factors (θ_λ), λ dimensional pAAC values are generated $(X_{20+1}, X_{20+2}, X_{20+3} \cdots X_{20+\lambda})$. Figure 12.6, demonstrates the first, second, and third order correlation modes along the protein sequence. The first tier correlation is between i to i + 1 positioned amino acid residues, the second tier correlation is between i to i + 2 positioned amino acid residues and the third tier correlation is between i to i + 3 positioned amino acid residues [21].

12.3.2 Physicochemical Properties

Protein features can be extracted based on the physical and chemical properties of the amino acids or proteins [22].

1. **Charge:** Based on the functional (-R) group present in the amino acids, amino acids are either positively or negatively charged. The property can be calculated at particular pH values. The net charge of the proteins is calculated using Lehninger's pK scale at pH 7.

$$netZ = \sum_{i=i}^{n} \left(\frac{Z_{i_d}}{10^{pK_i - pH} + 1} + \frac{Z_{i_u}}{10^{pH - pK_i} + 1} \right) \tag{12.4}$$

2. **Molecular Weight:** Molecular weight of the peptide sequence is the summation of individual residues' weights subtracted by the molecular weight of (N-1) number of water molecules.

$$MolWt = \left[\sum_{i=1}^{20} N_i \times wt_i \right] - (L-1) \times 18 \tag{12.5}$$

This is measured in Dalton. Where, N_i corresponds to the number of each amino acid residue i and wt_i is the molecular weight of amino acid i; L corresponds to the length of the polypeptide and 18 is the weight of one H2O molecule.

3. **Instability Index:** It is the measure of the ability of a protein to retain its primary, secondary and tertiary structural conformations, especially when it is taken out of cellular environment, i.e., in *in-vitro* condition when there is variation in the protein's environment (solvent or temperature). It is computed based on Instability weight values for consecutive or contiguous dipeptide residues. The higher the value is, the more unstable the protein is, and the lower the value is, the more stable the protein is in the solvent. Stable proteins can be easily crystallized, unlike unstable ones. The score of forty is considered as the cutoff value below which the proteins are said to be stable. It can be calculated using the following notation:

$$InstIndex = \frac{10}{L} \sum_{i=1}^{L-1} Diwt \left(X_{i,i+1} \right) \tag{12.6}$$

Where L is the length of the protein sequences, $Diwt$ is the instability weight values for amino acid dipeptide and $X_{i,i+1}$ where the value of i ranges from 1 to L-1

4. **Isoelectric Point:** It is well known as the point on a surface with a pH gradient where the net charge of an amino acid becomes zero. It is represented by:

$$pI \left(X_u \right) = \frac{pK_{a1} - pK_{a2}}{2} \tag{12.7}$$

The net pI of a protein, X would be the summation of the pI values of all amino acids constituting the protein from the position, $a = 1$ to L, the length of the sequence:

$$pI \left(X \right) = \sum_{a=1}^{L} pI \left(X_a \right) \tag{12.8}$$

5. **Amino Acid Composition of Particular Class:** Amino acids can be classified into nine to eleven classes, based on their physicochemical properties. These properties contribute to the difference in the functionality of the proteins. The difference in the amino acid properties is the outcome of the difference in the structure of the side chains (-R groups) attached to each amino acid residue. Nine numerical values were generated based on nine categories:. Aromatic, Aliphatic, Acidic, Basic, Charged, Non-polar, Polar, Small and tiny amino acid [17]. Each amino acid may fall into one or more than one category. It may be calculated using the following notation:,

$$AAcomp(t) = \frac{\sum N_t}{L} \times 100 \qquad (12.9)$$

Where N_t is the amino acid of a particular type or category and L is the length of the sequence. In the equation one hundred may or may not be multiplied to the fraction. Table 12.2 shows the distribution of amino acids into the nine categories mentioned here.

12.3.3 COMPOSITION TRANSITION DISTRIBUTION

CTD encompasses the set of descriptors where C defines the number of particular characteristics of amino acid, T describes the frequency percentage of amino acid and D measures the length of the chain of a particular property of amino acid [23].

1. **Hydrophobicity:** Based on the hydrophobicity profile or scale, the amino acids can be grouped into highly hydrophobic or less hydrophobic, hydrophilic, or highly hydrophilic [24]. The average hydrophobicity of the protein can be obtained by using the formula:

$$AvgHb = \frac{\sum N_i \times hp_i}{L} \qquad (12.10)$$

TABLE 12.2
Amino Acids Grouped Based on Nine
Physicochemical Properties Used in This Chapter

Subset of Amino Acid	Amino Acids
Aliphatic subset	I, L, V
Aromatic subset	F, H, W, Y
Acidic subset	D, E
Basic subset	H, K, R
Charged subset	D, E, H, R, K
Nonpolar subset	A, C, F, G, I, L, M, P, V, W, Y
Polar subset	D, E, H, K, N, Q, R, S, T
Small subset	A, C, D, G, N, P, S, T, V, W, Y
Tiny subset	A, C, G, S, T

TABLE 12.3
Hb1. Tanford-Nozaki Hydrophobicity Values Computed for Each Amino Acid Based on Relative Solubility in Water and Ethanol. Hb$_2$. Ponnuswamy Gormiha Hydrophobicity Computed Based on the Surrounding Amino Acid (Computationally) within a Specific Radius. C. Kyte Doolite (1982) Experimentally Predicted Values

Amino Acid	Hb$_1$	Hb$_2$	Hb$_3$
A	0.87	13.85	1.8
C	1.52	15.37	2.5
D	0.66	11.61	−3.5
E	0.67	11.38	−3.5
F	2.87	13.93	2.8
G	0.1	13.34	−0.4
H	0.87	13.82	−3.2
I	3.15	15.28	4.5
K	1.64	11.58	−3.9
L	2.17	14.13	3.8
M	1.67	13.86	1.9
N	0.09	13.02	−3.5
P	2.77	12.35	−1.6
Q	0	12.61	−3.5
R	0.85	13.1	−4.5
S	0.07	13.39	−0.8
T	0.07	12.7	−0.7
V	1.87	14.56	4.2
W	3.77	15.48	−0.9
Y	2.67	13.88	−1.3

α Helical membrane proteins have a much higher hydrophobicity value as compared to cytoplasmic proteins. In the mathematical notation, N_i represents the number of amino acid i and hp_i represent the hydrophobicity value of ith amino acid and L represents the length of the polypeptide. Hydrophobicity values of twenty amino acids are listed in Table 12.3 [25–27].

2. **Solvent accessibility:** Solvent accessible surface area can be computed by finding the number of times the solvent molecule is known as a probe molecule with known radius, can rollover the protein molecule to access the Vander Waals surface of the whole protein [18]. The surface area of a sphere of radius R, will depend on the radius of the sphere of radius r that will roll over it. So, ASA of a protein can be represented as:

$$ASA = n\left(4\pi r^2\right) \tag{12.11}$$

Where n is the number of times the solvent molecule with radius r rolls over the solute molecule. As depicted in Figure 12.7, a water molecule is generally considered as the solvent molecule whose a radius of 1.4 A^0. The solute

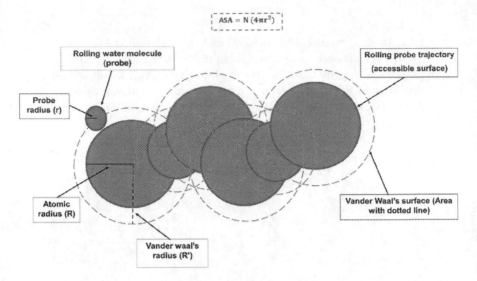

FIGURE 12.7 Rolling water molecule (probe) is shown to be moving around the peptide molecule, touching the surfaces of exposed amino acid residues.

molecule (protein) is represented by a set of interlocking spheres where each sphere represents the Vander Waals radii of the atoms like carbon, hydrogen, oxygen, nitrogen, sulfur, *etc.*, that are present within the protein. Buried molecules or residues of a solute cannot be accessed by the solvent, whereas exposed molecules are partially or fully accessible. For water as solvent, hydrophobic residues would be buried and hydrophilic would be exposed.

3. **Van der Waals interaction:** When two atoms are at an infinite distance apart, they cannot interact with each other. But after they are brought close enough so as to allow them to interact with each other by applying some amount of force, they will create an attractive force between them. This attractive force between the two entities tend to align them even closer, to attain a state of equilibrium also known as the lowest free energy state. At this point, the electrostatic forces between the atoms are perfectly balanced. This distance is known as Van der Waals distance. To bring the spheres even closer to make their Van der Waals radii intersect with each other will require additional energy from an outside source.

4. **Polarity and Polarizability:** Polarazibility is the tendency of any molecule to attain a state of electrical dipole where the center of positive charge and the center of negative charge are not properly balanced. Polarizable molecules are known as polar molecules, and the molecules that can polarize other molecules are known as polarizing molecules. Water has a polarizable as well as polarizing tendency [28]. Non-polar molecules make linear structures, while polar ones make nonlinear structures. In general, negatively charged large-sized ions are polarizable, whereas positively charged small-sized ions are polarizing in nature. Amino acids like Arg (R), Lys (K), Asp (D), Glu (E), Asn (N), and Gln (Q) are highly polar and amino acids like His

(H), Ala (A), Tyr (Y), Thr (T), Ser (S), Pro (P) and Gly (G) are less polar. The rest of the amino acids are non-polar in nature [29]. Polar amino acids are water lovers and are present towards the surface of the protein.

5. **Secondary structural elements:** Any polypeptide can fold to attain its functionally stable confirmation. Different kinds of secondary structural elements that can be found within any protein fall into the following categories: Helices are either α helix, π helix or three turn helix. Three turn helices are formed when hydrogen bonding is between i & i+3 positioned amino acids of the polypeptide. In α helix, the hydrogen bond is between i & i+4 positioned amino acid, and in π helix, that is between i & i+5 positioned amino acids. Strands can be parallel or antiparallel in orientation. Or these strands can bend, turn, coil, and have irregular structural patches. These subsequences generally join the helices and/or strands. The specific type of structure is predominant in certain protein families. So, finding the probability of the presence of any such structural elements plays an important role in protein-protein classification.

12.3.4 FASGAI Vectors

As many as more than 500 physicochemical indices can be generated from an amino acid sequence. Factor analysis of 335 physicochemical indices reduces the dataset into a much smaller dataset, with only six factors that describe the correlated variables: Hydropathy Index (F1), compositional characteristics Index (F2), local flexibility (F3), bulky properties (F4), alpha and turn propensities (F5), and electronic properties (F6) [30].

1. **Hydrophathy Index:** It denotes the hydrophobic profile of a polypeptide. The addition of the hydropathy indices of individual amino acids that are present in a polypeptide gives the GRAVY or Grand Average Hydropathy value of the polypeptide. Higher values indicate that the query protein is highly hydrophobic, and lower values indicate the query is highly hydrophilic in nature [24].

2. **Alpha and turn propensities:** Amphipathicity refers to the periodicity in the polar or non-polar character of amino acid sequences in a protein sequence resulting in one-dimensional function [18]. The average hydrophobicity of residues constituting the edge i (i = 1, 4) is given by:

$$\alpha_i = \frac{\sum_0^M h_{i+j}}{n} \tag{12.12}$$

Where, n is the total number of residues in the edge and j increases in the interval of four, from 0 to M, where M is the number of residues in the helix and h is the hydrophobic index of the residue 'i'. In helix, the periodicity of two consecutive amino acids being more hydrophobic in nature than that of the other two following consecutive amino acids being less hydrophobic in nature is being observed.

Similarly, the β strand is considered to have a periodicity of one amino acid with a high hydrophobicity value and the consecutive one amino acid with law. Hydrophobicity value is observed. The average hydrophobicity of residue constituting the face i ($i = 1, 2$) is given by

$$\beta_i = \frac{\sum_0^m h_{i+k}}{n} \qquad (12.13)$$

Where n is the total number of residues in the face. k Increases at an interval of two, from 0 to m, where m is the number of residues in the strand and h is the hydrophobic index of residue i.

3. **Local flexibility:** Within a protein structure, local flexibility, or rotations, is allowed at the peptide bonds but are restricted to some extent due to the presence of the different types of side chains (R) attached to the main chain. Rotations are represented by torsional angles Φ and Ψ. Φ is the angle between Cα and C atoms and Ψ is the angle between Cα and N. The portions of the polypeptide with Φ and Ψ angles indicating collision between atoms is in sterically disallowed conformation [31]. Unlike an angle that is formed between two lines, a dihedral angle is formed between two planes where 1, 2, and 3 atoms may form plane 1 and 2, 3 and 4 atoms may form plane 2, with atoms 2 and 3 common between the two planes. The torsional angles can be calculated if the coordinates (x, y, z) of the atoms forming the dihedral angles are available. Based on the number of rotamers and their orientation an occupancy, value can be calculated that falls between 0 to 1.

   Another factor known as the B factor also defines the flexibility of each atom in a protein structure [32]. If we can hold an atom rigidly fixed in one place, we could observe its distribution of electrons in an ideal situation. Usually, the electron has a wider distribution than its ideal form. This may be due to the vibration of the atoms or the differences between the many different molecules in the crystal lattice. B factor or temperature factor leading to disruption of the ideal structure can be calculated. If the B factor is less than fifty, the atom is highly rigid, and if it is more than fifty, then the atom is highly flexible.

4. **Bulky properties:** Bulky side chains of a protein promote Hydrophobic interaction within the protein. Bulky property enhances the proteolytic stability and folding process. The estimate of Amino acids with bulky side chains is Alanine, Valine, Leucine, Isoleucine; their interacting capacity is an important aspect in protein classification.

5. **Electronic properties:** The presence of free radicals within the proteins is known as the main reason behind the electronic property of a protein. Some proteins have exposed free radicals, and some have been buried. The presence of cofactor (metal ions) near the active sites of enzymatic proteins facilitates the tunneling of electron charge carriers. This property leads to the conduct of several signaling pathways within biological systems [33]. Nevertheless, research on the artificial retina, artificial auditory system,

prosthetic limbs, *etc.*, revolves around harnessing this electronic or electrical property of proteins. Such proteins are generally structural or enzymatic proteins of the cell. An electrical gradient may be generated facilitating the movement of electrons either by exposing it to an electrolyte solution or by immobilizing the protein onto a solid electrode, and thus the electrochemical properties can be measured. Electron flow through a protein molecule involves intramolecular charge transport and electron exchange with the surrounding or also within itself [34]. So, the electron may either be transferred (ET) or transported (ETp)

6. **Compositional characteristics Index:** As every amino acid is structurally different from each other, the physicochemical properties of each of the twenty amino acids differ from each other. This difference at the amino acid level leads to the variations in the properties of the protein molecule as a wholesome effect, since the majority of features of a protein depend on the composition of amino acids within the protein. As shown in Figure 12.8, the graphs showing molecular weight, Hydrophobicity, and Isoelectric point properties indexes of twenty amino acids placed at **A, B,** and **C** sections respectively. Section **D** of Figure 12.8, displays the graph plotted taking AAC values for a schizophrenia-associated protein (DRD2) and a protein not associated with schizophrenia (TNFSRF).

12.4 DEEP NEURAL NETWORK

In this Section, the implementation of Deep Neural Network (DNN) is discussed for classification purposes.

Deep Learning is a promising Machine Learning technology that attempts to replicate and recreate the neural network seen in the human brain in order to understand and interpret data. In general, DNN has four or more hidden layers, with the first layer receiving the input. The final layer provides the projected output. Each hidden layer's artificial neuron computes the weighted total, which is then passed through an activation function, which in this case is a rectified linear unit function (ReLU). The positive values are sent on while the negative values are blocked by this activation function. The training time is reduced by using the activation function. In addition, to speed up the training process, the cross-entropy function is used as the loss function [35, 36, 37]. The proposed DNN architecture is similar to the implementation used in [35, 38]; the training is performed using error back back-propagation with a stochastic gradient descent algorithm with momentum. In the proposed implementation, neurons used in the first, second, third, and fourth layers are sixty-four, thirty-two, sixteen, and eight respectively. The block diagram of the DNN implementation is shown in Figure 12.9. A total of 119 features are used for a three-class classification problem. The Softmax Activation function is employed at the last layer to normalize the output of a network to a probability distribution over expected output classes. The dataset was divided into two categories, training and testing. For training, 80% of the data is used and 20% of the dataset is used for testing and validation, using a five-fold cross-validation scheme. After simulation, the classification accuracy is 94%. Further, the performance is compared using two standard classifiers, Support Vector

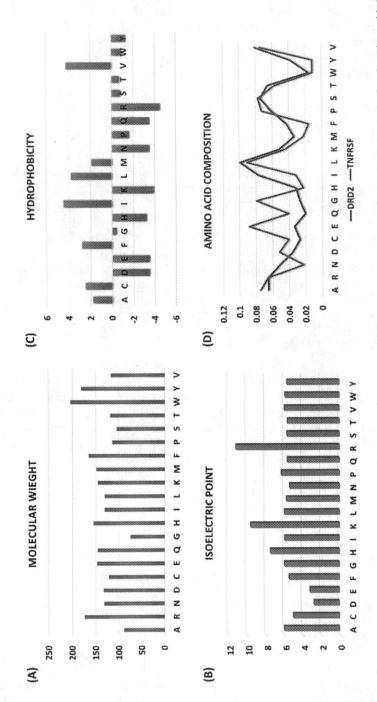

FIGURE 12.8 Variations in the Features of individual amino acids leading to differences at protein level: (a) Molecular weight of each amino acid. (b) Hydropathy index of each amino acid. (c) pH of amino acids at Isoelectric point. (d) Comparison between AAC descriptors of one from positive dataset (DRD2) and another from negative dataset (TNFRSF).

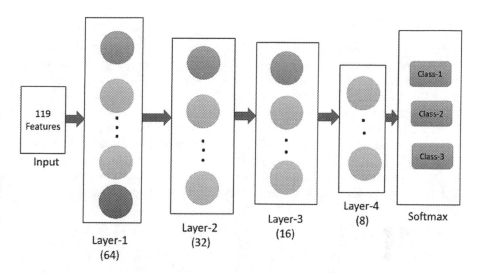

FIGURE 12.9   The Deep Neural Network Model.

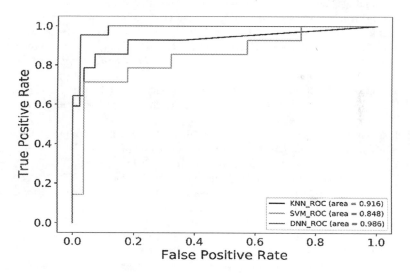

FIGURE 12.10   Comparison of ROC curves of different classifiers.

Machine [39], and K-NN algorithm [40]. For visualization of the results, the Receiver operator characteristic (ROC) curve is used. It is a two-dimensional plot between the true positive rate (TPR) and the false false-positive rate (FPR), which provides the relative tradeoffs between true positives and false positives. The ROC curves are shown in Figure 12.10. The proposed DNN based classifier has a better area under the curve (AUC) than the other classifiers, with an AUC of 0.985. This demonstrates the importance of the proposed features in the protein classification task.

12.5 CONCLUSION

In the detection of Schizophrenia disorder, the Amino Acid Descriptors-based features are used. Four broad types of features are extracted under the category of Amino Acid Composition, Physicochemical features, Composition Transition Description, and FASGAI vectors. Then the database is divided into three categories, such as "schizophrenia with high, moderate, and low intensity." A total of 119 features are used for a Deep Neural Network-based classifier with four hidden layers, which provides 94% classification accuracy. The experimental results demonstrate the efficacy of the proposed Amino Acid features and the DNN based classification of protein set. Although the proposed method performs better than the standard classifiers, the features can be evaluated in larger datasets and considering other diseases.

REFERENCES

1. Martha, S. R., D. Mallik, and M. Patri. 2017. Genome wide screening and analysis of *Homo sapiens* genes and proteins associated with schizophrenia. *Research Journal of Recent Sciences* 6(7): 63–70. http://www.isca.in/rjrs/archive/v6/i7/8.ISCA-RJRS-2017-063.php.

2. Bhadra, P., J. Yan, J. Li, S. Fong, and S. W. I. Siu. 2018. AmPEP: Sequence-based prediction of antimicrobial peptides using distribution patterns of amino acid properties and random forest. *Scientific Reports*, 8: 1697. doi: 10.1038/s41598-018-19752-w.

3. Martha, S. R., and M. Patri. 2020. Network driven discovery of dopamine, serotonin and glutamate receptors as key players in schizophrenia, *Journal of Scientific and Industrial Research*, 79 (5): 406–412.

4. Macarthur, J., E. Bowler, M. Cerezo, et al. 2017. The new NHGRI-EBI Catalog of published genome-wide association studies (GWAS Catalog). *Nucleic Acids Research*, 45: 896–901. doi: 10.1093/nar/gkw1133.

5. Jourquin, J., D. Duncan, Z. Shi, et al. 2012. GLAD4U: Deriving and prioritizing gene lists from PubMed literature. *BMC Genomics*, 201213(8):. doi: 10.1186/1471-2164-13-S8-S20.

6. Kevin, H. J., R. B. Jeffrey, S. Katrin, et al. 2015. Clinical Pharmacogenetics Implementation Consortium (CPIC) guideline for CYP2D6 and CYP2C19 genotypes and dosing of selective serotonin reuptake inhibitors. *Clinical Pharmacology and Therapeutics*, 98(2): 127–34. doi: 10.1002/cpt.147.

7. Bauer-Mehren, A., M. Bundschus, M. Rautschka, et al. 2010. DisGeNET: A Cytoscape plugin to visualize, integrate, search and analyze gene–disease networks. *Bioinformatics*, 26(22): 2924–2926. doi: 10.1093/bioinformatics/btq538.

8. Frezal, J. 1998. Genatlas database, genes and development defects. *Comptes Rendus de l'Académie des Sciences. Série III*, 321(10): 805–817.

9. Law, V., C. Knox, Y. Djoumbou, et al. 2014. DrugBank 4.0: Shedding new light on drug metabolism. *Nucleic Acids Research*. 42. doi: 10.1093/nar/gkt1068.

10. Stelzer, G., I. Dalah, S. T. Iny, et al. 2010. In-silico Human Genomics with GeneCards. *Human Genomics*, 5 (6): 709–717. doi: 10.1186/1479-7364-5-6-709.

11. Boutet, E., D. Lieberherr, M. Tognolli, et al. 2016. UniProtKB/Swiss-prot, the manually annotated section of the UniProt KnowledgeBase: How to use the entry view. *Methods in Molecular Biology*, 1374: 23–54. doi: 10.1007/978-1-4939-3167-5_2.

12. Szklarczyk, D., A. Franceschini, S. Wyder, et al. 2015. STRING v10: protein-protein interaction networks, integrated over the tree of life. *Nucleic Acids Research*, 43: 447–52. doi: 10.1093/nar/gku1003.

13. Goodsell, D. S., S. Dutta, C. Zardecki, et al. 2015. The RCSB PDB "Molecule of the month": Inspiring a molecular view of biology. *PLoS Biology* 13(5): e1002140. doi: 10.1371/journal.pbio.1002140.

14. Pndy-Szekeres, G., C. Munk, T. M. Tsonkov, et al. 2017. GPCRdb in 2018: adding GPCR structure models and ligands. *Nucleic Acids Research*, 16. doi: 10.1093/nar/gkx1109.

15. Laskowski, R. A., M. W. MacArthur, and J. M. Thornton. 2001. PROCHECK: validation of protein structure coordinates, in *International Tables of Crystallography, Volume F. Crystallography of Biological Macromolecules*, Eds. Rossmann, M. G., and E. Arnold, Dordrecht, Kluwer Academic Publishers. 722–725.

16. Dassault Systmes BIOVIA, 2016. Discovery Studio Modeling Environment. *Dassault Systmes*. https://www.3dsbiovia.com/products/collaborative-science/biovia-discovery-studio/.

17. Vishnoi, S., P. Garg, A. Pooja. 2020. Physicochemical n-Grams Tool: A tool for protein physicochemical descriptor generation via Chou's 5-step rule. *Chemical Biology & Drug Design*, 95: 79–86.

18. Gromiha, M. M. 2010. *Protein Bioinformatics*. Academic Press (Elsevier)

19. Zhang, L., B. Dong, Z. Teng, Y. Zhang, and L. Juan. 2020. Identification of human enzymes using amino acid composition and the composition of k-spaced amino acid pairs. *BioMed Research International*, 2020, 9235920, doi:10.1155/2020/9235920.

20. Chou, K. 2001. Prediction of protein cellular attributes using pseudo- amino acid composition. *Proteins: Structure, Function, and Genetics*, 43(3): 246–255.

21. Akbar, S., A. U. Rahman, H. Maqsood, S. Mohammad. 2020. cACP: Classifying anti-cancer peptides using discriminative intelligent model via Chou's 5-step rules and general pseudo components. *Chemometrics and Intelligent Laboratory Systems*, 196 103912.

22. Tyunina, E. Y., and V. G. Badelin. 2009. Molecular descriptors of amino acids for the evaluation of the physicochemical parameters and biological activity of peptides. *Russian Journal of Bioorganic Chemistry* 35(4): 453–460.

23. Wang, Y, Y. Guo, X. Pu, and M. Li. 2017. A sequence-based computational method for prediction of MoRFs. *RSC Advances*, 7(31): 18937–18945.

24. Simm, S., J. Einloft, O. Mirus, and E. Schleiff. 2016. 50 years of amino acid hydrophobicity scales: revisiting the capacity for peptide classification. *Biological Research*, 49: 31. doi:10.1186/s40659-016-0092-5.

25. Nozaki, Y., and C. Tanford. 1971. The solubility of amino acids and two glycine peptides in aqueous ethanol and dioxane solutions. Establishment of a hydrophobicity scale. *The Journal of Biological Chemistry*, 246(7): 2211–7.

26. Ponnuswamy, P. K., and M. M. Gromiha. 1993. Prediction of transmembrane helices from hydrophobic characteristics of proteins. *International Journal of Peptide and Protein Research*, doi:10.1111/j.1399-3010.1993.tb00502.x

27. Kyte, J., R. F. Doolittle. 1982. A simple method for displaying the hydropathic character of a protein. *Journal of Molecular Biology*, 157(1): 105–32.

28. Millefiori, S., A. Alparone, A. Millefiori, and A. Vanella. 2007. Electronic and vibrational polarizabilities of the twenty naturally occurring amino acids. *Biophysical Chemistry*, 132 (2–3): 139.

29. Swarta, M., J. G. Snijdersb, and P. T. Duijnenb. 2004. Polarizabilities of amino acid residues. *Journal of Computational Methods in Science and Engineering*, 4: 419–425.

30. Liang, G., L. Yang, Z. Chen, H. Mei, M. Shu, and Z. Li. 2009. A set of new amino acid descriptors applied in prediction of MHC class I binding peptides. *European Journal of Medicinal Chemistry*, 44(3): 1144–1154.

31. Hollingsworth, S. A., and P. A. Karplus 2010. A fresh look at the Ramachandran plot and the occurrence of standard structures in proteins. *Biomolecular Concepts*, 1(3–4): 271–283.
32. Berman, H. M., J. Westbrook, Z. Feng, G. Gilliland, T. N. Bhat, H. Weissig, I. N. Shindyalov, P. E. Bourne. 2000. The protein data bank. *Nucleic Acids Research*, 28: 235–242.
33. Marcus, R. A., N. Sutin. 1985. Electron transfers in chemistry and biology, *Biochimica et Biophysica Acta (BBA) - Reviews on Bioenergetics*. 811(3): 265–322. doi:10.1016/0304-4173(85)90014-X.
34. Winkler, J. R., and H. B. Gray. 2014. Electron flow through metalloproteins. *Chemical Reviews*, 114 (7), 3369–3380. doi: 10.1021/cr4004715.
35. Mahapatra, S., Gupta, V. R. R., Sahu, S. S., and Panda, G. (2021). Deep neural network and extreme gradient boosting based Hybrid classifier for improved prediction of Protein-Protein interaction. *IEEE/ACM Transactions on Computational Biology and Bioinformatics* (2021).
36. Dash, T.K. and Solanki, S.S., 2020. Speech intelligibility based enhancement system using modified deep neural network and adaptive multi-band spectral subtraction. *Wireless Personal Communications*, 111(2), pp. 1073–1087.
37. Mahapatra, S., and Sahu, S. S. (2020, February). Boosting predictions of Host-Pathogen protein interactions using Deep neural networks. In *2020 IEEE International Students' Conference on Electrical, Electronics and Computer Science (SCEECS)* (pp. 1–4). IEEE.
38. Zhang, L., Yu, G., Xia, D. and Wang, J., 2019. Protein–protein interactions prediction based on ensemble deep neural networks. *Neurocomputing*, 324, pp. 10–19.
39. Mohabatkar, H., Beigi, M.M. and Esmaeili, A., 2010. Prediction of GABAA receptor proteins using the concept of Chou's pseudo-amino acid composition and support vector machine. *Journal of Theoretical Biology*, 281(1), pp. 18–23.
40. Xiao, X., Wang, P. and Chou, K.C., 2010. GPCR-2L: predicting G protein-coupled receptors and their types by hybridizing two different modes of pseudo amino acid compositions. *Molecular BioSystems*, 7(3), pp. 911–919.

13 Deep Learning Architectures, Libraries and Frameworks in Healthcare

Nongmeikapam Brajabidhu Singh, and Moirangthem Marjit Singh

North Eastern Regional Institute of Science & Technology, Nirjuli, India

Arindam Sarkar

Ramakrishna Mission Vidyamandira, Howrah, India

CONTENTS

DOI: 10.1201/9780367548445-15

13.1 INTRODUCTION

Recently, Deep Learning has become quite prominent, effectively providing highly satisfactory results in solving many complex problems. The potentials and values of DL are unlimited. DL is able to slowly adapt itself to any problem domain, even though the task is quite challenging, because of DL's cognitive abilities and features as this DL architecture mimics the inner working of neurons in the central nervous systems of dissimilar animals. The problem of not having a large enough number of quality datasets has been a major issue in Deep Learning. But recently many Deep Learning techniques, like deep RL and GAN architectures, have emerged. These are quite successful in tackling the issues of insufficient datasets or not having any datasets at all. Every Deep Learning architecture needs a good hardware configuration in order to deliver a good performance. The heavy dependance of Deep Learning architectures on hardware is due to the fact that advanced DL architectures have complex hidden layers, requiring a huge amount of computational resources in order to train a very deep DL architecture. Working in the field of DL involves using many libraries, frameworks, advanced Machine Learning techniques, optimization and loss algorithms, high performance hardware, and huge labeled datasets. Without these resources, Deep Learning would not have been able to progressively develop so fast just in a few years. It is correct to say that we have been experiencing DL without knowing it [1]. Very recently, researchers have attempted using DL techniques to deal with the COVID-19 pandemic, as well [2–4].

This survey chapter has been organized as follows: (1) For the sake of convenience, the full forms of the abbreviations being used in this chapter are given in Table 13.1. (2) A detailed overview of DL, including its history, techniques, libraries, and frameworks has been given in Section 13.2. A detailed discussion on various major Deep Learning frameworks and libraries has been presented in Section 13.2.2. In Section 13.3, a detailed discussion is given on various basic and advanced DL architectures.

TABLE 13.1

Abbreviations in Deep Learning

ANN	Artificial Neural Network	RL	Reinforcement Learning
NTM	Neural Turing Machine	DRN	Deep Residual Network
GAN	Generative Adversarial Network	DCIGN	Deep Convolutational Inverse Graphics Network
DBN	Deep Belief Network	RBM	Restricted Boltzmann Machine
FNN	Fuzzy Neural Network	CE	Cross-Entropy
AE	Auto Encoder	SAE	Sparse Auto Encoder
DAE	Denoising Auto Encoder	VAE	Variational Auto Encoder
LSTM	Long Short-Term Memory	GRU	Gated Recurrent Unit
DFF	Deep Feed Forward	DN	Deconvolutional Network
SL	Supervised Learning	UL	Unsupervised Learning
BM	Boltzmann Machine	ANN	Artificial Neural Network
PTB	Penn Tree Bank	NMAE	Normalize Mean Absolute Error
NRMSE	Normalize Root Mean Square Error	NWP	Numerical Wealth Prediction
PNN	Probailistic Neural Network	IoU	Intersection over Union
FCN	Fully Connected Network	MAE	Mean Absolute Error
PeMS	Portfolio Measurement System	RL	Reinforcement Learning
Pixel Acc	Pixel accuracy	CRR	Character Recognition Rate
NMI	Normalized Mutual Information	FSIM	Feature Similarity Index Metric
RMSEc Edge	Root Mean Squared Error of the 3D Canny	SAC	Self-Adaptive Controller
BP	Back Propagation	MSLE	Mean Squared Log Error
DL	Deep Learning	RNN	Recurrent Neural Network
GAN	Generative Adversarial Network	DNN	Deep Neural Network
ML	Machine Learning	MLP	Multilayer Perceptron

13.2 DEEP LEARNING

13.2.1 OVERVIEW OF DEEP LEARNING

DL is a complex sub-field of a class of artificial intelligence called Machine Learning (as shown in Figure 13.1) where the DL model progressively extracts higher level features from raw input data or from interactions with the environment. Unlike Machine Learning, DL learns multiple levels of data representation using a hierarchical multiple layers architecture, with each layer applying its own transformation.

Every Deep Learning architecture uses Artificial Neural Network techniques in order to learn higher level features from raw or random input or through environmental interactions. ANNs (or NNs) are machine learning architectures that learn to do cognitive tasks (like pattern recognition, decision making) in a humanlike way by

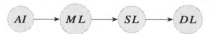

FIGURE 13.1 Deep Learning as a sub-field of AI.

mimicking the working concepts of biological neural networks in animal brains. In 1943, the first mathematical model of a NN, typically termed as McCulloch-Pitts neurons, had been developed by Walter Pitts and Warren McCulloch and was the foundation for all future NN models [5]. In 1957, Frank Rosen Blatt proposed a perceptron model called Cornell Photoperceptron [6]. This model is capable of pattern perception and generalization. The Cornell Photoperceptron model paved the way for all future DNN models. DNN is a feedforward ANN, with multiple hidden layers. The concept of feedforward is mathematically given by the Universal Theorem of Approximation of ANN Theory. The Universal Theorem of Approximation for width-n network with limited expressive power can be stated as follows:

Given any Lebesgue-integrable function $g: \mathbb{R}^m \to \mathbb{R}$ satisfying the condition: $\{z: g(z) \neq 0\}$ is a +ve measure set in Lebesgue measure, and if $\exists$ some function G_Λ represented by a fully connected ReLU net Λ with the width $d_m \leq m$, then,

$$\int_{\mathbb{R}_m} |g(z) - G_\Lambda(z)| da = +\infty \quad \text{or} \quad \int_{\mathbb{R}^m} |g(z)| dz \qquad (13.1)$$

Equation 13.1 holds true for a feedforward NN with hidden layer(s). ANNs such as CNNs, FCNs are classes of DNNs. In 1980, Kunihiko Fukushima introduced the CNN called necognitron [7]. Neocognitron is the first practical CNN model ever introduced. Basic layers such as convolutional layers and downsampling layers are present in neocognitron. In 1982, John Hopfield introduced Hopfield network, which is a class of RNN [8]. The subsequent RNNs are based on the ANN model, with backpropation by David Rumelhart in 1986.

Deep Learning as a whole is supervised in nature. Deep Learning architectures implemented for various major application domains may be supervised, semi-supervised, or unsupervised, or RL. Deep RL (such as Deep Q Network) belongs to the category of RL. In deep RL architecture, agents involved in the RL algorithm (like Q-algorithm) learn by interacting with the environment. We can choose our DL architecture to be supervised, semi-supervised, unsupervised, or RL. When the DL Model is being trained on the labeled dataset(s), having both input and output parameters, it is supervised learning. Training the DL Model on unlabeled dataset(s) not having output parameter(s) is unsupervised learning. Training with a small amount of unlabeled dataset and a large amount of labeled dataset will make it semi-supervised learning. No dataset is required for training a deep RL Model.

In Table 13.2, we have presented a brief history of Deep Learning. A summary of various major works and contributions by various researchers and scholars from the conception of neural net in 1943 and the conception of DL in 1957 to the state-of-the-art development of various advanced DL architectures in 2018 and 2019 has been presented. DL architectures belonging to the classes of ANNs, DNNs, and deep RL have been applied in fields like Healthcare, Climate Change, Biotechnology, Autonomous Driving, Biomedical Imaging, Gaming Technology, Cyber Security, NLP, and Forecasting, where researchers have been producing state-of-the-art results comparable to or better than the ability of human beings.

TABLE 13.2
A Brief Chronological History of Deep Learning

Year	Contributions
1943	Walter Pitts & Warren McCulloch developed the first mathematical model of an NN. This model typically termed as McCulloch-Pitts neurons had paved the foundation for the NN models invented later on [5].
1950	Alan Turing proposed the idea of Learning Machine and its development. He even hinted at genetic algorithms & neurons in computing. He created the Imitation Game which we called the Turing Test [9].
1952	Christopher Strachey had developed a program which could play the complete Draughts game on Manchester Machine at a fairly reasonable speed [10].
1952	Anthony G. Oettinger gave the earliest successful demonstration of Machine Learning. He created a Shopping Program running it on the EDSAC Computing Machine which mimicked the idea of a child sent on a shopping tour [11].
1957	Frank Rosen Blatt had paved the foundation for DNNs by proposing a perceptron model called Cornell Photoperceptron capable of pattern perception and generalization [6].
1959	David H.Hubel and Torsten Wiesel had discovered simple & complex cells in the cat primary visual cortex. This inspired the development of ANNs and heavily influenced the progress of DL [12].
1960	Henry J. Kelly proposed the idea of the behavior of systems with inputs and feedback mechanism in control theory. This idea has been applied to ANNs over the years. This has been used to develop the concept of backpropagation of errors used in training NNs [13].
1965	Alexecy Grigorevich Ivakhnenko proposed the first learning algorithms for DNNs by developing Group Method of Data Handling denoted by GMDH. He is considered the Father of Deep Learning by many people for this contribution [14].
1971	A. G. Ivakhnenko developed a multilayer perceptron network structure with 8 layers of selection for solving a complex problem. He presented a network implemented model of the British economy and the resuts forecasted by the model were also given. He implemented the deep learning model in a computer identification system called Alpha [15].
1980	Kunihiko Fukushima developed a multilayered neural network model called neocognitron for visual pattern recognition with unsupervised learning. The neocognitron has the ability of self-organization, stimulus pattern reconigtion based on the geometrical properties of th shape without being affected by any shift in position [7].
1982	J. J. Hopfield successfully developed a popular RNN called Hopfield Network with general and error correcting content addressable memory [8].
1986	D. E. Rumelhart, G. E. Hinton & R. J. Williams had proposed BP learning algorithm with the ability to create useful new features there by bringing significant improvement in the shape recognition and word prediction domain. G. E. Hinton is called the Godfather of Deep Learning by many [16].
1989	Yann LeCun proposed a neural network model using CNN and constrained BackPropagation producing state-of-the-art result in handwritten digit recognition task [17].
1989	Christopher J. C. H Watkins developed the improved reinforcement learning algorithm called the Q-learning algorithm where the agent involved tries to learn the optimcal policy from its own history of interaction with the environment under context [18].
1993	Jurgen Schmidhuber solved a very deep DL task using an RNN having more than 1000 layers [19].
1995	C. Cortes and V. Vapnik presented an improved SVM called Support Vector Network for 2-group classification and experimented with NIST and US Postal databases producing state-of-the-art results [20].
1997	Jurgen Schmidhuber and Sepp Hochreiter developed an improved RNN called LSTM with memory cells and gates solving the very problem of long time storage due to insufficient and decaying error back flow issues [21].
1998	Yann LeCun presented a better pattern recognition system using a Gradient Based approach called Graph Transformer Networks relying on automatic learning [22].

(Continued)

TABLE 13.2 (Continued)

Year	Contributions
2009	Jiang Dong developed a large scale 14 million plus labeled image dataset called ImageNet having the hierarchical structure of WordNet [23].
2012	Alex Krizhevsky developed a CNN called AlexNet and successfully implemented with ImageNet dataset achieving for the first time in history the state-of-the-art result with such a huge dataset [24].
2012	Google trained its deep NN to recognize cats in YouTube videos with unsupervised learning achieving more than 70% accuracy [25].
2014	Facebook AI Resource Team developed DeepFace System for face recognition trained with SFC, LFW, YTF datasets achieving 97.35% accuracy. DeepFace was built with 3D Model-base alignment and many large scale FeedForward Models [26].
2014	Ian J. Goodfellow created Generative Adversarial Networks. He is called the GAN father [27].
2015	Alec Radford, Luke Metz and Soumith Chintala developed deep convolutiional Generative Adversarial Network and trained with image datasets achieving state-of-the-art improvement and results [28].
2017	Phillip Isola of Berkeley AI Research developed conditional GAN for image translation tasks which involved highly structured graphical outputs [29].
2018	Google developed the state-of-the-art language representation model called BERT with 24 layers and 340M parameters for pre-training deep bidirectional representations from unlabeled text [30].
2019	Yikang Shen presented ordered neurons with ON-LSTM recurrent units thereby integrating tree structure into RNNs [31].
2019	DeepMind created a dataset of mathematical problems and analyzed the mathematical reasoning abilitites of NNs by training with state-of-the-art models of RNNs and attention/transformer [32].

13.2.2 DEEP LEARNING FRAMEWORKS AND LIBRARIES

In simple terms, a DL framework is the interface, library, or the tool for building DL models quickly and easily. DL frameworks provide some clear and concise ways for defining DL models using a well-defined collection of pre-built and optimized components [33]. DL frameworks offer building blocks for DL architectures. Popular DL frameworks, such as Google's Tensorflow, PyTorch, Sonnet, Keras, MXNet, DL4J and others, rely on some GPU-accelerated libraries, such as MAGMA, cuBLAS, cuSPARSE, LibSciACC, and cuRAND, to actually deliver high performance GPU accelerated training, which is several times faster than CPU accelerated training.

In order to build and deploy DL models to execute highly sophisticated tasks, we have numerous and diverse DL frameworks at our disposal, providing us deep learning tools. Such deep learning tools offer higher levels of abstraction and, in turn, simplify highly difficult programming challenges. When we choose a suitable DL framework, we can forget about wasting our time writing extra lines of code as we can use the framework to build our DL model very efficiently and quickly. A good DL framework is characterized by optimized performance, easy codes, community support, parallel computations, auto, or self-gradient computation. Let us examine some of the important DL frameworks to understand which framework would be the most perfect or useful in solving each of our programing challenges [33, 34]. Table 13.3 gives a detailed comparison of various DL libraries and frameworks

TABLE 13.3

Comparison of Deep Learning Libraries and Frameworks according to Github Data Collected on 30 March, 2020

Deep Learning Library/Framework	Written In	Stars(in Github)	Open Source	Contributors (in Github)	Used In	Used For
Tensorflow	C++	134502	Yes	2457	[35]	Classification, Perception, Understanding, Discovering, Prediction, Creation
Theano	Python	8923	Yes	332	[36]	Describing, Evaluating & Optimizing expressions (mathematics) by using Multi-dimensional arrays
CNTK	C++	16434	Yes	198	[37]	Describing DNNs and Other arbitrary learning machines
Eclipse DL4J	Java	11159	Yes	32	[38]	Integrating various DL models with Apache Hadoop & Spark
MXNet	Python	17729	Yes	786	[39]	Training & Deploying DNNs
PyTorch	C++	31755	Yes	1365	[40]	Used for DL applications including computer vision and NLP
Caffe	Shell	29100	Yes	256	[39]	Making research models, prototypes and industrial applications in multimedia, vision and speech
Caffe2(Integrated into PyTorch)	C++	8465	Yes	197	[41]	Fast prototype building, fast & less resource intensive training
BigDL	Scala	3163	Yes	65	[42]	Analyzing huge Hadoop Spark Cluster where data is stored in database such as HDF, HBase, Hive etc. adding DL functionalities to Spark programs
DeepLearnToolbox	Matlab	3365	Yes	12	[43]	Developing & training DL models in matlab
Pandas	Python	21384	Yes	1897	[44]	Real world data analysis in Python
NVIDIA DIGITS	HTML	3781	Yes	42	[45]	For efficient & fast training of DNNs in Image classification Segmentation & object detection
Lasagne	Python	3662	Yes	64	[46]	Developing & training NNs in Theano
Numpy	C	11814	Yes	896	[47]	Scientific computing with Python
matplotlib	Python	10069	Yes	901	[48]	Embedding plots into Python applications using GUI toolkits

(Continued)

TABLE 13.3 (Continued)

Deep Learning Library/Framework	Written In	Stars (in Github)	Open Source	Contributors (in Github)	Used In	Used For
scikit-learn	Python	37128	Yes	1638	[49]	Solving machine learning & data science problems
Sonnet	Python	7944	Yes	38	[50]	Constructing NNs
Gluon-NLP	Python	1773	Yes	66	[51]	Text processing, dataset loading and buidling NNs to speed up the research
ONNX	PureBasic	7096	Yes	156	[52]	Moving between DL frameworks and providing shared optimization
TFLearn	Python	9273	Yes	120	[53]	Providing a higher-level API to Tensorfow thereby facilitating and speeding up experimentations
nolearn	Python	929	Yes	15	[54]	Providing wrappers for NN libraries, and other utilities
fastai	Jupyter Notebook	15727	Yes	506	[55]	Training fast and accurate NNs using modern best practice with simplicity
Elephas	Python	1286	Yes	16	[56]	Running distributed DL Models at scale with Spark
DLib	C++	7977	Yes	134	[57]	Making ML and data analaysis application in C++
Blocks	Python	1146	Yes	48	[58]	Building and managing NN Models on using Theano
H2O 3	Java	4343	Yes	127	[59]	Building ML models trained with big data and providing easy production of these models in enterprise environments
ALiPy	Python	344	Yes	5	[60]	Active Learning
Lingvo	Python	1.8k	Yes	39	[61]	Sequence-to-sequence modeling
Keras	Python	44.6k	Yes	816	[62]	Fast prototyping with DNNs

according to GitHub data (like Tensorflow, Theano, CNTK, Eclipse DL4J, MXNet, DeepLearnToolBox, NVIDIA DIGITS, keras, and Lasagne).

13.2.2.1 Tensorflow

Google's TensorFlow is a free and open-source ML library for research and development, a platform for building ML models. It is the most commonly used software library in DL and is highly flexible, easy to handle, and offers a dynamic abundance of tools, libraries, and community resources. Tensorlfow's flexible architecture allows deployment of computation to multiple CPUs or GPUs in mobile devices, desktops, or servers with just a single API [63, 64]. TensorFlow is developed using C++ and provides open-source Python APIs developed by Google under the Apache 2.0 license. It has a very fast growing community. It can be used in the development of NNs, focusing on training and inference, and in other interesting fields, such as optimization, deep image processing, etc. TensorFlow provides reliable performance in diverse environments consisting of multicore GPU, CPU on desktops, mobile devices, Tensor Processing Unit (TPU) and server in production data centers [64]. Demirović et al. [64] have evaluated various image processing algorithms, such as Canny filter, Image deblurring, Gaussian filtering, Gradient filter, K-means segmentation, Matrix multiplication, Bicubic resize, Bilateral image resize, Image rotation, 2-D image convolution, 3-D image convolution, and PDE-Heat equations, on different parallel processing units using TensorFlow.

13.2.2.2 Keras

Keras is a free and open-source, high level NN API and library, purely written in Python, developed by a Google engineer called François Chollet. Keras handles low-level computations, using another library called the Backend. Actually, Keras is capable of running with the CNTK backend, Theano backend, or Tensorflow backend, and is a high level API wrapper for the low-level API. Capable of fast prototyping with the least possible delay, which is the key to better research, it enables fast experimentation [65]. Si et al. [66] have proposed Predictive Emissions Monitoring System (PEMS) for Nitrogen Oxides prediction called Nitrogen Oxides Predictive Models developed using Keras. Luttrell IV et al. [67] has proposed a model by using the GPU version of TensorFlow and Keras.

13.2.2.3 PyTorch

PyTorch is a free and open-source DL platform developed by Facebook's AI Research group to provide a smooth and continuous path from research prototyping to deployment of products. A number of ML tools and libraries extends PyTorch, supporting the various development works in NLP, computer vision, and more. Libraries and tools such as PyTorch Geometric, PyTorch Skorch, and Glow are available in PyTorch. PyTorch supports Python and provides smooth integration with the python data science stack. PyTorch presents a framework for building both simple and complex computational graphs with runtime debugging support. This is a powerful technique for building NNs [68]. Stančin et al. [69] have provided a comparison of various DL libraries, such as Tensorflow, Keras, PyTorch (developed and used by Facebook), and Caffe, with various supported methods.

13.2.2.4   Caffe

Caffe is a free and open-source DL framework introduced by Berkeley AI Research (BAIR) and supported by community contributors. During his PhD at UC Berkeley, Yangqing Jia created the Caffe project. It is developed in C++, released under the BSD 2-Clause license, and uses Nvidia CUDA for computation. Caffe provides Python and Matlab libraries for training and development purposes. It allows us to quickly explore different DL architectures and NN layers on either GPU or CPU [70, 71]. Turchenko et al. [72] have developed a deep convolutional AE in the Caffe framework and presented experimental evaluations. The proposed model doesn't contain pooling/unpooling layers. Aghdam et al. [73] have briefly discussed Caffee library, Caffe Installation, design, and data related issues in ML, using a framework like Caffe.

13.2.2.5   MXNet

MXNet is full-featured, highly flexible in programming, and highly-scalable. It is a DL framework supporting state-of-the-art DL models, including CNNs and LSTMs. The origin of this framework can be traced back to academia and was created through the collaboration and contributions of programmers, volunteers, and researchers at several top universities. The University of Washington and Carnegie Mellon University are a few of the founding institutions of MXNet [74]. MXNet includes Gluon library. Gluon library provides a high level interface for prototyping, training, and deploying DL models without compromising training speed. Gluon provides high-level abstractions for predefined layers, loss functions, and optimizers [75]. Li et al. [76] have proposed an MXNet design by implementing RDMA-based ps-lite, which is a low-level communication system, extending the MXNet framework to make it RDMA capable, which demonstrates great performance on their experimentations on both communication and overall training for MXNet.

13.2.2.6   Chainer

Written in Python, Chainer is powerful and flexible. This framework, used for neural networks, was developed by a Japanese based startup called Preferred Networks. Complex architectures can be written simply and intuitively when using this framework. It helps in bridging the gap between algorithms and DL implementations [77]. A free and open source DL framework, it is meant for efficient research into and development of DL algorithms [78]. It provides a comparatively easier and more straight forward way of implementing the more complex DL architectures currently being research and is characterized by the way in which a model's definition is closely related to its training. Chainer uses the Define-by-Run approach during model training [79].

13.2.2.7   Deeplearning4J

DL4J is an open-source, distributed DL library for JVM. It works with the advanced distributed computing frameworks, including Hadoop and Apache Spark, for accelerating training speed and performs comparatively equally when used across multi-GPUs [80]. As Weka is also implemented in Java, DL4J is the most suitable library for integration with Weka. DL4J supports standard DL architectures, such as CNNs, RNNs, and training on graphics processing units [81]. Revista de la et al. [82] have implemented a Deep Belief Network (DBN) architecture using the Deeplearning4j framework with Apache Spark for classifying the stream of Internet of Things (IOT) data.

13.3 BASIC DEEP LEARNING ARCHITECTURES

Basic DL architectures can be called not so deep DL architectures, serving as the building blocks for advanced DL architectures. They deliver average, normal, slightly better, or state-of-the-art performance. They may not have a continuous track record of successful performance. Basic DL architectures are very important on the basis that all advanced DL architectures have been built on top of those basic DL architectures. We can still consider some of the advanced DL architectures as the basic DL architectures, and vice-versa, since there is no definite criteria for a DL architecture to be basic or advanced. Some advanced DL architectures of today may become the basic DL architectures of tomorrow because of the rapid development of DL technology. For now, consider that this classification is being done for our convenience. A brief summary of various DL architectures used in solving problems in many major fields is presented in Table 13.4. We will discuss the important basic DL architectures, such as CNNs, RNNs, LSTM, GRU, DBN, MLP, GAN and FCNs, here.

13.3.1 CONVOLUTIONAL NEURAL NETWORKS

CNN is a class or a type of DNN which takes full benefit of the inherent hierarchical patterns in data, thereby assembling more complex patterns by using smaller and smaller patterns. A CNN uses convolution operation instead of using general multiplication operation in one or more of its many hidden layers. CNNs use very little pre-processing and is extremely suitable for use in image classification. It is also used in Video and Image recognition, Medical Image analysis, NLPs, and Recommender Systems. Fukushima and Kunihiko [7] have proposed a self-organizing unsupervised multi-layer CNN Model called "neocognitron". CNN Models proposed prior to this proposition are severely limited in performance in pattern recognition because of the positional shift and the shape distortion of the input patterns. The neocognitron model is not affected by the distortion in the position of pattern and/or small positional change and/or size of the stimulus pattern.

13.3.2 RECURRENT NEURAL NETWORKS

An RNN is a type or class of ANN where the connected nodes actually form a directed cyclic or acyclic graph along a temporal sequence. RNNs use feedback loops such as BP Through Time, connecting inputs together and enabling RNNs to process sequential and temporal data.

Let the superscript and subscript denote the layer and the timestamp respectively.

TABLE 13.4

Summary of DL Architectures and Their Implementations

Cited	Deep Learning Architecture Used	Task	Learning Method	Performance Metrics	Dataset(s)
[83]	Multi resolution foveated CNN + Transfer Learning	Large Scale Video Classification	Supervised	Mean Average Precision(mAP)	Sports-1M, UCM-101
[31]	ordered neuronsLSTM (ON-LSTM)	Language modelling, unsupervised constituency parsing, targeted syntactic evaluation, logical inference	Unsupervised	F1 Score, Accuracy	WSJ10
[84]	restricted-RNTN	To balance expressiveness & computational cost in word representation & conditional computing	Supervised	Model Perplexity (PPL)	PTB Corpus, text8 Corpus
[85]	DBN+PNN	To evaluate proposed ID Model with DoS, R2L, U2R and ProbClasses of attacks	Supervised	Running time, detection accuracy, detection rate, false alarm rate	KDD CUP 1999
[86]	deep PCA-LSTM	Forecasting the 24 hours power of Manchester Wind Farm on June, 2011	Supervised	NMAE, NRMSE	NWP data of Manchester Wind Farm
[87]	DCGAN + GAN's Discriminator + Multi-Scale heat map fuision strategy	Visual surface inspection, abnormal surface detection	Unsupervised	IoU, Pixel acc	WOOD, CRACK
[88]	Probilistic Deep AE	Outlier detection and reconstruction of power system measurement	Supervised	Outlier detection accuracy, Outlier detection sensitivity	Real observations, upper bound and lower bound of the current and past D time periods
[89]	Stacked Sparse AE + SVM	Efficiently classify the data	Supervised	Classification accuracy	Australian(Staglog), Diabetes(UCI), German Number(Staglog), W1a(JP98a), W4a(JP98a), a1a(UCI), Dna(Staglog), Satimage(Staglog)

Ref	Model	Application	Learning	Metrics	Dataset
[90]	DBN+Stacked AE	Classifying hyperspectral images with high accuracy	Supervised	Overall accuracy of BatchSize, Learning Rate, Number of Epochs	Pavia University Scene(PUS)
[91]	3-layered FCN(BP + Optimal Learning Rate + FNN)	Improviding Convergent Rate	Unsupervised	MSE	Normal distribution with mean 0 and SD 1 generated by normalrnd()
[92]	Deep Q-Network	Successfully implemented Deep RL Net on 8 Open AI gym environments		Average Loss per Episode, Duration per Episode, Average Max Q-value per Episode, Total reward per episode	
[93]	spatial temporal discriminative RBM	Detecting & analyzing ERP	Supervised	AUC, p values CRR	BCI Competition Databases
[94]	DCIGN + CBCT-CT DIR	cone-beam CT to deformable image registration(DIR)	Unsupervised	NMI, FSIM, RMSEc	CBCT images from 285 patients + 100 synthetic cases
[95]	Deep Care GRU	Diabetes disease diagnosis prediction	Supervised	Accuracy	Diabetes dataset
[96]	ResNets + CNN	Spectral-spatial classification	Supervised	Overall accuracy	HSI datasets
[97]	Runge-Kutta MLP+ SAC	Reduction of parametric uncertainties for non-linear MIMO System	Supervised	system output, control signal, uncertain outflow parameter	current states & inputs of non-linear three tank systems

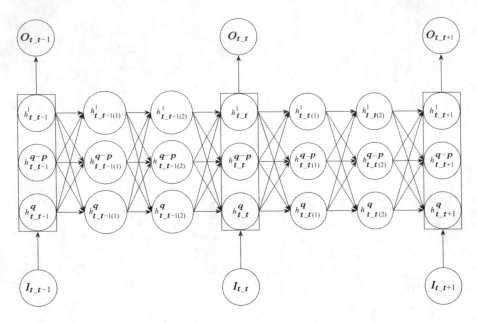

FIGURE 13.2   A simple RNN.

Asume that the states are k- dimensional. A hidden state in layer l in timestamp t_t is given by $h_{t\_t}^l \in \mathbb{R}^k$. Let $T_{k,j}: \mathbb{R}^k \longrightarrow \mathbb{R}^j$ be an affine transform ($Dx + b$ for some D and b). Let $\odot$ and $h_{t\_t}^0$ be element-wise multiplication and input word vector at timestep k respectively where t_t denotes the current timestamp.

The RNN can be described by a deterministic state transition function,

$$f : h_{t\_t}^{l-1}, h_{t\_t-1}^l \rightarrow h_{t\_t}^l$$

For classical RNNs as shown in Figure 13.2, this function is represented as

$$h_{t_t}^l = f\left(T_{k,k} h_{t_t}^{l-1} + T_{k,k} h_{t_t-1}^l\right), \text{where } f \in \{\tanh, \text{sigm}\}$$

13.3.2.1 Long-Short-Term-Memory

LSTM is a type or a class of RNN architecture used in DL capable of learning long-term depedencies. They are easily capable of memorizing information for an extended number of timestamps. As in Figure 13.3, a common LSTM unit consists of a memory part called Cell denoted by $c_{t\_c}^l \in \mathbb{R}^k$, Input Gate denoted by i_{t_c}, Output Gate denoted by o_{t_c}, and Forget Gate denoted by f_{t_c}. Let t_c denote the current timestamp. Let h_{t_c}, h_{t_c-1}, C_{t_c}, C_{t_c-1}, x_{t_c}, and U_{t_c} represent the current forget gate content, current cell output, previous cell output, current cell memory, previous cell memory, input vector and weight vector

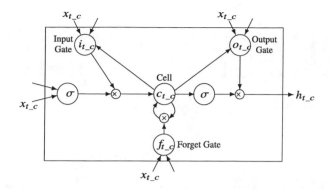

FIGURE 13.3   A simple graphical representation of LSTM unit.

respectively. The LSTM architecture shown in Figure 13.3 is given by the following equations:

$$f_{t\_c} = \sigma\left(U_f \odot \left[h_{t\_c-1} \odot x_{t\_c}\right] + b_f\right) \tag{13.2}$$

$$i_{t\_c} = \sigma\left(U_i - \left[h_{t\_c-1}, x_{t\_c}\right] + b_i\right) \tag{13.3}$$

$$\tilde{C}_{t\_c} = f_{t\_c} \odot C_{t\_c-1} + i_{t\_c} + C_{t\_c} \tag{13.4}$$

$$C_{t\_c} = f_{t\_c} \odot C_{t\_c-1} + i_{t\_c} \odot \tilde{C}_{t_c} \tag{13.5}$$

$$o_{t\_c} = \sigma\left(U_o\left[h_{t\_c-1}, x_{t\_c}\right] + b_o\right) \tag{13.6}$$

$$h_{t\_c} = o_{t\_c} \odot \tan h\left(C_{t\_c}\right) \tag{13.7}$$

In these equations, sigm and tan*h* are applied element-wise.

13.3.2.2   Gated Recurrent Unit

GRU is a gating mechanism in RNN having only two gate vectors, update gate and reset gate, unlike LSTM which is having three gate vectors (input gate, output gate and forget gate). The reset gate is meant for making the decision of forgetting some unnecessary information. The update gate vector is used to determine the necessary past information from previous time steps that needs to be passed along to the future. Let x_{t_c}, h_{t_c}, z_{t_c}, r_{t_c} and U be the input vector, output vector, update vector, reset vector and weight vector respectively. Let t_c denote the current timestamp. A simple GRU unit is shown in Figure 13.4 having input gate, reset gate and memory cell.

The GRU is represented by the following equations:

$$z_{t\_c} = \sigma \odot \left(U_z \odot \left[h_{t\_c}, x_{t\_c}\right]\right) \tag{13.8}$$

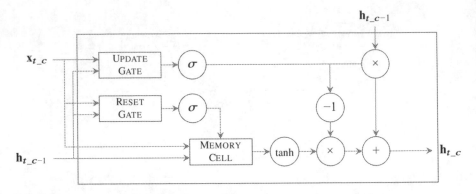

FIGURE 13.4   Simple Gated Recurrent Unit Layer.

$$r_{t\_c} = \sigma \odot \left(U_r \odot \left[h_{t\_c-1}, x_{t\_c} \right] \right) \tag{13.9}$$

$$\tilde{h}_{t\_c} = \tan h \left(U_o \odot \left[r_{t\_c} * h_{t\_c-1}, x_{t\_c} \right] \right) \tag{13.10}$$

$$h_{t\_c} = \left(1 - z_{t\_c} \right) * h_{t\_c-1} + \left(z_{t\_c} \right) * \tilde{h}_{t\_c} \tag{13.11}$$

13.3.3 Deep Belief Networks

DBN is a type or a class of DNNs having multiple layers of hidden units. There are no connectivity between the units in any NN layer. It can be considered as a stack of nets such as Autoencoder or RBM where each sub-net's hidden layer serves as a visible layer for the next sub-net. The stack of RBMs in DBN can be trained one at a time. Once the current RBM has been trained, next RBM must be stacked on the top of it taking its input from the final trained NN layer of the current RBM. Hinton and Geoffrey E [98] have proposed a training method for RBM called Contrastive Divergence. Let t_c denote the current timestamp and let U, x_x, α be the weight vector, visible vector, learning rate respectively. In training a single RBM net, weight update can be represented by the equation: $U_{ij}\left(t\_c+1 \right) = U_{ij}\left(t\_c \right) + \alpha \dfrac{\delta \log p\left(x\_x \right)}{\delta u_{ij}}$ where $p(x\_x)$ is the probability of a visible vector and is given by $p\left(x\_x \right) = \dfrac{1}{Q} \sum_h -e^{-E\left(x\_x, h \right)}$ where Q is the partition function used in normalization and $E(x\_x, h)$ is some energy function assigned to the state of the RBM net. The gradient $\dfrac{\delta log\left(p\left(x\_x \right) \right)}{\delta u_{ij}}$ is given the form $\langle x\_x_i h_j \rangle_{data} - \langle x\_x_i h_j \rangle_{model}$ where $\langle .... \rangle_p$ represents averages w.r.t. distribution p.

Zhao et al. [99] have developed a Deep Belief Network (DBN) to establish the terminal replacement prediction model for learning the characteristics which can deeply affect the terminal replacement, thereby achieving better prediction accuracy than shallow learning models like 1NN, SVM, and NN. Zhao et al. [85] have proposed a DBN-PNN DL based IDS model for Intrusion Detection on the KDD CUP 1999 dataset, where using DNN shortens the training and testing duration of a PNN network by actually converting raw data into low-dimensional data and a PSA algorithm is used for optimizing the number of nodes in the hidden layers of Deep Belief Network. Zhang et al. [100] have presented a hybrid model by introducing an improved Genetic Algorithm into a Deep Belief Network (DBN). The GA is used to optimize the network structure for further use by a DBN for network attack classification on the NSL-KDD dataset.

13.3.4 GENERATIVE ADVERSARIAL NETWORKS

Generative Adversarial Net is a type or class of DNN which uses two adversarial nets – discriminative net and generative net. The generative net generates synthetic data, while the discriminative net distinguishes synthetic data generated by the generator from the true data in the data distribution. The generator net tries to fool the discriminator net by increasing the error rate of the discriminator net. As shown in Figure 13.5, a minibatch of known dataset $\mathbb{P}_r$ serves as the initial training data for the discrminator. The generator net is seeded with randomized input $p(z)$ sampled from some predefined latent space. The discriminator evaluates the synthetic data synthesized by the generator net and the BP is used in the training process.

[101] has developed a new approach using a parallel data paradigm with generative adversarial networks (GANs) to impute traffic flow by performing experimentations on traffic flow datasets obtained from Caltrans Performance Measurements System (PeMS). Synthetic Data is generated using GAN by taking real time data from Caltrans PeMS dataset. Then, the parallel data, which is a mixed set of real data and synthetic data, is used to train the imputation model. [102] has discussed in detail about the recent progress and developments on GANs, basic concepts of GANs, comparison amongst various GAN models, potential GAN applications, and future research directions in GANs.

13.3.5 MULTILAYER PERCEPTRON

A multilayer perceptron is a type of feedforward ANN having three or more layers including at least one hidden layer. It can be represented by a finite directed acyclic graph. MLP utilizes non-linear activation functions and backpropagation method in training. In deep learning applications, ReLU activation function is more frequently

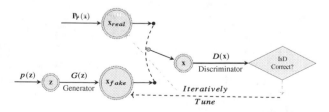

FIGURE 13.5 An abstract representation of GAN architecture.

used as compared to other activation functions such as sigmoids (like $\tanh$ and $(1 + e^{-V_i})$) in MLP models.

A 3-layers MLP is represented by the function notation, $f: \mathbb{R}^A \longrightarrow \mathbb{R}^B$, where A is the size of the input vector and B is the size of the output vector. $f(x\_a)$ can be expressed as:

$$f\left(x\_a\right) = softmax\left(c\_b^{(2)} + U^{(2)}\left(\tan h\left(c\_b^{(1)} + U^{(1)}\left(x\_a\right)\right)\right)\right) \quad (13.12)$$

where $c\_b^{(1)}$ and $c\_b^{(2)}$ are bias vectors, $U^{(1)}$ and $U^{(2)}$ are weight vectors, softmax and tanh are activation functions.

Let $U^{(1)} \epsilon R^{A \times A_h}$ be the weight matrix. Then the hidden layer can be given as:

$$h\left(x\_a\right) = \phi\left(x\_a\right) = \tan h\left(c\_b^{(1)} + U^{(1)}x\_a\right) \quad (13.13)$$

The output vector is given by:

$$o\left(x\_a\right) = softmax\left(c\_b^{(2)} + U^{(2)}h\left(x\_a\right)\right) \quad (13.14)$$

13.3.6 FULLY CONNECTED NEURAL NETWORKS

A Fully Connected Neural Net is a net having a series of some fully connected layers where each neuron in each layer is connected to every neuron in its adjacent layer. FCNs are structure agnostic and no special assumptions for the input are necessary. FCN is being used in most of the DL domains as it has the capability of learning any function having the ability of universal approximators. A fully connected layer is a function from R^c to R^d.

Let $x \in R^c$ be the input to a fully connected layer in FCN. Let $o_i \in R$ be the ith output from the fully connected layer. The $o_i \in R$ is computed as:

$$o_i = \sigma\left(u_1 x_1 + .. + u_c x_c\right) \quad (13.15)$$

where $\in$ is a non-linear activation function and u_i is linear parameter in the network. The full output o is then given by:

$$o = \begin{pmatrix} \sigma\left(u_{1,1}x_1 + .. + u_{1,c}x_c\right) \\ \sigma\left(u_{d,1}x_1 + .. + u_{d,c}x_c\right) \end{pmatrix} \quad (13.16)$$

Wang et al. [103] have developed a new approach in which the four layer FNN gets transformed into a three layer Equivalent FCNN by combining a systematic fuzzy rule generation for FNN. Liu et al. [104] have developed a hybrid model of CNN, LSTM and Fully-Connected Conditional Random Field (FCCRF) for group activity recognition, where the CNN based spatial-temporal model and LSTM are used to learn the position and moving features of each person in the group activity and predict its behavior, and FCCRF is used to learn the interactions between the individuals in the group. Wang et al. [91] have proposed a BP training algorithm implemented with learning rates (with near optimal values) for tuning premise as well as the consequent parameters for FCNN and for improving convergence speed, by transforming FNN into an equivalent fully connected three layer feed-forward neural network.

13.4   ADVANCED DEEP LEARNING ARCHITECTURES

Advanced Deep Learning architectures are those DL architectures proven to deliver state-of-the-art performances and are considered to be quite successful in completing the desired tasks. A typical advanced DL architecture can be very deep. Some of them may have more than 1,000 layers. But it is not compulsory for an advanced DL architecture to be extremely deep. Some of them are not that deep but still deliver state-of-the-art performance well. The performance of the advanced DL architectures heavily depends on advanced hardware resources and can be sped up by supplying more hardware resources. Some of the important advanced DL architectures, such as AlexNet, VCGNet, GoogLeNet, ResNet, YOLO, ResNeXt, Deep Recurrent CNN, SegNet, SqueezeNet, Mask Scoring R-CNN, ordered neurons LSTM, and Spherical CNN, are discussed here.

13.4.1   ALEXNET

Alex Krizhevsky has developed an eight layer Convolutional Neural Net called AlexNet in [24] as a part of an ImageNet LSVRC-2010 Contest. AlexNet has five convolutional layers and three fully connected layers, along with ReLU activation function. AlexNet is a variant of the CNN designs developed by Yann Lecun in [17].

13.4.2   VCG NET

Simonyan et al. [105] have proposed a ConvNet architecture called VCGNet, consisting of 3 × 3 convolutional layers, maxpooling layers, and fully-connected layers. VCGNet has a maximum depth of nineteen weight layers (sixteen convolutional layers + three FC) and a minimum depth of sixteen weight layers (thirteen convolutional layers + three FC) in its configuration. It uses ReLU activation function. VCGNet has been developed as a part of the ILSVRC 2014 Contest, achieving top-1 accuracy (76.3%) and top-5 accuracy (93.2%) with the validation set. Mazumdar et al. [106] have proposed a hybrid model of CNN (based on the VCG network and having five max pooling layers and thirteen convolutional layers) and Fully Connected Neural Network (FCNN) for object recognition in videos. The proposed model is being trained on the Imagenet dataset.

13.4.3   GOOGLENET

Christian Szegedy et al. [107] proposed a twenty-two layers deep ConvNet called GoogLeNet for performing classification and detection tasks in ILSVRC 2014

Contest. GoogLeNet has twenty-seven layers, if pooling layers are also included. It has a 5×5 pooling layer, a 1×1 convolutional layer, a fully-connected layer with ReLU, a dropout layer with 70% dropout o/p ratio, and a linear layer with softmax loss. GoogLeNet has been trained with the DistBelief distributed Machine Learning system. It receives a top-5 error rate of 6.67% in classification.

13.4.3.1 ResNet

Residual Neural Net (ResNet) is a type of ANN implemented by utilizing skip connections, or short-cuts to jump over some layers, and is inspired by constructs known from pyramid cells in the cerebral cortex. ResNet is developed by Kaiming He [108]. The authors have developed ResNet with up to 152 layers, with lower complexity for classification tasks for the ILSVRC 2015 Challenge, winning first place. They have also attempted using ResNet with over 1,000 layers on CIFAR-10.

13.4.4 Deep Recurrent CNN

Sourour Brahimi et al. [109] have developed an improvised deep recurrent CNN called Improved Very Deep Recurrent Convolutional Neural Network, also called IVD-RCNN for object recognition task. In this network, VCG-19 methods are integrated with RCL layers. It utilizes an improved generalized pooling function and an SPP layer.

13.4.5 Mask Scoring R-CNN

Huang et al. [110] have developed a Mask R-CNN called Mask Scoring R-CNN with MaskIoU head module for the task of scoring instance segmentation masks. MaskIoU head module is meant to learn the Mask-IoU aligned mask score. The net is implemented with the COCOC dataset having eighty object categories. Mask Scoring R-CNN uses the backbone ResNet in the architecture.

13.4.6 Ordered Neurons LSTM

Shen et al. [31] have developed an LSTM architecture called ordered neurons LSTM (also called ON-LSTM). The authors have introduced a new activation function called *cumax*( ) and a new gating mechanism with a Master Forget Gate and a Master Input Gate to serve high level control for the update operations of cell states. ON-LSTM is the state-of-the-art in inducing the latent structure of natural language achieving greater performance on language modeling, logical inference tasks, and long-term dependency.

13.4.7 Spherical CNNs

Taco Cohen et al. [111] have developed an improved Spherical CNN (S^2 – CNN) by implementing it with a Generalized FFT-based correlation method based on S^2

and $SO(3)$ cross-correlations. This net is developed for 3D Model Recognition and Molecular Energy Regression tasks achieving near state-of-the-art performance.

13.4.8 ResNeXt

Saining Xie et al. [112] have developed a practical modular network architecture called ResNeXt for image classification task. The authors introduce a new dimension called cardinality besides the dimensions of depth and width. With the increasing net capacity, increasing cardinality is more effective than going deeper or wider. ResNeXt, when experimented with ImageNet-5k and COCO datasets, produces better results than ResNet.

13.4.9 YOLO

Redmon et al. [113] have developed a state-of-the-art unified model for object detection called You Only Look Once, also called YOLO. The proposed YOLO net has twenty-four convolutional layers and two FC layers. The authors also experiment with a fast YOLO design for fast object detection but with fewer convolutional layers, i.e., nine convolutional layers instead of twenty-four convolutional layers.

13.4.10 SegNet

Badrinarayanan et al. [114] have developed a deep ConvNet for semantic segmentation called SegNet. This net consists of an encoder net, a corresponding decoder net and a final pixelwise classification layer with softmax classification. The thirteen convolutional layers of encoder correspond to those thirteen convolutional layers in VCG16 Net. Each encoder layer in the Encoder net has a corresponding decoder layer in the Decoder net.

13.4.11 SqueezeNet

Iandola et al. [115] have developed a smaller state-of-the-art ConvNet architecture called SqueezeNet, achieving the same accuracy level as that of AlexNet but with fifty times fewer parameters. SqueezeNet is compatible with frameworks like Caffe, MXNet, Chainer, Keras and Torch.

13.5 CONCLUSION

Deep Learning architectures and other deep learning techniques are quite successful in solving many complex real world problems. In this chapter, we have presented a discussion on various major Deep Learning frameworks and libraries, major Deep Learning architectures. Major Deep Learning architectures have been categorized into basic and advanced architectures. Categorization of Deep Learning architectures is a complex issue, and further survey and analysis is required for more refined and better classifications of Deep Learning architectures that will be useful for applications in healthcare systems.

REFERENCES

1. Techlabs M (2019) 8 problems that can be easily solved by machine learning. URL https://www.marutitech.com/problems-solved-machine-learning/, accessed October 7, 2019.
2. Shan F, Gao Y, Wang J, Shi W, Shi N, Han M, Xue Z, Shen D, Shi Y (2020) Lung infection quantification of covid-19 in ct images with deep learning. arXiv preprint arXiv:200304655.
3. Wang L, Wong A (2020) Covid-net: A tailored deep convolutional neural network design for detection of covid-19 cases from chest radiography images. arXiv preprint arXiv:200309871.
4. Wang S, Kang B, Ma J, Zeng X, Xiao M, Guo J, Cai M, Yang J, Li Y, Meng X, et al. (2020) A deep learning algorithm using ct images to screen for corona virus disease (covid-19). medRxiv.
5. McCulloch WS, Pitts W (1943) A logical calculus of the ideas immanent in nervous activity. *The Bulletin of Mathematical Biophysics* 5(4):115–133.
6. Rosenbaltt F (1957) The perceptron–a perciving and recognizing automation. Report 85-460-1 Cornell Aeronautical Laboratory, Ithaca, Tech Rep.
7. Fukushima K (1980) Neocognitron: A self-organizing neural network model for a mechanism of pattern recognition unaffected by shift in position. *Biological Cybernetics* 36(4):193–202.
8. Hopfield JJ (1982) Neural networks and physical systems with emergent collective computational abilities. *Proceedings of the National Academy of Sciences* 79(8):2554–2558.
9. Turing AM (2009) Computing machinery and intelligence. In: *Parsing the Turing Test*, Springer, pp. 23–65.
10. Strachey CS (1952) Logical or non-mathematical programmes. In: *Proceedings of the 1952 ACM national meeting (Toronto)*, ACM, pp. 46–49.
11. Oettinger AG (1952) Cxxiv. programming a digital computer to learn. *The London, Edinburgh, and Dublin Philosophical Magazine and Journal of Science* 43(347):1243–1263.
12. Hubel DH, Wiesel TN (1959) Receptive fields of single neurones in the cat's striate cortex. *The Journal of Physiology* 148(3):574–591.
13. Kelley HJ (1960) Gradient theory of optimal flight paths. *Ars Journal* 30(10):947–954.
14. Schmidhuber J (2015) Critique of paper by deep learning conspiracy. *Nature* 521 p. 436. Accessed October 5, 2019.
15. Ivakhnenko AG (1971) Polynomial theory of complex systems. *IEEE Transactions on Systems, Man, and Cybernetics* (4):364–378.
16. Rumelhart DE, Hinton GE, Williams RJ, et al. (1988) Learning representations by back-propagating errors. *Cognitive Modeling* 5(3):1.
17. LeCun Y, Boser B, Denker JS, Henderson D, Howard RE, Hubbard W, Jackel LD (1989) Backpropagation applied to handwritten zip code recognition. *Neural Computation* 1(4):541–551.
18. Watkins CJCH (1989) Learning from delayed rewards.
19. Schmidhuber J, Mozer MC, Prelinger D (1993) Continuous history compression. In: *Proc. of intl. workshop on neural networks*, pp. 87–95.
20. Cortes C, Vapnik V (1995) Support-vector networks. *Machine Learning*, 20(3):273–297.
21. Hochreiter S, Schmidhuber J (1997) Long short-term memory. *Neural Computation* 9(8):1735–1780.
22. LeCun Y, Bottou L, Bengio Y, Haffner P, et al. (1998) Gradient-based learning applied to document recognition. *Proceedings of the IEEE* 86(11):2278–2324.

23. Deng J, Dong W, Socher R, Li LJ, Li K, Fei-Fei L (2009) Imagenet: A large scale hierarchical image database. In: 2009 IEEE conference on computer vision and pattern recognition, IEEE, pp. 248–255.

24. Krizhevsky A, Sutskever I, Hinton GE (2012) Imagenet classification with deep convolutional neural networks. In: *Advances in neural information processing systems*, pp. 1097–1105.

25. Markoffjune J (2019) How many computers to identify a cat? 16,000. URL https://www.nytimes.com/2012/06/26/technology/in-a-big-network-of-computers-evidence-of-machine-learning.html, accessed October 6, 2019.

26. Taigman Y, Yang M, Ranzato M, Wolf L (2014) Deepface: Closing the gap to human-level performance in face verification. In: *Proceedings of the IEEE conference on computer vision and pattern recognition*, pp. 1701–1708.

27. Goodfellow I, Pouget-Abadie J, Mirza M, Xu B, Warde-Farley D, Ozair S, Courville A, Bengio Y (2014) Generative adversarial nets. In: *Advances in neural information processing systems 27*, Curran Associates, pp. 2672–2680.

28. Radford A, Metz L, Chintala S (2015) Unsupervised representation learning with deep convolutional generative adversarial networks. arXiv preprint arXiv:151106434.

29. Isola P, Zhu JY, Zhou T, Efros AA (2017) Image-to-image translation with conditional adversarial networks. In: *Proceedings of the IEEE conference on computer vision and pattern recognition*, pp. 1125–1134.

30. Devlin J, Chang MW, Lee K, Toutanova K (2018) Bert: Pre-training of deep bidirectional transformers for language understanding. arXiv preprint arXiv:181004805.

31. Shen Y, Tan S, Sordoni A, Courville A (2018) Ordered neurons: Integrating tree structures into recurrent neural networks. arXiv preprint arXiv:181009536.

32. Saxton D, Grefenstette E, Hill F, Kohli P (2019) Analysing mathematical reasoning abilities of neural models. arXiv preprint arXiv:190401557.

33. Sharma P (2019) 5 amazing deep learning frameworks every data scientist must know! (with illustrated infographic). URL https://www.analyticsvidhya.com/blog/2019/03/deep-learning-frameworks-comparison/, accessed October 17, 2019.

34. Techlabs M (2019) Top 8 deep learning frameworks. URL https://www.marutitech.com/top-8-deep-learning-frameworks/, accessed September 25, 2019.

35. Shao T, Guo Y, Chen H, Hao Z (2019) Transformer-based neural network for answer selection in question answering. IEEE Access 7:26146–2615626.

36. Ding W, Wang R, Mao F, Taylor G (2014) Theano-based large-scale visual recognition with multiple gpus. arXiv preprint arXiv:14122302.

37. Jithesh V, Sagayaraj MJ, Srinivasa K (2017) Lstm recurrent neural networks for high resolution range profile based radar target classification. In: *2017 3rd International Conference on Computational Intelligence & Communication Technology (CICT)*, IEEE, pp 1–6DL

38. Karim MR (2018) Java Deep Learning Projects: Implement 10 real-world deep learning applications using Deep learning 4j and open source APIs. Packt Publishing Ltd.

39. Zhang Z, Xu W, Gaffney N, Stanzione D (2017) Early results of deep learning on the stampede2 supercomputer. IXPUGFall 2017 pp. 1–3.

40. Fey M, Lenssen JE (2019) Fast graph representation learning with pytorch geometric. arXiv preprint arXiv:190302428.

41. Hazelwood K, Bird S, Brooks D, Chintala S, Diril U, Dzhulgakov D, Fawzy M, Jia B, Jia Y, Kalro A, et al. (2018) Applied machine learning at facebook: A datacenter infrastructure perspective. In: *2018 IEEE International Symposium on High Performance Computer Architecture (HPCA)*, IEEE, pp. 620–629.

42. Dai J, Liu X, Wang Z (2017) Building large-scale image feature extraction with bigdl at jd. com. Honeywell launches UAV industrial inspection service, teams with Intel on innovative offering" Sept.

43. Zhou Y, Jiang J (2015) An fpga-based accelerator implementation for deep convolutional neural networks. In: *2015 4th International Conference on Computer Science and Network Technology (ICCSNT)*, IEEE, vol 1, pp. 829–832.

44. Konda P, Das S, Doan A, Ardalan A, Ballard JR, Li H, Panahi F, Zhang H, Naughton J, Prasad S, et al. (2016) Magellan: toward building entity matching management systems over data science stacks. *Proceedings of the VLDB Endowment* 9(13):1581–1584.

45. Gurghian A, Koduri T, Bailur SV, Carey KJ, Murali VN (2016) Deeplanes: End-to-end lane position estimation using deep neural networksa. In: *Proceedings of the IEEE Conference on Computer Vision and Pattern Recognition Workshops*, pp 38–45.

46. Devooght R, Bersini H (2017) Long and short-term recommendations with recurrent neural networks. In: *Proceedings of the 25th Conference on User Modeling, Adaptation and Personalization, ACM*, pp. 13–21.

47. Blum T, Kristensen MR, Vinter B (2014) Transparent gpu execution of numpy applications. In: 2014 IEEE International Parallel & Distributed Processing Symposium Workshops, IEEE, pp. 1002–1010.

48. Lou X, van der Lee S, Lloyd S (2013) Aimbat: A python/matplotlib tool for measuring teleseismic arrival times. *Seismological Research Letters* 84(1):85–93.

49. Gouillart E, Nunez-Iglesias J, Van Der Walt S (2017) Analyzing microtomography data with python and the scikit-image library. *Advanced Structural and Chemical Imaging* 2(1):18.

50. Jonschkowski R, Rastogi D, Brock O (2018) Differentiable particle filters: End-to-end learning with algorithmic priors. arXiv preprint arXiv:180511122.

51. Jahan R (2019) Vulgar and spam comment identification using gluon natural language processing and convolution neural networks. PhD thesis, Brac University.

52. Lin WF, Tsai DY, Tang L, Hsieh CT, Chou CY, Chang PH, Hsu L (2019) ONNC: A compilation framework connecting onnx to proprietary deep learning accelerators. In: *2019 IEEE International Conference on Artificial Intelligence Circuits and Systems (AICAS)*, IEEE, pp. 214–218, doi:10.1109/AICAS.2019.8771510

53. Tang Y (2016) Tf. learn: Tensorflow's high-level module for distributed machine learning. arXiv preprint arXiv:161204251.

54. Beckham CJ (2015) Classification and regression algorithms for weka imple-mented in python. (Working paper 02/2015). Hamilton, New Zealand: University of Waikato, Department of Computer Science.

55. Dietle J (2019) How i used deep learning to classify medical images with fast.ai. URL https://www.freecodecamp.org/news/how-i-used-deep-learning-to-classify-medical-images-with-fast-ai-cc4cfd64173c/, accessed September 20. 2019.

56. MIT (2019) Distributed deep learning with keras & spark. https://github.com/max-pumperla/elephas, accessed October 18, 2019.

57. King DE (2009) Dlib-ml: A machine learning toolkit. *Journal of Machine Learning Research* 10(Jul):1755–1758.

58. Van Merriënboer B, Bahdanau D, Dumoulin V, Serdyuk D, Warde-Farley D, Chorowski J, Bengio Y (2015) Blocks and fuel: Frameworks for deep learning. arXiv preprint arXiv:150600619.

59. Torres JF, Galicia A, Troncoso A, Martínez-Álvarez F (2018) A scalable approach based on deep learning for big data time series forecasting. Integrated Computer-Aided Engineering (Preprint):1–14.

60. Tang YP, Li GX, Huang SJ (2019) Alipy: Active learning in python. ArXiv abs/ 1901.03802.

61. Shen J, Nguyen P, Wu Y, Chen Z, Chen MX, Jia Y, Kannan A, Sainath T, Cao Y, Chiu CC, et al. (2019) Lingvo: a modular and scalable framework for sequence-to-sequence modeling. arXiv preprint arXiv:190208295.

62. Nagisetty A, Gupta GP (2019) Framework for detection of malicious activities in iot networks using keras deep learning library. In: *2019 3rd International Conference on Computing Methodologies and Communication (ICCMC)*, IEEE, pp. 633–637.

63. Brain G (2019) Tensorflow. URL https://www.tensorflow.org/, accessed September 10, 2019.

64. Demirović D, Skejić E, Šerifović-Trbalić A (2018) Performance of some image processing algorithms in tensorflow. In: *2018 25th International Conference on Systems, Signals and Image Processing (IWSSIP)*, IEEE, pp. 1–4.

65. Google (2019) Keras: The python deep learning library. URL https://keras.io/, accessed August 27, 2019.

66. Si M, Tarnoczi TJ, Wiens BM, Du K (2019) Development of predictive emissions monitoring system using open source machine learning library–keras: A case study on a cogeneration unit. *IEEE Access* 7:113463–113475.

67. Luttrell IV JB, Zhou Z, Zhang C, Gong P, Zhang Y (2017) Facial recognition via transfer learning: Fine-tuning keras_vggface. In: *2017 International Conference on Computational Science and Computational Intelligence (CSCI)*, IEEE, pp. 576–579.

68. Facebook (2019) Pytorch. URL https://pytorch.org/, accessed August 27, 2019.

69. Stančin I, Jović A (2019) An overview and comparison of free python libraries for data mining and big data analysis. In: *42nd International Convention MIPRO 2019*.

70. Bahrampour S, Ramakrishnan N, Schott L, Shah M (2019) Com-parativestudy of caffe, neon, theano, andtorch fordeeplearning at work-shop track - iclr 2016. URL https:// openreview.net/pdf?id=q7kEN7WoXU8LEkD3t7BQ, accessed September 25, 2019.

71. Jia Y (2019) Deep learning framework by bair. URL https://caffe.berkeleyvision.org/, accessed October 20, 2019.

72. Turchenko V, Luczak A (2017) Creation of a deep convolutional auto-encoder in caffe. In: *2017 9th IEEE International Conference on Intelligent Data Acquisition and Advanced Computing Systems: Technology and Applications (IDAACS)*, IEEE, vol 2, pp. 651–659.

73. Aghdam HH, Heravi EJ (2017) *Guide to convolutional neural networks*. New York, NY: Springer 10:(978–973), 51.

74. Vogels W (2019) Mxnet - deep learning framework of choice at aws. URL https://www. allthingsdistributed.com/2016/11/mxnet-default-framework-deep-learning-aws.html, accessed August 10, 2019.

75. Amazon (2019) Apache mxnet on aws. URL https://aws.amazon.com/mxnet/, accessed August 15, 2019.

76. Li M, Wen K, Lin H, Jin X, Wu Z, An H, Chi M (2019) Improving the performance of distributed mxnet with rdma. *International Journal of Parallel Programming* 47(3):467–48024.

77. quintagroup (2019) Chainer - a python-based framework for neural networks. URL https://www.quintagroup.com/cms/python/chainer, accessed October 4, 2019.

78. Reilly Media O (2019) Complex neural networks made easy by chainer. URL https:// www.oreilly.com/learning/complex-neural-networks-made-easy-by-chainer, accessed August 23, 2019.

79. Tokui S, Oono K, Hido S, Clayton J (2015) Chainer: a next-generation open source framework for deep learning. In: *Proceedings of workshop on machine learning systems (LearningSys) in the twenty-ninth annual conference on neural information processing systems (NIPS)*, vol 5, pp. 1–6.

80. Black AD, Gibson A, Kokorin V, Patterson J (2019) Deep learning for java. URL https://deeplearning4j.org/, accessed September 17, 2019.

81. Lang S, Bravo-Marquez F, Beckham C, Hall M, Frank E (2019) Wekadeeplearning4j: A deep learning package for weka based on deeplearning4j. *Knowledge-Based Systems* 178:48–50.

82. Manimegalai MP (2017) 10. a deep belief network (dbn) for classifying iot data. *Revista de la Facultad de Agronomia de la Universidad del Zulia* 34(1).

83. Karpathy A, Toderici G, Shetty S, Leung T, Sukthankar R, Fei-Fei L (2014) Large-scale video classification with convolutional neural networks. In: *Proceedings of the IEEE conference on Computer Vision and Pattern Recognition*, pp. 1725–1732.

84. Salle A, Villavicencio A (2017) Restricted recurrent neural tensor networks: Exploiting word frequency and compositionality. arXiv preprint arXiv:170400774.

85. Zhao G, Zhang C, Zheng L (2017) Intrusion detection using deep belief network and probabilistic neural network. In: *2017 IEEE International Conference on Computational Science and Engineering (CSE) and IEEE International Conference on Embedded and Ubiquitous Computing (EUC)*, IEEE, vol. 1, pp. 639–642.

86. Xiaoyun Q, Xiaoning K, Chao Z, Shuai J, Xiuda M (2016) Short-term prediction of wind power based on deep long short-term memory. In: *2016 IEEE PES Asia-Pacific Power and Energy Engineering Conference (APPEEC)*, IEEE, pp. 1148–1152.

87. Zhai W, Zhu J, Cao Y, Wang Z (2018) A generative adversarial network based framework for unsupervised visual surface inspection. In: *2018 IEEE International Conference on Acoustics, Speech and Signal Processing (ICASSP)*, IEEE, pp. 1283–1287.

88. Lin Y, Wang J (2019) Probabilistic deep autoencoder for power system measurement outlier detection and reconstruction. *IEEE Transactions on Smart Grid* pp. 1–1, doi: 10.1109/TSG. 2019.2937043

89. Ju Y, Guo J, Liu S (2015) A deep learning method combined sparse auto encoder with svm. In: *2015 International Conference on Cyber-Enabled Distributed Computing and Knowledge Discovery*, IEEE, pp. 257–260.

90. Özdemir AOB, Gedik BE, Çetin CYY (2014) Hyperspectral classification using stacked autoencoders with deep learning. In: *2014 6th Workshop on Hyperspectral Image and Signal Processing: Evolution in Remote Sensing (WHISPERS)*, IEEE, pp. 1–4DL Architectures, Libraries and Frameworks in Healthcare 25.

91. Wang J, Chen CP, Wang CH (2012) Finding the near optimal learning rates of fuzzy neural networks (fnns) via its equivalent fully connected neural networks (ffnns). In: *2012 International Conference on System Science and Engineering (ICSSE)*, IEEE, pp. 137–142.

92. Rao PA, Kumar BN, Cadabam S, Praveena T (2017) Distributed deep reinforcement learning using tensorflow. In: *2017 International Conference on Current Trends in Computer, Electrical, Electronics and Communication (CTCEEC)*, IEEE, pp. 171–174.

93. Li J, Yu ZL, Gu Z, Tan M, Wang Y, Li Y (2019) Spatial–temporal discriminative restricted boltzmann machine for event-related potential detection and analysis. *IEEE Transactions on Neural Systems and Rehabilitation Engineering* 27(2):139–151.

94. Kearney V, Haaf S, Sudhyadhom A, Valdes G, Solberg TD (2018) An unsupervised convolutional neural network-based algorithm for deformable image registration. *Physics in Medicine & Biology* 63(18):185017.

95. Pavithra M, Saruladha K, Sathyabama K (2019) Gru based deep learning model for prognosis prediction of disease progression. In: *2019 3rd International Conference on Computing Methodologies and Communication (IC-CMC)*, IEEE, pp. 840–844.

96. Zhong Z, Li J, Ma L, Jiang H, Zhao H (2017) Deep residual networks for hyperspectral image classification. In: *2017 IEEE International Geoscience and Remote Sensing Symposium (IGARSS)*, IEEE, pp. 1824–1827.

97. Uçak K (2019) A runge-kutta mlp neural network based control method for nonlinear mimo systems. In: *2019 6th International Conference on Electrical and Electronics Engineering (ICEEE)*, IEEE, pp. 186–192.

98. Hinton GE (2002) Training products of experts by minimizing contrastive divergence. *Neural Computation* 14(8):1771–1800.

99. Zhao Z, Guo J, Ding E, Zhu Z, Zhao D (2015) Terminal replacement prediction based on deep belief networks. In: *2015 International Conference on Network and Information Systems for Computers, IEEE*, pp. 255–258.

100. Zhang Y, Li P, Wang X (2019) Intrusion detection for iot based on improved genetic algorithm and deep belief network. IEEE Access 7:31711–3172228.

101. Chen Y, Lv Y, Wang FY (2019) Traffic flow imputation using parallel data and generative adversarial networks. *IEEE Transactions on Intelligent Transportation Systems, 21*(4), 1624–1630.

102. Pan Z, Yu W, Yi X, Khan A, Yuan F, Zheng Y (2019) Recent progress on generative adversarial networks (gans): A survey. *IEEE Access* 7:36322–36333.

103. Wang J, Wang CH, Chen CP (2011) Finding the capacity of fuzzy neural networks (fnns) via its equivalent fully connected neural networks (ffnns). In: *2011 IEEE International Conference on Fuzzy Systems (FUZZ-IEEE 2011)*, IEEE, pp. 2193–2198.

104. Liu J, Wang C, Gong Y, Hao X (2019) e fully connected model for collective activity recognition. IEEE Access.

105. Simonyan K, Zisserman A (2014) Very deep convolutional networks for large scale image recognition. arXiv preprint arXiv:14091556.

106. Mazumdar M, Sarasvathi V, Kumar A (2017) Object recognition in videos by sequential frame extraction using convolutional neural networks and fully connected neural networks. In: *2017 International Conference on Energy, Communication, Data Analytics and Soft Computing (ICECDS)*, IEEE, pp. 1485–1488.

107. Szegedy C, Liu W, Jia Y, Sermanet P, Reed SE, Anguelov D, Erhan D, Vanhoucke V, Rabinovich A (2014) Going deeper with convolutions. *2015 IEEE Conference on Computer Vision and Pattern Recognition (CVPR)* pp. 1–9.

108. Reddy ASB, Juliet DS (2019) Transfer learning with resnet-50 for malaria cell image classification. In: *2019 International Conference on Communication and Signal Processing (ICCSP)*, IEEE, pp. 0945–0949.

109. Brahimi S, Aoun NB, Amar CB (2018) Improved very deep recurrent convo-lutional neural network for object recognition. In: SMC.

110. Huang Z, Huang L, Gong Y, Huang C, Wang X (2019) Mask scoring r-cnn. In: *Proceedings of the IEEE Conference on Computer Vision and Pattern Recognition*, pp. 6409–6418.

111. Cohen T, Geiger M, Köhler J, Welling M (2018) Spherical cnns. ArXiv abs/1801.10130.

112. Xie S, Girshick RB, Dollár P, Tu Z, He K (2016) Aggregated residual transformations for deep neural networks. *2017 IEEE Conference on Computer Vision and Pattern Recognition (CVPR)* pp. 5987–5995.

113. Redmon J, Divvala S, Girshick R, Farhadi A (2016) You only look once: Unified, real-time object detection. In: *Proceedings of the IEEE conference on computer vision and pattern recognition*, pp. 779–788.

114. Badrinarayanan V, Kendall A, Cipolla R (2017) Segnet: A deep convolutional encoder-decoder architecture for image segmentation. *IEEE Transactions on Pattern Analysis and Machine Intelligence* 39(12):2481–2495.

115. Iandola FN, Han S, Moskewicz MW, Ashraf K, Dally WJ, Keutzer K (2016) Squeezenet: Alexnet-level accuracy with 50x fewer parameters and <0.5 mb model size. arXiv preprint arXiv:160207360.

14 Designing Low-Cost and Easy-to-Access Skin Cancer Detector Using Neural Network Followed by Deep Learning

Utkarsh Umarye, Vishal Rathod, and Trilochan Panigrahi

National Institute of Technology Goa, Goa, India

Samrat L. Sabat

School of Physics, University of Hyderabad, Hyderabad, India

CONTENTS

DOI: 10.1201/9780367548445-16

14.1 INTRODUCTION

Melanoma, is a skin cancer where the growth of pigment-producing cells becomes abnormal, resulting in skin color change [1]. Early detection and necessary intervention stop this cancer from spreading to nearby tissues and to the rest of the body parts. Different stages of skin cancer and the surviving rate are shown in Figure 14.1.

At the first stage of this disease, any change in skin color should be physically examined by the practitioner. The color of skin helps to categorize the melanoma [2]. But the diagnosis of melanoma requires expert knowledge because natural skins have similar physical characteristics [3]. Thus, the only option for dermatologists is to perform a laboratory medical procedure known as a biopsy. In fact, a biopsy helps to know whether a skin tumor is benign or malignant. But this testing is expensive and guidelines are very rigid. Therefore, early detection methods are sought for rapid and convenient screening. Melanoma can be screened by processing the dermoscopy image of the skin using image processing techniques for initial screening [4]. In this chapter, color images of skin tumors are trained using a reduced complexity convolution neural network to classify the tumor as melanoma or not.

The goals of the proposed work are as follows:

14.1.1 LOCAL AND OFFLINE DEPLOYMENT

Doctor Hazel (developed at the TechCrunch Disrupt Hackathon in September 2017 [5], for melanoma detection) uses Machine Learning (ML) and adopts Intel's cloud computing service. It utilizes up to 192 GB of RAM and Xeon processors. It provides real time detection accuracy close to 85%, at the same it uses huge resources for large data sets training and validating. It is not possible to provide these resources

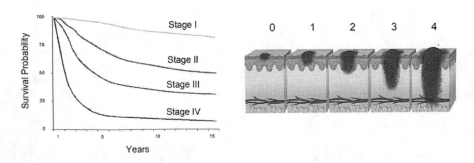

FIGURE 14.1 Surviving rate of skin cancer with stages (Source: cancerindia.org.in).

for local and offline deployment. Moreover, the automatic classifier should act as a supplementary tool to the practitioner and not replace the practitioners; therefore, a lighter model with slightly less accuracy (less false negative) is sufficient if it can be deployed locally. It is feasible and manageable locally on regular computers with substantially low processing power and RAM for a lighter model. This reduction in complexity also helps to decrease latency, enabling it to be used for transfer learning locally.

14.1.2 ELIMINATING CUSTOM HARDWARE REQUIREMENTS

Intel's implementation makes use of expensive custom hardware, such as the UP Board and Movidius Neural Compute stick [6]. The UPÂCÂš board houses Intel's Apollo Lake line of processors, while the Movidius Stick is used for edge computation for deploying and optimizing ML algorithms. Our goal is to eliminate this custom hardware and deploy the model on any hardware setup, such as in Raspberry PI [7, 8]. The main advantage of universalizing hardware and software is to ensure ease of access and availability across the globe, even in the remotest village dispensaries and hospitals.

14.1.3 DIVERSIFYING CLASSIFICATION

Doctor Hazel helps to classify melanoma from among nevus/mole, and seborrheic keratosis/warts [5, 6]. Since this form of cancer presents itself externally and is visible on the skin, it causes a stir among people as it can easily be confused with a rash or burn. Thus in our present approach, we have expanded the data set to cover burns of several degrees and various types of rashes, thereby further expanding its use case to help prevent false alarms.

An integrated system can be designed to pave the way for systematic hospital appointments. The detection system aims to give fairly accurate results; thus those cases having a higher probability of having melanoma can be shortlisted for the first slot of appointments to nearby oncologists, while the lower probability cases can be appointed later slots, thereby ensuring an efficient and systematic appointment system.

14.2 COMPUTER-AIDED SYSTEM FOR SKIN CANCER DIAGNOSIS

Automated image classification using Machine Learning (ML) is a useful tool for early skin cancer diagnosis [8, 9]. There exists some limitations of ML for general usability in current clinical practice in rural areas [3]. Developing a robust ML melanoma classifier requires a large number of trainings of the images specific to the demographical area. Developing a computer-aided system that can play a role of support in the diagnosis of skin cancer may require the following steps [10] given in Figure 14.2. The primary step of classification begins with processing a digital image. It includes removing the structures from the digital image which are not relevant or can affect the main subject. The background information, due to skin hairs, bubbles, or gel, causes noise in the image. Adding to it, if we use images from dermoscopy,

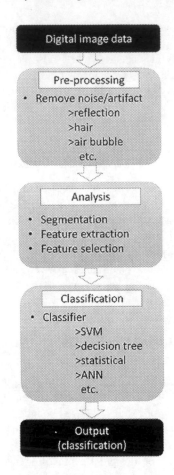

FIGURE 14.2   Flow diagram of a Machine Learning classifier.

the ink marks or any other noises such as salt pepper add graininess to the image. Different types of filters are used to remove such noise. This step is known as image preprocessing; without it, the efficiency of the model deteriorates.

The next step in the process is to analyze the image. It includes the feature extraction [11]. Before extracting features, one must segregate the main subject: the tumor area, from the rest of the image. This process is known as the segmentation of the image. It is a challenging part. Many segmentation algorithms are listed in [12].

After performing the image preprocessing, feature extraction is carried out. The basic features a medical practitioner uses are ABCDE of a melanoma skin cancer, where A refers to the Asymmetry nature of a tumor, B refers to the border of the lesion, C refers to the Color feature, D stands for the diameter of it, whereas, E refers to the Evolution over a period of time. These features are used to classify the image as benign or malignant. In a computerized classifier, the classification accuracy depends on the feature set.

In image processing, Histogram Of Oriented Gradients features (HOG) are the prominent features being used, along with the support vector machine (SVM) for image classification [13]. It captures the shape features of the image. In literature, different feature extraction techniques are used to extract texture features of a dermoscopy image for melanoma detection [14, 15]. Further, these features are used in an SVM classifier. Since all these references have experimented on a smaller and private data set, it is difficult to compare and improve the model. The smaller data set training in SVM leads to degraded accuracy. Many algorithms are reported to perform the skin cancer classifications, for instance, support vector machine, statistical [logistic regression], decision tree, or artificial neural network (ANN) [16]. The schematic structure of an ANN is shown in Figure 14.3. Each node of ANN behaves like a neuron of the human brain, and ANN has a similar biological neural network structure to that in the brain.

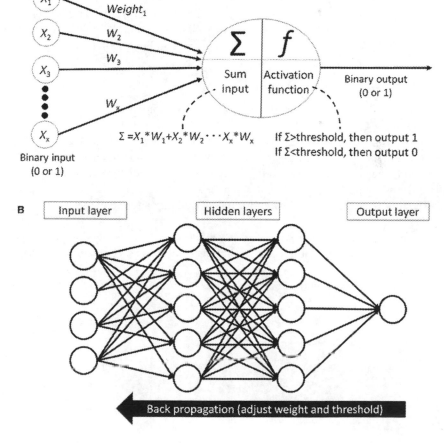

FIGURE 14.3 Schematic diagram of artificial neural network. (a) Model with single perceptron. (b) ANN with hidden layers.

Developing a melanoma classifier has several caveats which need to be tackled [9]. As opposed to the conventional approach of visiting a doctor, there is a lack of experience and intuition involved. With intuition and experience, doctors assess several factors, such as the asymmetrical shape, border, color profile, diameter, and long-term evolution. When designing a model, the evolution and experience parameters are discarded, as it is imperceptible by artificial intelligence (AI).

Given the widespread nature of this issue, one needs to create an improved solution for melanoma classification. The cost, ease of access, and offline training are the primary requirements of a simple Machine Learning-based classifier. The conventional ANN model's accuracy depends on the feature set; consequently, we chose CNN in place of ANN. The appropriate structure and the learning of CNN lead to an increase in classification accuracy [17].

14.3 PROPOSED METHOD

In this section, skin cancer detection from the color image by using a deep learning method is explained. The proposed methodology discriminates an image as either melanoma or benign. It is achieved by designing a simple Convolution Neural Network (CNN) model. A structure of CNN is given in Figure 14.4.

14.3.1 FLOW OF A CNN MODEL

We detect two different classes in the proposed work, i.e., melanoma and non-melanoma, using Convolution Neural Network (CNN) with keras Tensor Flow. Non-melanoma includes burns, rashes, seborrheic keratosis, and nevus [2]. Then, the proposed model is analyzed to see how it can be useful in a practical scenario with constrained resources. The general architecture of the prediction model using CNN is shown in Figure 14.4.

The following steps are followed for model building and evaluation:

1. Import all the libraries which are essential.
2. Obtaining images and creating a dictionary of these pictures, and labeling them.
3. Labels are based on the image category.

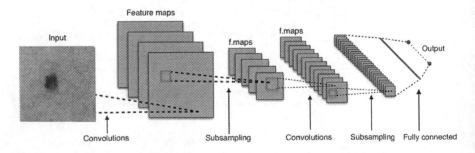

FIGURE 14.4 General architecture of CNN.

4. Normalization of the image set.
5. Splitting data into training and testing sets.
6. Building an architecture of the model.
7. Cross-validating the model.
8. Testing the model.

14.3.2   BUILDING CNN

The application programming interface (API) used for building the model is Keras Sequential API. It allows building the model systematically, one layer at a time, beginning from the input layer and followed by other deep layers, as shown in Figure 14.5. The functionality of each layer used in the proposed model is explained as follows:

14.3.2.1   Layer 1

The input layer in the CNN model is the convolution (Conv2D) layer. It acts as a set of filters that are learnable. At the first layer, we have selected a set 32 filters. Filters are used for the transformation of the image. The part of the image is selected using kernel size, which in our case is 3 × 3. The kernel filter matrix is multiplied by the complete image.

14.3.2.2   Layer 2

The layer immediately following the convolution layer is the pooling layer. This layer helps to downsample the image. The filter used in our case is the Maxpool filter, size 2 × 2. It looks for the two neighboring pixels and selects the maximal value among them. These filers are used to reduce the computational cost and avoids overfitting. The downsampling rate directly depends on the size of the kernel. The combination of convolutional and pooling layers helps CNN to learn from local as well as global features.

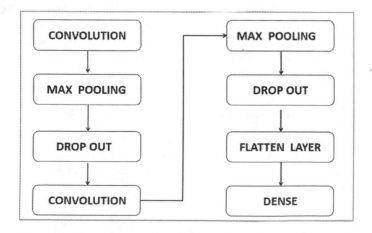

FIGURE 14.5   Layer-wise block diagram of basic CNN Model.

14.3.2.3 Layer 3

Another important layer of CNN is the Dropout layer. In this layer, some randomly selected weights are set as zero. That means it drops a few nodes and allows the CNN model to learn the parameters from this new distributed set. It helps to overcome the problem of over-fitting of the model and also makes the network more generalized.

14.3.2.4 Layer 4

The layer used after the dropout layer is the flattened layer. It is used to convert the final feature maps into a single 1D vector. This step of flattening is essential to use the fully connected network after the layers of convolutional or maxpool. It further combines all the local features found from the previous layers, i.e., convolution or pooling layer.

14.3.2.5 Layer 5

The outer layer of the CNN model is the dense layer, which can be seen as a fully connected layer. This layer is nothing but an ANN classifier.

14.3.3 FEATURE EXTRACTION

As opposed to typical Machine Learning algorithms, CNN, in particular, is extensively more comprehensive and has a significantly lower error rate, which is absolutely vital when dealing with medical-related applications [18, 19]. It has an innate ability to extract and extrapolate several informative parameters from even a limited data set by virtue of convolution, backpropagation, and intuitive repeated convolution. The ability of backpropagation helps the algorithm automatically adjust weights of convolutional kernels [20]. In doing so, from a relatively limited data set, we can achieve a significantly large number of parameters. All these parameters may not be tangible to humans or even intuitive for them to interpret; however, these weights and biases in a congregation add immense improvement to the ability of a CNN classifier and increase the accuracy with an increase in the epochs.

In the beginning, the algorithm either randomly assigns weights to the kernel or uses a couple of basic conventional filters, such as a high pass filter, low pass filter, band pass filter, etc., as tabulated in Table 14.1. These extracted features deviate from the known tangible elements with the backpropagation algorithm and defer towards seemingly random weights and biases.

14.3.4 ACTIVATION FUNCTIONS

In a ANN structure, every node in each layer is associated with an activation function, which is essentially a non-linear function. This activation function transforms the output of the node. In our model, we have used two activation functions. In the convolution layer, Rectified linear unit (ReLU) activation function is used, which is given by:

$$f(x) = \max(0, x)$$

TABLE 14.1
List of Common Kernels
Used for Feature Extraction

Operation	Kernels
Sharpen	$\begin{bmatrix} 0 & -1 & 0 \\ -1 & 5 & -1 \\ 0 & -1 & 0 \end{bmatrix}$
Gaussian blur	$\dfrac{1}{16} \times \begin{bmatrix} 1 & 2 & 1 \\ 2 & 4 & 2 \\ 1 & 2 & 1 \end{bmatrix}$
Top Sobel	$\begin{bmatrix} 1 & 2 & 1 \\ 0 & 0 & 0 \\ -1 & -2 & -1 \end{bmatrix}$
Left Sobel	$\begin{bmatrix} 1 & 0 & -1 \\ 2 & 0 & -2 \\ 1 & 0 & -1 \end{bmatrix}$
Box Blur	$\dfrac{1}{9} \times \begin{bmatrix} 1 & 1 & 1 \\ 1 & 1 & 1 \\ 1 & 1 & 1 \end{bmatrix}$
Outline	$\begin{bmatrix} -1 & -1 & -1 \\ -1 & 8 & -1 \\ -1 & -1 & -1 \end{bmatrix}$

Softmax is another activation function that is used at the final layer of the CNN model. The input for the Softmax function is a vector of K real numbers. These are then normalized into a probability distribution, which consists of K probabilities directly proportional to the exponential of the input numbers. The function is defined by:

$$f(x)_i = \frac{e^{x_i}}{\sum_{j=1}^{K} e^{x_j}} \text{ for } i = 0,1,2,\ldots,K$$

A Softmax function helps to convert each node's output value between 0 and 1, which further allows to sum the outputs up to 1 and gives the final output result as a certain probability (0–100% probability).

14.4 RESULTS AND DISCUSSIONS

This section presents the classification results of a dermoscopy image as either melanoma or non-melanoma from the images using CNN. The CNN is designed with nine layers. The performance of the CNN model is evaluated using accuracy and cross-entropy. Further sensitivity and selectivity were also evaluated based on the confusion matrix. The proposed simple CNN model performance is compared with different existing CNN models.

We used the publicly available dataset from an archive of the International Skin Image Collaboration (ISIC) (www.isic-archive.com) for training and validation in the current work. The dataset consists of 1,780 pictures of benign moles, including rash and burn samples, and 1,781 pictures classified as melanoma moles. All the pictures are resized to a lower resolution of (224 × 224 × 3) RGB format. Figure 14.6 shows the first fifteen images from the test case and how they are classified. The data included two new test cases and images. In our case, these are hand-picked so that they have a maximum impact on the model's output.

The CNN model is designed with nine convolution layers, followed by a fully connected network. The model's features across each convolution layer are provided in Table 14.2 by using python command model. summary(). The total of trainable parameters is 2,133,826.

Cross entropy is considered as a loss function for model training. Cross entropy is defined as:

$$\text{Cross} - \text{entropy} = -\sum_{l=1}^{n}\sum_{k=1}^{m} y_{l,k} \log\left(p_{l,k}\right)$$

where, $y_{l,k}$ stands for the true value and is equal to 1 if sample l belongs to class k, and in otherwise condition, it is zero and $p_{l,k}$ stands for the probability, which is predicted

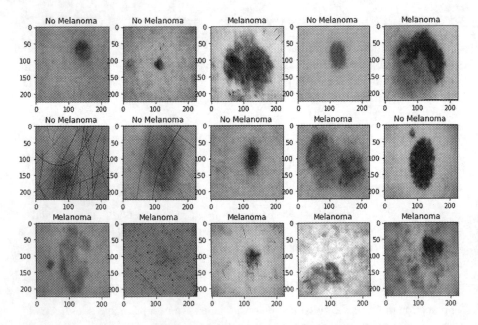

FIGURE 14.6 First fifteen images from test set and how they are classified.

TABLE 14.2
Model Summary of CNN Using Python
Command model.summary()

Layers(type)	Shape of Output	Parameters
2D convolution 1	$112 \times 112 \times 32$	864
2D convolution 2	$112 \times 112 \times 64$	2048
2D convolution 3	$56 \times 56 \times 128$	8192
2D convolution 4	$28 \times 28 \times 526$	32768
2D convolution 5	$28 \times 28 \times 256$	65536
2D convolution 6	$14 \times 14 \times 512$	131072
2D convolution 7	$14 \times 14 \times 512$	262144
2D convolution 8	$14 \times 14 \times 1024$	524288
2D convolution 9	$14 \times 14 \times 1024$	1048576
Global Average Pooling	1024	0
Dense	2	2050

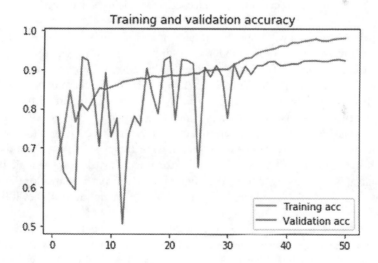

FIGURE 14.7 Loss curve varying with epochs for training and testing.

by CNN model of i sample which belongs to j class. Figure 14.7 provides the graph of loss of proposed model versus epoch number.

The accuracy of true detection is considered as one of the main performance indicators of the model.

The accuracy is calculated as:

$$Accuracy = \frac{\text{Number of correct predictions}}{\text{Total number of predictions}}$$

We obtain the model training accuracy as 95%, while testing the achieved training accuracy was 92.78%, whereas [5] reported real time skin cancer detection up to

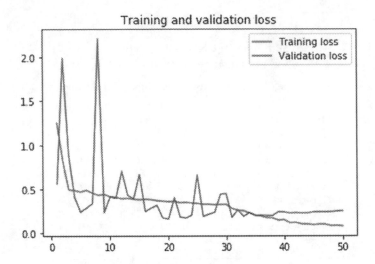

FIGURE 14.8 Accuracy curve varying with epochs for training and testing.

85% validation accuracy. Although the later model is a real time implementation result compared to the proposed model, it requires a larger infrastructure for training and has a huge number of trainable parameters, making it unfit for deploying in a resource-constrained, embedded platform. Figure 14.8 shows the plot between accuracy versus epochs. The x-axis refers to epochs, and the y-axis represents accuracy.

The confusion matrix is another parameter to define the efficiency of a CNN model for classification. It is also referred to as an error matrix. It is in the form of a table and used to visually show the performance of the classifier on a test data set, provided we know the true values for it. Error matrix also allows visualizing the performance of a CNN algorithm. The confusion matrix performance of the model is shown in Figure 14.9. From Figure 14.9, we came across the following observations.

The confusion matrix makes it easy to evaluate the sensitivity, selectivity, and positive predictive for determining the classification accuracy. These parameters are derived from True Positive Rate (TPR), True Negative Rate (TNR), Positive Predictive Value (PPV) as defined below.

True Positive Rate (TPR): It gives the hit point of the model. It is also called the sensitivity of CNN model and given by:

$$TPR = \frac{TP}{P} = \frac{TP}{TP + FN} = 1 - FNR$$

True Negative Rate (TNR): It gives selectivity of the model and is given by:

$$TNR = \frac{TN}{N} = \frac{TN}{TN + FP} = 1 - FPR$$

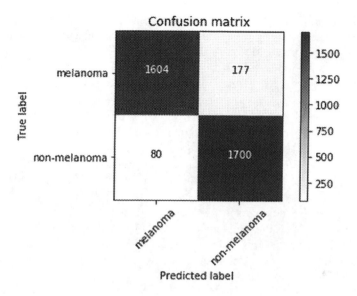

FIGURE 14.9   Confusion matrix of CNN model.

Positive Predictive Value(PPV): It helps to find the precision of correctly predicting positive case. It is given by:

$$PPV = \frac{TP}{TP + FP}$$

Based on the above parameters, sensitivity, selectivity, and positive predictive values are evaluated as:

- Sensitivity = 0.9525
- Selectivity = 0.9057
- PPV = 0.9

These performance parameters are better than the models reported in the literature [17].

The trained model file of the proposed CNN is 121.54MB. It is small enough to be installed in any embedded device, like an android smartphone.

14.4.1   COMPARISON WITH MACHINE LEARNING APPROACH

Compared to typical Machine Learning algorithms or classifiers like Support Vector Machine (SVM) or K Nearest Neighbors (KNN), CNN consistently outperforms and outranks these every time. The ability of CNN to comprehensively and intuitively extract, extrapolate and infer parameters from a data set is unarguably unmatched.

KNN and SVM Machine Learning methods are both clustering-based algorithms used for classification [13]. KNN adopts lazy learning while training; there is absolutely no information extracted from the data set, whereas, while testing, a lot of processing and analysis needs to be done to decide in which cluster the input image belongs for its classification. It requires a significantly more extensive data set to be even comparable to CNN, and the longer testing time implies it is not suited for real-time applications. Its error rate is also much higher, in the range of 20% to 30%.

From the standpoint of this work, we aimed to design a power efficient, easily computable, real-time melanoma detection model. In the proposed work, we use CNN because, as opposed to KNN or even SVM, it can work with a much smaller data set to give significantly higher accuracy, due to its convolutional and back propagational techniques that extract a plethora of features, significantly more than any other ML classifier.

To visualize this comparison, we performed melanoma detection on a KNN model. As this involves the hand-engineered feature extraction part, few feature values are shown in Table 14.3, while Figure 14.10 shows the image output using extracted features. The extracted features are listed in Table 14.3. These features are used in KNN for classification.

14.4.1.1   Mean

The input image is in RGB format. Therefore, we have calculated the mean value for these to get color information about the skin lesions.

14.4.1.2   Area

Effected area is calculated using Otsu's image thresholding, which acts as a binary mask. The image in the second row of Figure 14.10 shows binary mask output. From this, we calculate the number of pixels that are black to get the area.

14.4.1.3   Border

To get the information about the border, we apply edge detection to the binary masked image that gives the boundary of the lesions. Now the white pixels are calculated to get the perimeter.

TABLE 14.3

Extracted Values of Features for Melanoma and Non-melanoma Images

	Benign	Malignant
R Mean	197.1230	196.5412
G Mean	148.5121	137.0165
B Mean	161.7338	150.8907
Effected area	1222	10628
Perimeter	1063	1328

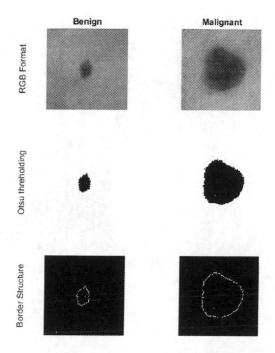

FIGURE 14.10   Features extractions.

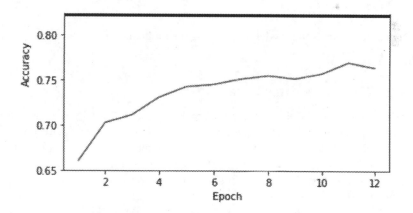

FIGURE 14.11   Validation accuracy of KNN model.

Results of the KNN model is plotted in Figure 14.11. The highest accuracy achieved is 76.27%. The confusion matrix parameters, i.e., sensitivity, selectivity, and precision are calculated using Figure 14.12 as:

- Sensitivity = 0.7799
- Selectivity = 0.7475

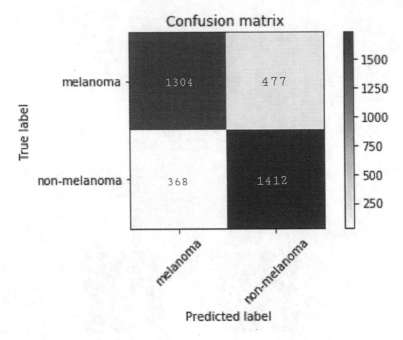

FIGURE 14.12　Confusion Matrix of KNN model.

- Precision = 0.7322

CNN decimates the KNN model by showing a significant improvement in accuracy, lower error rates, and a much lower testing time.

14.4.2　Comparison with Other CNN Models

There are a plethora of CNN architectures that are used in the field of medical imaging. Few popular such networks are LeNet, AlexNet [21], VGGNet [22], ZFNet [23], ResNet [24], SENet [25], and GoogLeNet [26], In the yearly competition on computer vision held by ImageNet Large Scale Visual Recognition (ILSVR), these CNN architectures were tested. All these models are used for melanoma classification for the same dataset.

The results of the challenge are shown in Table 14.4. From this table, it is observed that the proposed net gives an error rate of 7.2%. It gives an acceptable accuracy level with a reduced number of trainable parameters.

14.5　FUTURE SCOPE

The current implementation of the model achieves 92.78% accuracy, with testing times under two minutes, making this package an extremely effective, power-efficient, real-time melanoma detection model. After an initial one-time training, the

TABLE 14.4

List of CNN Architectures

Architecture	Year	Error rate	Layers	Parameters
LeNet	1998	NA[a]	5	60 thousand
AlexNet	2012	15.3%	7	60 million
ZFNet	2013	14.8%	8	NA
GoogLeNet	2014	6.67%	22	4 million
Proposed Net	2020	**7.22%**	9	2 million
VGG Net	2014	7.3%	16	138 million
ResNet-50	2015	3.6%	152	26 million
SENet	2017	2.3%	NA	NA

[a] NA: Not Available

parametric weights and biases are stored as a Hierarchical Data Format file occupying 121.54 MB of storage space.

In order to implement this model in a Raspberry Pi or any other Android or iOS device, we need to parse an image through this HDF file to get the probabilistic account of whether the input image is a case of melanoma or not. However, before parsing, we require some initial preprocessing of the image pertaining to dermoscopic photography standards. This preprocessing involves dropping the resolution of the image down to 224 × 224 in accordance with appropriate cropping into the area under consideration, namely, the suspected melanoma site.

In a Raspberry Pi, we can capture input images through the natively supported Pi Cam and utilize a python script for the preprocessing, while for mobile platforms, we need to bundle the script along with the HDF into a user-friendly application interface for which app development is required. Consequently, the experience could be enriched by an intuitive platform for hospital appointment bookings and useful information, links, and stern advisories regarding seeking the opinion of a licensed medical practitioner or oncologist.

14.6 CONCLUSION

The proposed work demonstrated the use of a simpler CNN model for melanoma detection. The designed model is deployable in a hardware-constrained embedded platform. Since the healthcare screening devices are not meant to replace the medical practitioners, a compromised accuracy of 8% is sufficient to screen the patient. The proposed CNN model provides better accuracy of 92.78%, with less data size. It also provides better sensitivity and selectivity over KNN and SVM, and other existing CNN models. Given these parameters, it is extremely suited for adoption as a mobile-based application and used in many smart wearables. The future work is to run this code on a hardware model and test it with clinical data.

REFERENCES

1. Skincancer.org, Melanoma – SkinCancer.org, 2016. Available: http://www.skincancer.org/skin-cancer-information/.
2. Patil Rashmi, Bellary Sreepathi, "Machine learning approach in melanoma cancer stage detection," *Journal of King Saud University – Computer and Information Sciences*, 2020.
3. Kadampur Mohammad Ali, Al Riyaee Sulaiman, "Skin cancer detection: Applying a deep learning-based model driven architecture in the cloud for classifying dermal cell images," *Informatics in Medicine Unlocked*. (2020) 18.
4. Wong A, Stalling S, Wei M, Hadley D, Bhattacharya A, Young A, "Precision diagnosis of melanoma and other skin lesions from digital images," *AMIA Summits on Translational Science Proceedings*. (2017) 2017:220–226.
5. Online Article: https://software.intel.com/en-us/articles/doctor-hazel-a-real-time-aidevice-for-skin-cancer-detection
6. Intel shows off the movidius myriad x, a computer vision chip with deep learning baked-in. Online Article https://techcrunch.com/2017/08/28/intel-shows-off-hemovidius-myriad-x-a-computer-version-chip-with-some-serious-brains.
7. Pasolini G, Bazzi A, Zabini F, "A Raspberry Pi-based platform for signal processing education [SP Education]," *IEEE Signal Processing Magazine*. (2017) 34(4):151–158.
8. Qin H, Sahu P, Yu D, Lightweight deep learning on smartphone for early detection of skin cancer, in *Symposium on Data Science for Healthcare (DaSH)*, Ridgefield, Connecticut, October 19–20, 2017.
9. González-Cruz C, Jofre MA, Podlipnik S, Combalia M, Gareau D, Gamboa M, Vallone MG, Faride Barragán-Estudillo Z, Tamez-Peása AL, Montoya J, Amlrica Jess-Silva M, Carrera C, Malvehy J, Puig S, "Machine learning in melanoma diagnosis limitations about to be overcome," *Actas Dermo-Sifiliogr Āaficas* (English Edition). (2020) 111(4):313–316.
10. Masood A, Al-Jumaily AA. "Computer aided diagnostic support system for skin cancer: a review of techniques and algorithms," *International Journal of Biomedical Imaging*. (2013) 2013:323268.
11. Winger DG, Ferris LK, Harkes JA, Gilbert B, Golubets K, Akilov O, "Computer-aided classification of melanocytic lesions using dermoscopic images," *Journal of the American Academy of Dermatology*. (2015) 73:769–776.
12. Youn Hyun J, Jinman K, Changyang L, Fulham M, Euijoon A, Lei B, "Automated saliency-based lesion segmentation in dermoscopic images," *Conference Proceedings of IEEE Engineering in Medicine and Biology Society*. (2015) 2015:3009–3012.
13. Bakheet S, "An SVM frame work for malignant melanoma detection based on optimized hog features," 2017.
14. Satheesha TY, Satyanarayana D, Prasad MNG, Dhruve KD, "Melanoma is skin deep: A 3D reconstruction technique for computerized dermoscopic skin lesion classification," *IEEE Journal of Translational Engineering in Health and Medicine*. (2017) 5.
15. Sheha MA, Mabrouk S, Sharawy A, "Automatic detection of melanoma Skin Cancer using Texture Analysis", *International Journal of Computer Applications*. (2012) 42.
16. Ripley BD. *Pattern Recognition and Neural Networks*. Cambridge: Cambridge University Press (1996). doi: 10.1017/CBO9780511812651
17. Winkler Julia K, Sies Katharina, Fink Christine, Toberer Ferdinand, Enk Alexander, Deinlein Teresa, Hofmann-Wellenhof Rainer, Thomas Luc, Lallas Aimilios, Blum Andreas, Stolz Wilhelm, Abassi Mohamed S, Fuchs Tobias, Rosenberger Albert, Haenssle Holger A, "Melanoma recognition by a deep learning convolutional neural network-Performance in different melanoma subtypes and localisations," *European Journal of Cancer*. (2020) 127:21–29, ISSN 0959-8049.

18. Iwashima Y, Wang J, Yashima Y, "Full reference image quality assessment by CNN feature maps and visual saliency," *2019 IEEE 8th Global Conference on Consumer Electronics (GCCE)*, 2019, pp. 203–207.

19. Indolia Sakshi, Goswami Anil Kumar, Mishra S.P., Asopa Pooja, "Conceptual understanding of convolutional neural network – A deep learning approach," *Procedia Computer Science.* (2018) 132:679–688.

20. Hekler Achim, Utikal Jochen S, Enk Alexander H, Solass Wiebke, Schmitt Max, Klode Joachim, Schadendorf Dirk, Sondermann Wiebke, Franklin Cindy, Bestvater Felix, Flaig Michael J, Krahl Dieter, von Kalle Christof, Fráuhling Stefan, Brinker Titus J., "Deep learning outperformed 11 pathologists in the classification of histopathological melanoma images," *European Journal of Cancer.* (2019) 118:91–96.

21. Yap MH, Pons G, Marti J, Ganau S, Sentis M, Zwiggelaar R, et al. Automated breast ultrasound lesions detection using convolutional neural networks. *IEEE Journal of Biomedical and Health Informatics.* (2018) 22:1218–1226.

22. Malhi HS, Cheng PM, Transfer learning with convolutional neural networks for classification of abdominal ultrasound images. *The Journal of Digital Imaging.* (2017) 30:234–243. doi: 10.1007/s10278-016-9929-2

23. Zeiler MD, Fergus R. Visualizing and understanding convolutional networks 2013. Available online at: arXiv[Preprint].arXiv:1311.2901 (accessed 25 July 2019).

24. Baltruschat IM, Nickisch H, Grass M, Knopp T, Saalbach A. Comparison of deep learning approaches for multi-label chest X-ray classification. *Scientific Reports.* (2019) 9:6381.

25. Hu J, Shen L, Albanie S, Sun G, Wu E. Squeeze-and-excitation networks 2017. arXiv[Preprint].arXiv:1709.01507.

26. Fujisawa Y, Otomo Y, Ogata Y, Nakamura Y, Fujita R, Ishitsuka Y, et al. Deep-learning based, computer-aided classifier developed with a small dataset of clinical images surpasses board-certified dermatologists in skin tumour diagnosis. *British Journal of Dermatology.* (2019) 180:373–381. doi: 10.1111/bjd.16924

Part III

---

*Internet of Things (IoT)
in Biomedical and Health
Informatics*

15 Application of Artificial Intelligence in IoT-Based Healthcare Systems

Ruby Dhiman and Riya Mukherjee
SRM University, Delhi, India

Gajala Deethamvali Ghousepeer, and Anjali Priyadarshini
SRM University, Delhi, India

CONTENTS

DOI: 10.1201/9780367548445-18

15.1   INTRODUCTION

Internet of Things has grabbed enormous attention in the field of health care in the last decade. It has manifested extensive practicality and application in healthcare devices, reducing the cost of medical expenses, better operational proficiency and patient's safety, expanding functional capability in the healthcare sector. This technology helps to spread medical knowledge to the isolated and secluded places. IoT has made a diagnosis of abnormality and disease easy with various monitoring miniatures, like a thermometer, glucometers, PO2 level measurement, blood pressure measurement, and other electronic devices connected with the Internet [1]. In straightforward terms, IoT enables the amalgamation and fruitful interchange of data between the utilizer and provider with a fully equipped IoT system containing operator devices, Network components, electronics, and data storage and analyzers [2]. The recent automation in healthcare, like the Internet of Things and artificial intelligence, and big data have shown their effectiveness in the COVID-19 pandemic. In addition to monitoring and analyzing the patient's health, one can prevent disease by early diagnosis and beginning therapy. It helps to address significant health issues and lightens the doctors' workload of physical examination [3]. In 2020 it was estimated that over 16.1 million of search medical devices had been used worldwide [4] (Figure 15.1).

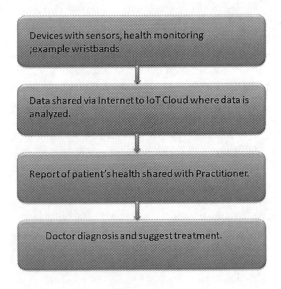

Devices with sensors, health monitoring ;example wristbands

Data shared via Internet to IoT Cloud where data is analyzed.

Report of patient's health shared with Practitioner.

Doctor diagnosis and suggest treatment.

FIGURE 15.1   IoT overview.

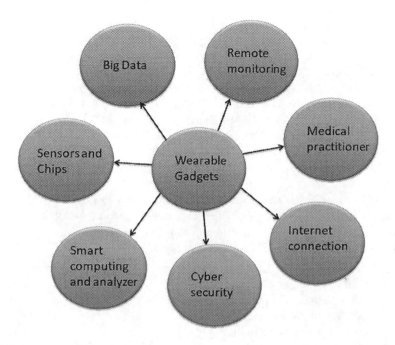

FIGURE 15.2 Components of IoT.

Internet of Things system requires a vast set-up with a wide range of components as shown in the illustration above [5] (Figure 15.2).

15.2 FUZZY LOGIC AND FUZZY MODELS OF HEALTH CARE

In the year 1960, a theory named "fuzzy logic theory" was established, which evolved from the theory of fuzzy sets. It means the participation of components is not absolutely true or absolutely false.

This theory supports concluding imprecise statistics or data. So, these theory-based designs can exhibit vast possibilities in the healthcare sector, where a rough fact can provide significant knowledge and construct excellent results, such as detecting any disease and proposing treatment for it. Another example, this fuzzy logic has been used in diabetes detection and cure, where a chip senses the blood glucose level changes. This data was used to analyze whether the insulin quantity is sufficient to tackle the usual response to food and exercise. Another fuzzy system works on tracking insulin levels, and then maintaining the insulin level using fuzzy logic algorithms. It also predicts the type of food to be taken by the patient. These algorithms were also used to detect heart-related ailments, breast cancer diagnosis, treatment in initial stages, the dengue epidemic, etc. [6].

As evident, the medical and clinical fields extensively utilize these prediction models and logics because we still do not have the proper system to conduct such studies from available data. One such system, called EWS (early warning scores), is employed in hospitals to observe and record patients' health. It can detect declining health and alert health care workers to take intensive care of the patient. This system is obligatory in hospitals of the United Kingdom, and other countries like the United States and Australia usually use it [7]. There are few new modifications of EWS that are under investigation. One of the latest versions is NEWS, a national early warning score, established in The United Kingdom. it monitors oxygen saturation, heart rate, pulse rate, respiratory rate, blood pressure, body temperature, consciousness level, etc. [8].

The fuzzy logic and models have delivered satisfactory results in the clinical field by helping medical practitioners improve pre-diagnosis and treatment accuracy. These models can be developed in different steps, as following:

- *Denote the problem*: the essential symptoms, indicators, and normal limits are examined, defining fuzzy logic. Then, conclusions are made concerning any irregularities highlighted through the symptoms.
- *Talk with multiple specialists*: The specialists will establish a relevant classification using the description.
- *Fuzzification*: In this step, the fuzzy sets will be defined. The variables are assigned to identify the fuzzy set they belong to (e.g., if two inputs connectively cooperate or functions are relevant, they are precisely assigned different fuzzy sets.
  MBP(mean blood pressure) has a fuzzy set.
  If low MBP i.e, < 80; L{MBP} = {(0,1),(60,1),(80,0)}
  For Normal MBP i.e, 80-130; N{MBP} = {(75,0),(105,1),(130,0)}
  For High MBP i.e, > 130; H{MBP} = {(126,0),(138.7,1),(200,1)}
- *Fuzzy rules are defined*: Here standardization is accomplished by defining the set of rules, set of vital signs, and set of possible diagnosis. At this step, it is crucial that the number of rules defined can include all feasible amalgamations of inputs and outputs.
  If low MBP and low POS2; instability.
  If normal MBP and low POS2; Hypoxemia.
  POS2 – Partial oxygen saturation classification.
- *Interpretation step*: For fuzzy sets, describe, declare, and access each case for every fuzzy rule, then combine the knowledge from the assigned rules.
- *Defuzzification step*: In this step, a single numerical value is given to get results. For example, if we get a value>8, linguistically suggest instability. Then, doctors can see an alarm signal on the device and focus on stabilizing the patient [9, 10].

This model is labor-saving and time-effective in real-life-saving circumstances, devoid of compromising standards and outcomes [11] (Figure 15.3).

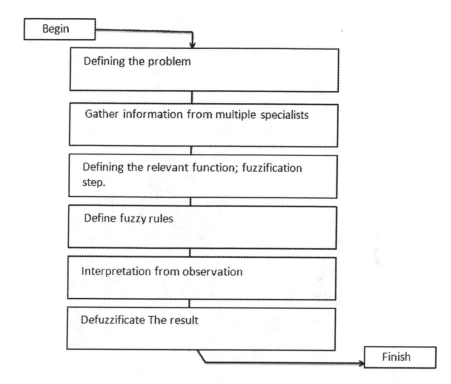

FIGURE 15.3 Flow chart for generating fuzzy medical system.

15.3 EVOLUTIONARY COMPUTING OF HEALTH CARE

Evolutionary computing is an application used for predictive analysis. It uses multi-dimensional data for predicting what diseases might occur in the future. The algorithms employed in evolutionary computing imitate biological activities to answer complicated questions as it allows a flexible structure that can accommodate a variety of problems or variants of the same problems [12]. It applies to theories of natural evolution, survival of the fittest, and natural selection for finding a solution to actual problems [13]. Evolutionary computing or algorithms is one of the branches of artificial intelligence. With the expanding applications of innovative health technologies, there is an genuine need for updated and relevant evolutionary computation algorithms to improve health care facilities in the developed countries [14] (Figure 15.4).

Evolutionary computations are employed to solve the electrical impedance tomography EIT related problems. It is an imaging and monitoring technique and noninvasive technique like MRI and CT scans [15] (Figure 15.5). This field of computer science studies a class of population-based, stochastic search algorithms known as evolutionary algorithms.

There are four main types of evolutionary algorithms [16].

1. *The genetic algorithm* is an excellent framework for studying biological evolution, which solves complex problems. Its main features are:

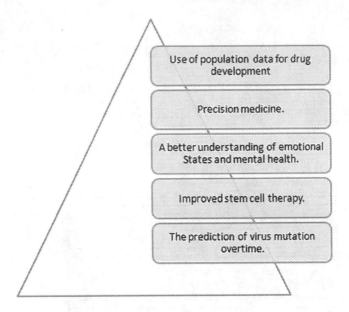

FIGURE 15.4   Applications of evolutionary computing in healthcare.

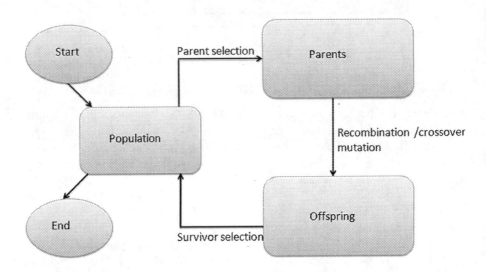

FIGURE 15.5   Evolutionary cycle.

- It performs with the coding of variable sets, not the variable itself.
- It starts the hunt from a population of different locations, not a single location or site.
- It involves only original formation, not secondary information.
- It uses selective conversion rules, not preordained rules [17].
2. *Genetic programming*: GP solves the problem using repetitive operations. It employs selective options of relevant and appropriate solutions and their

alternative by employing a set of genetic operators, generally mutation and crossover [18].

3. *Evolutionary programming*: This framework offers a rapid and economic problem-solving approach. It is utilized in hierarchical systems, evolutionary and combinatorial optimization, designing neural networks, and automatic control. This concept follows the evolutionary program body. It establishes the connection between mutation magnitude and phenotypic effects [19].

4. *Evolutionary strategies*: It belongs to the class of evolutionary algorithms which employ mutation, recombination based on the objective of evolution [20].

This approach is constructive in clinical diagnosis. So far, these algorithms have helped in studying the diagnosis of fatal ailments like breast cancer and hepatitis by collecting datasets from esteemed universities. It is so efficient in analyzing datasets that it can predict whether the patient with hepatitis will survive or not with the highest predictive accuracy [13]. It is also used to diagnose and monitor genetic disorders like Parkinson's disease, Alzheimer's disease, and other neurodegenerative diseases [21].

15.4   ARTIFICIAL NEURAL NETWORK FOR HEALTH CARE

The artificial neural network is an algorithm influenced by the design and functioning of the neural network present in our brain [22]. Deep Learning is a specialized Machine Learning system and a unique artificial neural network [23]. It is a type of network where numerous neurons form a multi-layer network comprising a minimum of three configuration layers in a sequence, an input layer, a hidden layer, and an output layer [24] (Figure 15.6) the input layer has five neurons in agreement, with each of the five prime anatomical characteristics recognized on preoperative MRI. The output of the first layer is linked to the middle layer, called the hidden layer.

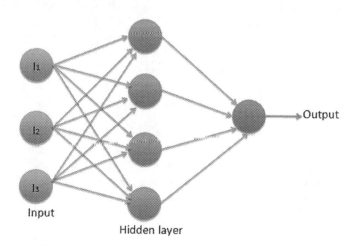

FIGURE 15.6   Artificial neural networks model.

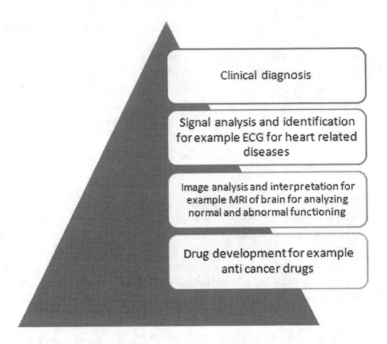

FIGURE 15.7   Applications of neural network in healthcare.

Finally, the network is linked to the final layer to provide the result it was built to produce, called the output layer [25].

Neural networks are carried out in two different manners called learning and evaluation or testing. It starts with ingesting new information, called the training phase, and giving out the output based on learning. With time and extensive learning, the networks refine and adjust to produce desired and acceptable outputs [26] (Figure 15.7).

It holds enormous potential in solving clinical problems and is easily achievable in the medical sector. It can measure the enzymes functional for heart and efficiency, and diagnose acute myocardial infarction with a 100% accuracy rate. Neural networks are trained in analyzing heart enzyme figures, symptoms, and variations after drug administration. It effectively analyzes the ECG signals and detects abnormalities. This proves that artificial network neurons are highly effective in cardiology. Hospitals have well-equipped ICUs with this technology for checking all normal parameters and predicting heart health deterioration of patients under intensive care. It has shown in multiple clinical surveys that this technology helps in the prognosis of oncology and gastroenterology with 100% accuracy. These networks can diagnose pulmonary embolisms with better accuracy than two experts working together. It detects tumors and evaluates the receptors' size and status with detailed cytology. It is also employed in Ophthalmology for detecting various abnormalities in the functioning of the cornea. In recent years, radiologists have established a neural network that can use SPECT images and diagnose liver diseases, coronary artery disease, and prostate cancer with utmost accuracy and better than experts. It has helped in other fields like pathology for tubular carcinoma, prostate cancer spread, and breast carcinoma stages, with the help of DNA flow cytometric histogram. It helps in studying different patterns of genetics, stages of embryogenesis, and changes or abnormalities

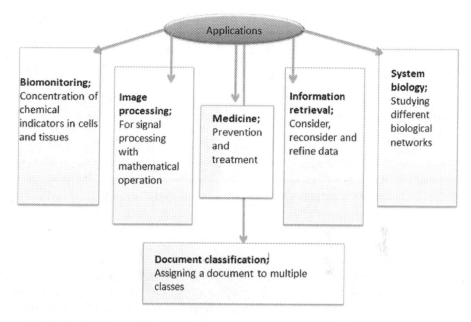

FIGURE 15.8 Applications in the healthcare flow chart.

in the shape and size of chromosomes for studying genetics. This history proves it has immense capability to solve medical problems, enhance accuracy, and make diagnosis an easy task for the experts [26, 27] (Figure 15.8).

15.5 A PROBABILISTIC MODEL FOR HEALTH CARE

As we have seen, various Machine Learning algorithms are serving their purpose in healthcare by easing the diagnosis; however, it demands enormous datasets for learning. If there is insufficient data related to that clinical pattern, we need to establish new models. One such model is the probabilistic model [28].

This model employs these statistics and random variables for diagnostic testing of patients in the hospitals. It lowers the cost without hampering the diagnostic quality and accuracy. This calculation-based method is economically advantageous as well as effective in testing and produces probability distribution as a solution [29].

15.5.1 Risks of Probability Model in Healthcare

Although producing these probabilistic models for clinical use is a very challenging task, there are always some missing values in healthcare data. For instance, if a diabetic patient gets his checkup done, and all the biomarkers were not considered for evaluation, the omission could lead to incomplete observation. This missing value hampers the quality of the results. Another issue developed with the patient's labels remaining unobserved, called censoring. This is an issue because probabilistic models are trained to produce worthy outcomes with the help of these labels. Another issue is calibration, and probability models should be using calibrated risk scores in clinical settings. Uncertainties and data shift are also considered significant issues [30].

15.6 BIG DATA IN HEALTHCARE

Big data can provide several advantages in the healthcare sector, like electronic health records; outcome reports can significantly improve the quality of research in medicine and relieve patient suffering. Big data implies that the data, with the current limitations of range, difficulty, and variety, need new algorithms and techniques to analyze, manage, and retrieve applicable and unknown, concealed information. Its features are quantity of data, submission and generation of data, processing, data analysis, non-uniformity and diversity of data, and calculable data to rely on [31].

Hospitals produce large quantities of data related to patients, such as medical imaging data and clinical data. These massive records consist of all the medical and health-related information that can be utilized to solve unknown answers.

The big data in the clinical healthcare field can be classified into three primary data: electronic health records (EHR), electronic medical record (EMR), personal health records, and medical images. These records provide complete information about health, CT scans, medicine, treatment, surgery, anesthesia, physical examination, MRI, etc., are recorded in standard databases, like the continuity of care documents(CCD), continuity of care records(CCR), etc.

Whereas, PHR reveals lifelong health-related data, for example: daily living observations, laboratory test results, vaccinations, allergic reactions, drug effects, and hospitalization-related data are recorded in electronic devices, web applications, computer-based software, etc.

Data related to medical images is the visual information of the human body. It contains X- Rays, PET, positron emission tomography, CT scans, MRI, nuclear medicine, echocardiogram recording, DICOM, and digital imaging communication or similar software.

This will help extract valuable information and discard irrelevant information from big data, providing significant advantages to the healthcare industry [32] (Figure 15.9).

15.6.1 APPLICATIONS OF BIG DATA IN THE HEALTHCARE SECTOR

- The design of the disease can be understood.
- Discover the disease early.
- Treating the disease effectively.
- The outbreak of a disease can be analyzed.
- Familiarizing the public and controlling disease.
- Development of treatment and prevention by the scientists, like vaccines.
- Developing a predictive system to comprehend requirements and avoid health disasters [33, 34].

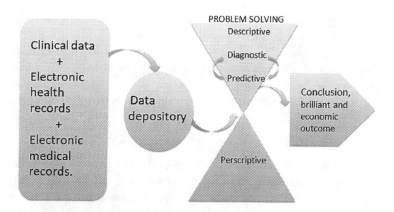

FIGURE 15.9 Workflow of big data in the healthcare system.

15.7 DATA MINING IN HEALTHCARE

As the data is increasing day by day, some valuable data gets invisible in this vast data. Therefore, to retrieve valuable and worthy data, different techniques are required. For example, CBIS, a computer-based information system, is used to retrieve helpful information. Data mining refers to finding patterns and withdrawing them from the massive bulk of data [35]. Extracting useful information from large data is done with the help of Knowledge Discovery in Databases (KDD). Data mining is the evaluating step of the KDD process. The aim of this step is the withdrawal of patterns from the data available. It has multiple steps starting with data understanding, then selecting it, followed by transformation and reduction [36]. Data mining technologies that have been used for many purposes, including research in the healthcare field, and specifically the biomedical sector [37]. There is a probability of broadening healthcare research to control and advantage patient data with advanced data mining techniques. Data mining provides advantages like fraud detection, chronic disease management, and general healthcare decision-making. There is much mental wandering in data mining in healthcare. More research is needed to provide a testament of its benefits in facilitating healthcare decision-making for healthcare providers, managers, and policymakers, and more evidence is needed on data mining's overall impact on healthcare services and patient care [38, 39].

15.8 CI APPLICATIONS IN HEALTHCARE

For running integration builds and automated tests on the code committed by the developers are involved by Continuous integration (CI). Continuous Integration Involves various Integration and automated tests. These tests involves various codes for easy understanding of their tests results. While profiling the application code, the CI combines development and testing by enabling unit and functional tests [40, 41]. Most digital transactions claimed in Healthcare involve transferring messages, digital invoices, or money between healthcare providers and insurers. When an individual receives some form of healthcare from a provider, for instance surgery, the provider makes several transactions that need to be appraised by the insurer. Here CI acts as a source developer in between the patient and doctor [42].

15.8.1 Increases Patient Engagement and Satisfaction

By practicing Continuous Integration (CI), solutions for patients can be deployed more quickly, for instance, when the patient is unwell and undergoing feelings of anxiety, sadness, pain, etc., a new check-in system to facilitate faster and more timely processing of inpatients [43]or implementing a new feature on a linked mobile patient portal that allows patients to be notified immediately for the test results, or perhaps a solution to monitor an appointment availability with specialists that automatically backfills them from a waiting list when cancellations occur.

Healthcare organizations should speed up automation without compromising patient security on more technology that either improves operational efficiency or directly addresses a patient need, increases the quality of care that patients receive, and increase the satisfaction and engagement of the patient [43].

15.9 ORGANIZATION OF DEEP LEARNING APPLICATIONS FOR IoT IN HEALTHCARE

The IoMT has been mentioned as the "Smart Healthcare" Paradigm [44]. It is like connecting the physical world to the digital world [45]. Designing small devices with sensing, processing, and communication capabilities using the new Internet of Things (IoT) enables the development of sensors, embedded devices, and other items capable of understanding the environment [46]. Many regular activities that involve physical exertion can be monitored by human biomedical signals [47]. IoT uses the term 3A to mean anytime, anywhere, and any media, which results in a sustained ratio between radio and man around 1:1" [48]Here are a few of the IoT applications that are used in healthcare:

15.9.1 Internet of Healthy Things

To monitoring vital signs in hospital wards, IoHT was introduced for monitoring blood pressure, body temperature, heart rate, respiratory rate, and oxygen saturation from its perspective. IoHT consists of interconnected objects [49], that have the capacity of exchanging and processing data to improve patient health. Hospitals have started using "smart beds" [50–53] which detect when they are occupied and when a patient is trying to get up, without the manual interaction of nurses, these IoHT hospitals with the modified beds can adjust themselves and ensure appropriate pressure and support is applied to the patient.

15.9.2   MEDICAL DIAGNOSIS AND DIFFERENTIATION APPLICATIONS

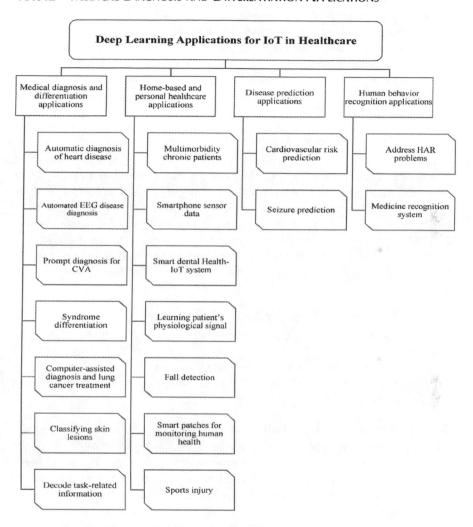

15.9.2.1   Automatic Diagnosis of Heart Disease

HealthFog for automatic diagnosis of heart diseases by employing Deep Learning and IoT, has been proposed. HealthFog is a lightweight fog service that provides effective data management from various IoT devices related to patients suffering from heart diseases. There are many advantages of a new pattern of fog and edge computing which are recommended to save energy. As for any fog computing weaknesses in medical applications, it is crucial to know the response and latency time and Quality of Service (QoS) parameters in real-time fog ambiances [54].

15.9.2.2 Automated EEG Disease Diagnosis

By the Deep Learning methods, neoteric technological breakthroughs in the automated EEG disease diagnosis and detection systems have occurred, as stated by Sarraft et al. 2016 [55]. It has an automatic feature extraction facility, which has been considered a positive point in ECG decoding performance. By assessing the EEG of the patient, the detection for unusual health conditions is possible. Although this EEG pathology dataset will lead to a challenge, few can be accessed online, but many of them are not suitable for some Deep Learning [56–59].

15.9.2.3 Cerebral Vascular Accidents (CVA) Diagnosis [60]

Generally, a stroke is due to ischemia or bleeding that stops the functioning of the brain, which can cause disability [61]. Many of the cases can lead to death. To tackle this issue, Prompt diagnoses like CT and MRI imaging are primarily used to diagnose. IoT framework can be utilized through a Conventional Neuron Network to classify the stroke through CT images to recognize if the brain is healthy or not. More minor human-dependent areas, leading to fewer human errors, is the advantage of introducing IoT into healthcare. Since there is no expansion of these systems, other medical imaging is not possible, which is considered as the limitation of this study.

15.9.2.4 Detection of Atrial Fibrillation (AF)

By using Heart Rate (HR) signals, and with a Deep Learning model based on Long Short-Term Memory (LSTM), Atrial Fibrillation (AF) episodes are detected [62]. It was piloted on twenty subjects and labeled HR signal data sourced from Physio Net's Atrial Fibrillation Database (AFDB). This Deep Learning model is less limited when compared to Machine Learning approaches.

15.9.2.5 Syndrome Differentiation

Chinese medication is used to cure infectious fever, i.e., syndrome differentiation [63]. Due to its complication, Chinese medication can make a difference between the infectious fever syndromes. Here Deep Learning acts as a promising model for differentiating computer-assisted syndromes of infectious fever through integration. Accuracy of classification is considered the strength of this study. However, the limitation is that the model cannot differentiate infectious fevers in many clinical cases.

15.9.2.6 Diagnosis and the Treatment for Lung Cancer

Deep Reinforcement learning models are explored for computer-assisted diagnosis and treatment of cancer, as lung cancer poses a severe threat to humans [64]. Benign tumors and malignant tumors are the two types of tumors. This Deep Reinforcement learning model can spot lung tumors and gives a good outcome. The biggest challenge is defining a proper valid function to update each action's Q-value.

15.9.2.7 Classifying Melanoma Diseases

This is a severe type of skin cancer, which undergoes metastasis. There are three types of Melanocytic lesions: common nevi, atypical nevi, and melanomas. An IoT technology-based system is used to classify lesions of skin [65]. This method applied CNN models to the ImageNet dataset to provide images. Merits of this method are accessible usage in different remote areas of the world and convenience. This study is more related to internet access. The application's connection to API (Application Programming Interface) is LINDA, and sending images requires a good connection.

15.10   HOME-BASED AND PERSONAL HEALTHCARE

In this technique for patients with chronic multimorbidity, the systems are provided a training set with a labeled set of input-output pairs, which maps from input to another output [66]. It provides a better quality of life for the patients who are suffering from comorbidities and controls costs. However, there are no data to confirm the effectiveness and functioning of this work. Smart care systems for Health-IoT of tooth images found on smartphones and hardware, Deep Learning, and the mobile terminal may solve issues about dental diseases in intelligent home care dental checkups [67]. Predicting one's own conditions, like heart attacks, chronic fever, and care for the elderly through Body Sensor Network (BSN) was addressed [68]. An exact detection model in cheap devices should be available. Smart chips should be developed to monitor human health conditions by employing IoT sensors [69].

15.10.1   DISEASE PREDICTION APPLICATIONS

Cardiovascular risk prediction through the input of sensors indicates the heart rate of the individual [70]. The epilepsy brain-computer interface (BCI) is designed to predict seizure [71]. BCI illustrates the brain's spontaneous electrical activity by recording utilizing several electrodes placed all over the scalp [72].

15.10.2   IoMT MONITORING SOLUTIONS

The **Internet of Medical Things (IoMT)** is one of the IoT applications used for medical and health-related purposes, data collection and analysis for research, and monitoring. It is conceptualized in various forms of intelligent wearable devices, home-use medical devices, point-of-care kits, and mobile healthcare applications and can communicate with medical experts remotely. Apart from their utility in managing regular health statuses, they have also been used for disease prevention, fitness promotion, and remote intervention in emergencies. In chronic disease management, IoMT-enabled devices, promising alternatives in conditions like hypertension, cardiac failure, and diabetes [73–75]. IoMT-enabled devices also facilitate health supervision by monitoring individual diet systems and physical activities, with a quality of life tracking of continuous data on patient activity and related vital changes, checked through the help of wearable devices, implantable chips, and embedded systems in biomedical devices. These devices allow us to analyze and correlate various vital events with health conditions at the local level. IoMT-based RFID (Radio-frequency identification) tags manage drug availability problems and supply cost. The FDA has suggested guidelines for RFID (Radio-frequency identification) and drug supply chain management [76].

15.11 MEDICAL INTERNET OF THINGS

This **one possibility** is considered the most important with great impact on people's lives. IoT has opened up a world of possibilities in the healthcare sector. This could be the solution to many problems. Medical IoT will bring about great opportunities for telemedicine, remote monitoring of patients' condition, and much more [77]. For example, it illustrates a view in a typical hospital that is practicing IoT. A patient with rheumatoid arthritis will have an ID card that, when scanned, links to a secure cloud that stores their electronic health record vitals and lab results, medical and prescription histories. These are Electronic Health Records (EHRs) [78]. Physicians and nurses would easily access the record on a mobile, tablet, or desktop computer. IoT-based medical services would reduce expenses and enhance the quality of life.

IoT healthcare tools are kept on slow processors. Moreover, these devices cannot perform computationally expensive operations [49]. In living spaces, specialized sensors are equipped for the senior citizens to monitor health and general well-being by ensuring that proper treatment is being administered and assisting people via therapy [79]. These sensors can create a network of intelligent sensors that collect, process, transfer, and analyze valuable information at different platforms in the environment, similar to connecting in-home monitoring devices to hospital-based systems [44] (Figure 15.10).

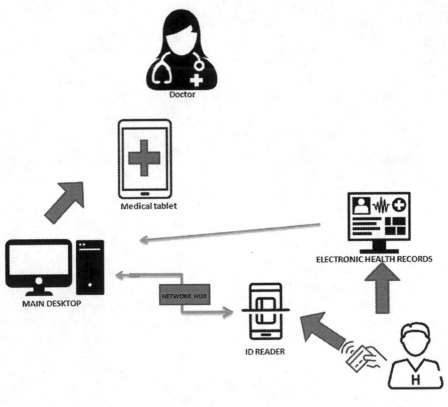

FIGURE 15.10 Complete set-up of IoT hospitals.

Laboratories in the future could be influenced by more robotics and more connectivity, in order to take advantage of the benefits of the Internet of Things and artificial intelligence (AI)-based systems.

15.11.1 ANALYSIS OF THE PHYSIOLOGICAL PARAMETERS

The healthcare sector has a vital role in assessing the patient's physiological parameters, and IoT applications have a huge role in assessing these parameters for better diagnosis. Optimization of healthcare is one of the important aspects that is increasing extensively for a better way of helping patients. The main objective of implementing AI in healthcare services is to monitor the health status of patients in a real-time manner. Real-time analysis with the help of CI has effectively increased the ease of the diagnosis pattern [80]. The vital physiological parameters in healthcare are temperature, heart rate, blood pressure, respiratory rate, and blood oxygen saturation. These vital parameters can be examined and analyzed through many non-invasive techniques with the help of the IoT [81]. These tests are primarily conducted through wearable devices capable of monitoring and recording real-time information about an individual's activities and physical condition. These wearable devices work with the help of sensors that can measure various physiological signs, thus easily obtaining electrocardiograms (ECGs), photoplethysmograms (PPGs), ballistocardiograms (BCGs),electromyograms (EMGs), and seismocardiograms (SCGs), along with body temperature, blood pressure, respiration rate (RR), oxygen saturation (SpO2), and heart rate (HR) [82] (Figure 15.11).

- Monitoring Cardiovascular Status
  There are various non-invasive techniques used for monitoring the cardiac activity of the patient. ECG monitoring measures the heart's electrical activity, and this method uses wet electrodes, which are less reliable for long-term monitoring. Hence, an alternative was developed consisting of dry textile electrodes.

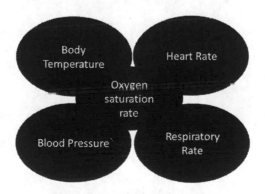

Five vital Physiological Parameters

FIGURE 15.11   Physiological parameters in healthcare.

These dry textile electrodes include a motion sensor capable of recording the ECG and all the motion signals to help diagnose any heart disease [83].

Another very effective system is a smart textile system that works on fiber optic sensors to monitor respiratory and cardiac activity. Through this system, small chest motions that begin with the beating of the heart can be recorded by placing the sensor at three different chest positions.

Apart from ECG, PPG photoplethysmography is the technique shown to have great results for HR and SpO2. It is a photodetector that uses light and contact of the skin's surface to measure the blood circulation variation in the veins and arteries [84].

- Monitoring Body Temperature
  Body temperature is one of the important parameters in diagnosing any health issue. Several wearable systems aid in providing real-time information, both for the skin and core temperature of the patient. A wireless sensing technology develops to monitor the patient's body temperature by using an ear-worn temperature sensor [85].
- Monitoring Daily Activities along with Patient's Body Motion
  Accelerometers and gyroscopes help to analyze gait dynamics with the help of wearable sensors. Several techniques are used to get the data of the patient's movement. Different wearables like wrist bands, when worn, can recognize the activity of the patient. Moreover, technologies are being developed so that they combine all data of the physiological parameters for achieving more concrete and accurate results of the patient's health status, for better and immediate medication [52, 53, 86].

15.12   REHABILITATION SYSTEMS

These sorts of computing systems within the home and or rehabilitation center help the person by giving his complete health status through the wearable devices in a real-time basis. There are different systems, such as the Fall detection system. This system applies to rehabilitation, which uses sensors placed in the patient's environment or placed in their wearable devices. The main aim of this system is to detect the fall or generate an alert to the health care service people for immediate support for the patient.

There are several mobile applications [74, 75] used in rehabilitation that are available for the health care system. These applications help monitor the risks that could arise in the patient's environment along with the medical status of the patient [41, 51, 87] (Figure 15.12).

15.13   SKIN PATHOLOGIES AND DIETARY ASSESSMENT

Skin pathology or dermatology is another sector that has started the use of AI in a progressive way.

- Keratinocyte Carcinomas and Melanoma

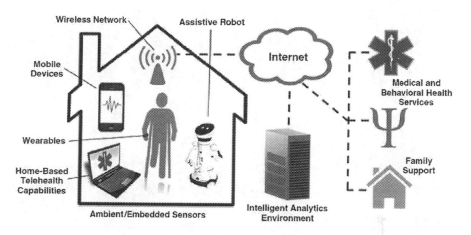

FIGURE 15.12 Rehabilitation system and IOT work process.

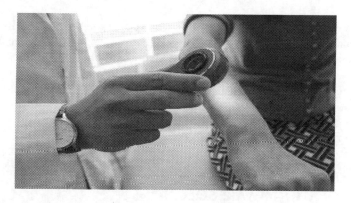

FIGURE 15.13 Deep learning used in skin cancer.

This application helps in breaking the dermatoscopic or non-dermatoscopic images of lesions into individual pixels for further analysis. An AI model development can easily differentiate between the normal skin and actinic keratosis, with almost 91.7% sensitivity (Figure 15.13) [88].

- Ulcer Assessment
  A particular AI application was developed that can segment ulcers with a specificity of 95.7% and sensitivity of 87.3%. Also, detecting necrosis, slough, and granulation can be remarkably accurate with AI's application.
- Inflammatory Skin Diseases

There are several smartphone AI tools that aid in the classification of acne lesions and also help in detecting any inflammation on the skin. The AI tool also helps in differentiating several diseases, such as psoriasis, lichen planus, chronic dermatitis.

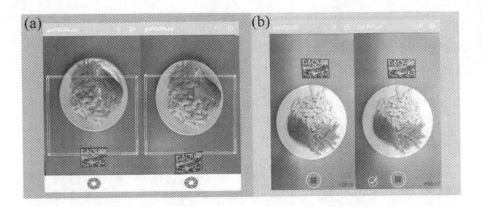

FIGURE 15.14   goFOOD™ analyzing.

The AI system analyzes the histopathological images, which helps classify basal cell carcinoma with almost 98.1 % accuracy and serves as a better option for analyzing benign tumors [89].

Dietary assessment is another crucial aspect for the patients that results in better clinical outcomes. A very novel dietary assessment system that is completely based on AI has been implemented to estimate the calorie and macronutrient content in the meal. The food images are being captured via the smartphone for detection. This whole system for dietary assessment is termed as goFOOD™. This system requires an input of meal images or videos. The network then helps detect the food volume. Therefore, the calorie and macronutrient content of each meal is further calculated [90, 91] (Figure 15.14).

15.14   EPIDEMIC DISEASES TREATMENT AND LOCATION-AWARE SOLUTIONS

We may now apply Machine Learning systems as the first line of defense against epidemics after the resources have been obtained and the comorbidities have been recognized. Using an AI system to combat an outbreak appears to be the most effective and logical path to take, assuming we can develop that system. The intrusion of intelligent devices into one's personal space can be used to people's advantage. AI systems are trained to recognize symptoms by recording and evaluating changes in voice patterns, breathing patterns and intervals, cough strength and rhythm, and dyspnea levels, which can all be used to diagnose symptoms. It can be utilized as a powerful screening tool for patients. The benefit of one of the advantages of these AI programs is that they have a baseline recording of when the user was normal and could detect small changes, changes in the voice, etc. [92, 93].

REFERENCES

1. Pradhan B, Bhattacharyya S, Pal K. IoT-Based applications in healthcare devices. *Journal of Healthcare Engineering*. 2021; 2021: 6632599. doi: 10.1155/2021/6632599. PMID: 33791084; PMCID: PMC7997744.

2. Singh RP, Javaid M, Haleem A, Suman R. Internet of things (IoT) applications to fight against COVID-19 pandemic. *Diabetes and Metabolic Syndrome: Clinical Research and Reviews*. 2020; 14(4): 521–524. doi: 10.1016/j.dsx.2020.04.041. Epub 2020 May 5. PMID: 32388333; PMCID: PMC7198990.

3. Vaishya R, Javaid M, Khan IH, Haleem A. Artificial Intelligence (AI) applications for COVID-19 pandemic. *Diabetes and Metabolic Syndrome: Clinical Research and Reviews*. 2020;14(4):337–339. doi: 10.1016/j.dsx.2020.04.012. Epub 2020 Apr 14. PMID: 32305024; PMCID: PMC7195043.

4. Mavrogiorgou A, Kiourtis A, Perakis K, Pitsios S, Kyriazis D. IoT in healthcare: Achieving interoperability of high-quality data acquired by IoT medical devices. *Sensors* (Basel). 2019;19(9):1978. doi: 10.3390/s19091978. PMID: 31035612; PMCID: PMC6539021.

5. Kashif Hameed, Imran Sarwar Bajwa, Shabana Ramzan, Waheed Anwar, Akmal Khan, An Intelligent IoT Based Healthcare System Using Fuzzy Neural Networks", *Scientific Programming*. 2020;8836927. doi:10.1155/2020/8836927

6. Al-Dmour JA, Sagahyroon A, Al-Ali AR, Abusnana S. A fuzzy logic-based warning system for patients classification. *Health Informatics Journal*. 2019;25(3):1004–1024. doi: 10.1177/1460458217735674. Epub 2017 Nov 6. PMID: 29108458.

7. Gerry S, Bonnici T, Birks J, Kirtley S, Virdee PS, Watkinson PJ, Collins GS. Early warning scores for detecting deterioration in adult hospital patients: systematic review and critical appraisal of methodology. *BMJ*. 2020. doi: 10.1136/bmj.m1501. PMID: 32434791; PMCID: PMC7238890.

8. Van Velthoven MH, Adjei F, Vavoulis D, Wells G, Brindley D, Kardos A. ChroniSense National Early Warning Score Study (CHESS): A wearable wrist device to measure vital signs in hospitalised patients-protocol and study design. *BMJ Open*. 2019;20(9):e028219. doi: 10.1136/bmjopen-2018-028219. PMID: 31542738; PMCID: PMC6756348.

9. Leite CR, Sizilio GR, Neto AD, Valentim RA, Guerreiro AM. A fuzzy model for processing and monitoring vital signs in ICU patients. *Biomedical Engineering Online*. 2011;10(68). doi: 10.1186/1475-925X-10-68. PMID: 21810277; PMCID: PMC3162941.

10. Liu S, Triantis KP, Zhao L, Wang Y. Capturing multi-stage fuzzy uncertainties in hybrid system dynamics and agent-based models for enhancing policy implementation in health systems research. *PLoS One*. 2018;25(4):e0194687. doi: 10.1371/journal.pone.0194687. PMID: 29694364; PMCID: PMC5918643

11. Achkoski J, Koceski S, Bogatinov D, Temelkovski B, Stevanovski G, Kocev I. Remote triage support algorithm based on fuzzy logic. *Journal of the Royal Army Medical Corps*. 2017;163(3):164–170. doi: 10.1136/jramc-2015-000616. Epub 2016 Jul 14. PMID: 27418264.

12. Sean Barnes, Suchi Saria, and Scott Levin, An evolutionary computation approach for optimizing multilevel data to predict patient outcomes. 2018.

13. K.C. Tan, Q. Yu, C.M. Heng, T.H. Lee, Evolutionary computing for knowledge discovery in medical diagnosis department of electrical and computer engineering, National University of Singapore, 10 December 2002.

14. Sai Ho Ling, and Hak Keung Lam, Evolutionary algorithms in health technologies, school of biomedical engineering, University of Technology Sydney, Ultimo, NSW 2007, Australia Published: 24 September 2019.

15. Khan T.A.; Ling, S.H. Review on electrical impedance tomography: Artificial intelligence methods and its applications. *Algorithms*. 2019;12:88.

16. Kumar, S., and Bentley, P. J. An introduction to computational development. *On Growth, Form and Computers*. 2003;1–43.doi:10.1016/b978-012428765-5/50034-7

17. Wilfried Roetzel, Xing Lou, Dezhen Chen (2020) *Design and Operation of Heat Exchangers and their Networks*, Elsevier, The Institute of thermodynamics of the Helmut Schmidt University: Hamburg, Germany.

18. Vanneschi L., Poli R. (2012) Genetic Programming — Introduction, Applications, Theory and Open Issues. In: Rozenberg G., Bäck T., Kok J.N. (eds) *Handbook of Natural Computing*. Springer, Berlin, Heidelberg. doi:10.1007/978-3-540-92910-9_24

19. Christian Jacob, *Illustrating evolutionary computation with mathematica*, A volume in the Morgan Kaufmann Series in Artificial intelligence, 2001.

20. Rudolph G. (2012) Evolutionary Strategies. In: Rozenberg G., Bäck T., Kok J.N. (eds) *Handbook of Natural Computing*. Springer, Berlin, Heidelberg. doi:10.1007/978-3-540-92910-9_22

21. Stephen L. Smith.2015. Medical Applications of Evolutionary Computation. In *Proceedings of the Companion Publication of the 2015 Annual Conference on Genetic and Evolutionary Computation GECCO Companion '15 Association for Computing Machinery*, New York, NY, 651–679. doi:10.1145/2739482.2756567

22. Choi J, Shin K, Jung J, Bae HJ, Kim DH, Byeon JS, Kim N. Convolutional neural network technology in endoscopic imaging: Artificial intelligence for endoscopy. *Clinical Endoscopy*. 2020;53(2):117–126. doi: 10.5946/ce.2020.054. Epub 2020 Mar 30. PMID: 32252504; PMCID: PMC7137563

23. Lee JG, Jun S, Cho YW, Lee H, Kim GB, Seo JB, Kim N. Deep learning in medical imaging: General overview. *Korean Journal of Radiology*. 2017;18(4):570–584. doi: 10.3348/kjr.2017.18.4.570. Epub 2017 May 19. PMID: 28670152; PMCID: PMC5447633.

24. Marcus AP, Marcus HJ, Camp SJ, Nandi D, Kitchen N, Thorne L. Improved prediction of surgical resectability in patients with glioblastoma using an artificial neural network. *Scientific Reports*. 2020;10(1):5143. doi: 10.1038/s41598-020-62160-2. PMID: 32198487; PMCID: PMC7083861.

25. Cunningham P, Carney J and Jacob S. Stability problems with artificial neural networks and the ensemble solution. *Artificial Intelligence in Medicine*. 2000;20:217–225, doi:10.1016/S0933-3657(00)00065-8

26. Sordo M. Introduction to neural networks in healthcare. Open Clinical: Knowledge Management for Medical Care. 2002.

27. Kumar SS, Kumar KA. Neural networks in medical and healthcare. *International Journal of Innovative Research and Development*. 2013;31(8);2.

28. Chishti S, Jaggi KR, Saini A, Agarwal G, Ranjan A. Artificial intelligence-based differential diagnosis: development and validation of a probabilistic model to address lack of large-scale clinical datasets. *Journal of Medical Internet Research*. 2020;22(4):e17550. doi: 10.2196/17550. PMID: 32343256; PMCID: PMC7218591.

29. Sameer Prakash, Tyler Hamby, Van Leung-Pineda,Don P Wilson, FNLA, probabilistic modeling approach to reducing healthcare costs with reflex testing, *Laboratory Medicine*. 2017;48(4):384–387, doi:10.1093/labmed/lmx049

30. Irene Y. Chen, Shalmali Joshi, Marzyeh Ghassemi, and Rajesh Ranganath. Probabilistic machine learning for healthcare. *Mastering Machine Learning*. doi:10.1146/2009

31. Gaitanou P., GaroufallouE., BalatsoukasP. (2014) The effectiveness of big data in health care: A systematic review. In: Closs S., Studer R., Garoufallou E., Sicilia M.A. (eds) *Metadata and Semantics Research*. MTSR 2014. Communications in Computer and Information Science, vol 478. Springer, Cham.doi:10.1007/978-3-319-13674-5_14

32. Hong Liang, Luo Mengqi, Wang Ruixue, Lu Peixin, Lu Wei and Lu Long. Big data in health care: Applications and challenges. *Data and Information Management*. 2018;2(3:175–197. doi: 10.2478/dim-2018-0014

33. Dash, S., Shakyawar, S.K., Sharma, M. et al. Big data in healthcare: Management, analysis and future prospects. *Journal of Big Data*. 2019;6:54. doi:10.1186/s40537-019-0217-0

34. P. Kamakshi, Importance of Big data in Healthcare System-A Survey, *International Journal of Applied Engineering Research*. 2018; 12184–12187. ISSN 0973-4562 13(15) © Research India Publications. http://www.ripublication.com 12184 Importance of Big data in Healthcare System-A Survey

35. Neesha Jothi, NurAini Abdul Rashid, Wahidah Husain, Data mining in healthcare – a review, *Procedia Computer Science*. 2015;72:306–313. ISSN 1877-0509, doi:10.1016/j.procs.2015.12.145.

36. Aljawarneh, S., Anguera, A., Atwood, J.W.et al. Particularities of data mining in medicine: lessons learned from patient medical time series data analysis. *Journal of Wireless Communication Network*. 2019;2019:260. doi:10.1186/s13638-019-1582-2

37. Yoo, I., Alafaireet, P., Marinov, M. et al. Data mining in healthcare and biomedicine: a survey of the literature. *Journal of Medical Systems*. 2012;36:2431–2448. doi:10.1007/s10916-011-9710-5

38. Swenson ER, Bastian ND, Nembhard HB. Healthcare market segmentation and data mining: A systematic review. *Health Marketing Quarterly*. 2018;35(3):186–208. doi:10.1080/07359683.2018.1514734. Epub 2018 Nov 23. PMID: 30470165.

39. Househ M, Aldosari B. The hazards of data mining in healthcare. *Studies in Health Technology and Informatics*. 2017;238:80–83. PMID: 28679892.

40. P.M. Duvall, S. Matyas, and A. Glover, *Continuous integration: Improving software quality and reducing risk*, Boston, MA: Addison-Wesley Professional, 2007.

41. Rahman, A., Dash, S., Luhach, A.K., Chilamkurti, N., Baek, S., and Nam, Y., A neuro-fuzzy approach for user behavior classification and prediction, *Journal of Cloud Computing: Advances, Systems and Applications*. 2019;8(17). doi:10.1186/s13677-019-0144-9

42. Amrit C, Meijberg Y. Effectiveness of test-driven development and continuous integration: A case study. *IT Professional*. 2018;20(1):27–35.

43. Greg Sienkiewicz how to implement DevOps in health care june 16, 2019: https://www.macadamian.com/learn/how-to-implement-devops-in-healthcare/

44. Dey, Nilanjan; Hassanien, Aboul Ella; Bhatt, Chintan; Ashour, Amira S.; Satapathy, Suresh Chandra (2018). *Internet of things and big data analytics toward next-generation intelligence*. Springer International Publishing. ISBN 978-3-319-60434-3

45. Ray PP. A survey on Internet of Things architectures. *Journal of King Saud University-Computer and Information Sciences*. 2018;30(3):291–319.

46. Mora H, Gil D, Terol RM, Azorín J, Szymanski J. An IoT-based computational framework for healthcare monitoring in mobile environments. *Sensors*. 2017;17(10):2302.

47. Zanella A, Bui N, Castellani A, Vangelista L, Zorzi M. Internet of things for smart cities. *IEEE Internet of Things Journal*. 2014;1(1):22–32.

48. Srivastava L. Pervasive, ambient, ubiquitous: the magic of radio. InEuropean Commission Conference "From RFID to the Internet of Things", Bruxelles, Belgium 2006 March.

49. Riazul Islam SM, Kwak D, Humaun Kabir M, Hossain M, Kwak K-S. The internet of things for health care: A comprehensive survey. *IEEE Access* 2015;3:678–708.

50. da Costa CA, Pasluosta CF, Eskofier B, da Silva DB, da Rosa Righi R. Internet of health things: Toward intelligent vital signs monitoring in hospital wards. *Artificial Intelligence in Medicine*. 2018;89:61–9.

51. S. Dash, S. Biswas, D. Banerjee and A. Rahman, Edge and fog computing in healthcare – A review. *Scalable Computing: Practice and Experience*. 2019;20(2):191–205. http://www.scpe.org, doi: 10.12694/scpe.v20i2.1504, ISSN 1895-1767

52. Atta-Ur-Rahman, Sujata Dash, Mahi Kamaleldin, Areej Abed, Atheer Alshaikhhussain, Heba Motawei, Nadeen Al Amoudi, Wejdan Abahussain, *A Comprehensive study of mobile computing in Telemedicine*, A. K. Luhach et al. (Eds):ICAICR2018,CCIS 956, pp. 413–425, 2019. Springer.

53. Atta-Ur-Rehman, Sujata Dash, Kiran Sultan, Muhammad Aftab Khan, Management of Resource Usage in Mobile Cloud Computing, *International Journal of Pure and Applied Mathematics*. 2018;119(16):255–261. ISSN: 1314-3395 (on-line version).

54. Mäkitalo N, Ometov A, Kannisto J, Andreev S, Koucheryavy Y, Mikkonen T. Safe, secure executions at the network edge: coordinating cloud, edge, and fog computing. *IEEE Software*. 2017;35(1):30–7.

55. Sarraf S, Tofighi G. Classification of alzheimer's disease using fmri data and deep learning convolutional neural networks. arXiv preprint arXiv:1603.08631. 2016.

56. Kamnitsas K, Ledig C, Newcombe VF, Simpson JP, Kane AD, Menon DK, Rueckert D, Glocker B. Efficient multi-scale 3D CNN with fully connected CRF for accurate brain lesion segmentation. *Medical Image Analysis*. 2017;36:61–78.

57. Hossain MS, Amin SU, Alsulaiman M, Muhammad G. Applying deep learning for epilepsy seizure detection and brain mapping visualization. *ACM Transactions on Multimedia Computing, Communications, and Applications (TOMM)*. 2019;15(1):1–7.

58. Acharya UR, Oh SL, Hagiwara Y, Tan JH, Adeli H, Subha DP. Automated EEG-based screening of depression using deep convolutional neural network. *Computer Methods and Programs in Biomedicine*. 2018;161:103–13.

59. Albert B, Zhang J, Noyvirt A, Setchi R, Sjaaheim H, Velikova S, Strisland F. Automatic EEG processing for the early diagnosis of traumatic brain injury. *Procedia Computer Science*. 2016;96:703–12.

60. Masoumi H, Behrad A, Pourmina MA, Roosta A. Automatic liver segmentation in MRI images using an iterative watershed algorithm and artificial neural network. *Biomedical Signal Processing and Control*. 2012;7(5):429–37.

61. Rebouças Filho PP, Sarmento RM, Holanda GB, de Alencar Lima D. New approach to detect and classify stroke in skull CT images via analysis of brain tissue densities. *Computer Methods and Programs in Biomedicine*. 2017;148:27–43.

62. Faust O, Shenfield A, Kareem M, San TR, Fujita H, Acharya UR. Automated detection of atrial fibrillation using long short-term memory network with RR interval signals. *Computers in Biology and Medicine*. 2018;102:327–35.

63. Jiang M, Lu C, Zhang C, Yang J, Tan Y, Lu A, Chan K. Syndrome differentiation in modern research of traditional Chinese medicine. *Journal of Ethnopharmacology*. 2012;140(3):634–42.

64. Bray F, Ferlay J, Soerjomataram I, Siegel RL, Torre LA, Jemal A. Global cancer statistics 2018: GLOBOCAN estimates of incidence and mortality worldwide for 36 cancers in 185 countries. *CA: A Cancer Journal for Clinicians*. 2018;68(6):394–424.

65. Ma Z, Tavares JM. A novel approach to segment skin lesions in dermoscopic images based on a deformable model. *IEEE Journal of Biomedical and Health Informatics*. 2015;20(2):615–23.

66. Fonseca C, Mendes D, Lopes M, Romao A, Parreira P. Deep learning and IoT to assist multimorbidity home based healthcare. *Journal of Health & Medical Informatics*. 2017;8(3):1–4.

67. Liu L, Xu J, Huan Y, Zou Z, Yeh SC, Zheng LR. A smart dental health-IoT platform based on intelligent hardware, deep learning, and mobile terminal. *IEEE Journal of Biomedical and Health Informatics*. 2019;24(3):898–906.

68. Sharma S, Chen K, Sheth A. Toward practical privacy-preserving analytics for IoT and cloud-based healthcare systems. *IEEE Internet Computing*. 2018;22(2):42–51.

69. Malasinghe LP, Ramzan N, Dahal K. Remote patient monitoring: A comprehensive study. *Journal of Ambient Intelligence and Humanized Computing.* 2019;10(1):57–76.

70. Ballinger B, Hsieh J, Singh A, Sohoni N, Wang J, Tison G, Marcus G, Sanchez J, Maguire C, Olgin J, Pletcher M. DeepHeart: semi-supervised sequence learning for cardiovascular risk prediction. In *Proceedings of the AAAI Conference on Artificial Intelligence* 2018;32, (1).

71. He W, Zhao Y, Tang H, Sun C, Fu W. A wireless BCI and BMI system for wearable robots. *IEEE Transactions on Systems, Man, and Cybernetics: Systems.* 2015;46(7):936–46.

72. Gandhi V, Prasad G, Coyle D, Behera L, McGinnity TM. EEG-based mobile robot control through an adaptive brain–robot interface. *IEEE Transactions on Systems, Man, and Cybernetics: Systems.* 2014;44(9):1278–85.

73. Omboni S, Campolo L, Panzeri E. Telehealth in chronic disease management and the role of the Internet-of-Medical-Things: the Tholomeus® experience. *Expert Review of Medical Devices.* 2020;17(7):659–70.

74. A. Rahman, S. Dash, Ahmad M., Iqbal T. (2021), Mobile cloud computing: A green perspective, In: Udgata S. K., Sethi S., Sriram S. N. (eds), *Intelligent notes in networks and systems*, vol. 185, pp: 523–533, Springer, Singapore. doi:10.1007/978-981-33-6081-5-46.

75. K. Rupbanta Singh, Sujata Dash, B. Deka, S. Biswas, (2020), Mobile Technology Solution for COVID-19. In Fadi Al-Turjaman et al., (Eds), *Emerging Technologies for battling COVID-19 applications and innovations*, Springer, pp: 271–294. ISBN: 978-030-60038-9

76. Shelke Y, Sharma A. Internet of medical things. Technology intelligence &ip research, aranca. 2016.

77. Aldahiri A, Alrashed B, Hussain W. Trends in using IoT with machine learning in health prediction system. *Forecast.* 2021;3(1):181–206.

78. Kruse CS, Kothman K, Anerobi K, Abanaka L. Adoption factors of the electronic health record: A systematic review. *JMIR Medical Informatics.* 2016;4(2):e19.

79. Istepanian R, Hu S, Philip N, Sungoor A (2011). "The potential of Internet of m-health Things "m-Ioᵗ" for non-invasive glucose level sensing". *2011 Annual International Conference of the IEEE Engineering in Medicine and Biology Society. Annual International Conference of the IEEE Engineering in Medicine and Biology Society. IEEE Engineering in Medicine and Biology Society.* Annual International Conference. 2011 pp. 5264–5266

80. Dohr, A.; Modre-Opsrian, R.; Drobics, M.; Hayn, D.; Schreier, G. The Internet of Things for Ambient Assisted Living. In *Proceedings of the 2010 Seventh International Conference on Information Technology: New Generations*, Las Vegas, NV, 12–14 April 2010.

81. Elliott, M.; Coventry, A. Critical care: The eight vital signs of patient monitoring. *The British Journal of Nursing.* 2012;21:621–625.

82. Jacob Rodrigues M, PostolacheO, CercasF. Physiological and behavior monitoring systems for smart healthcare environments: A review. *Sensors.* (Basel). 2020;20(8):2186. doi: 10.3390/s20082186. PMID: 32290639; PMCID: PMC7218909

83. Zhou, Y.; Ding, X.; Zhang, J.; Duan, Y.; Hu, J.; Yang, X. Fabrication of conductive fabric astextile electrode for ECG monitoring. *Fibers and Polymers.* 2014;15:2260–2264.

84. Castaneda, D., Esparza, A., Ghamari, M., Soltanpur, C., Nazeran, H. A review on wearable photoplethysmography sensors and their potential future applications in health care. *International Journal of Biosensors Bioelectronics.* 2018;4.

85. Popovic, Z.; Momenroodaki, P.; Scheeler, R. Toward wearable wireless thermometers for internal bodytemperature measurements. *IEEE Communications Magazine.* 2014;52:118–125.

86. Majumder, S.; Mondal, T.; Deen, M. Wearable sensors for remote health monitoring. *Sensors*. 2017;17:130.
87. Luxton, D. D., and Riek, L. D. (2019). *Artificial intelligence and robotics in rehabilitation*. In L. A. Brenner, S. A. Reid-Arndt, T. R. Elliott, R. G. Frank, & B. Caplan (Eds.), *Handbook of rehabilitation psychology* (p. 507–520). American Psychological Association. doi:10.1037/0000129-031
88. Taylor Kubota, 2017, Deep learning algorithm does as well as dermatologists in identifying skin cancer, Standford| News Service; https://news.stanford.edu/press/view/12239
89. Gomolin A, Netchiporouk E, Gniadecki R, Litvinov IV. Artificial intelligence applications in dermatology: Where do we stand? *Frontiers in Medicine* (Lausanne). 2020;7:100. doi: 10.3389/fmed.2020.00100. PMID: 32296706; PMCID: PMC7136423.
90. Lu Y, Stathopoulou T, Vasiloglou MF, Pinault LF, Kiley C, Spanakis EK, Mougiakakou S. goFOOD™: An artificial intelligence system for dietary assessment. *Sensors* (Basel). 2020;20(15):4283. doi: 10.3390/s20154283. PMID: 32752007; PMCID: PMC743610285.
91. Dehais J., Anthimopoulos M., Shevchik S., Mougiakakou S. Two-view 3D reconstruction for food volume estimation. *IEEE Transactions on Multimedia*. 2017;19:1090–1099. doi: 10.1109/TMM.2016.2642792.
92. Malik, Y. S., Sircar, S., Bhat, S., Ansari, M. I., Pande, T., Kumar, P., …Dhama, K. (2020). How artificial intelligence may help the Covid-19 pandemic: Pitfalls and lessons for the future. *Reviews in Medical Virology*. doi:10.1002/rmv.2205
93. Said Agrebi and Anis Larbi, Use of artificial intelligence in infectious diseases. *Artificial Intelligence in Precision Health*. 2020;415–438. doi:10.1016/B978-0-12-817133-2.00018-5

16 Computational Intelligence in IoT Healthcare

Olugbemi T. Olaniyan, Charles O. Adetunji, and Mayowa J. Adeniyi
Edo State University Uzairue, Iyamho, Nigeria

Daniel Ingo Hefft
University of Birmingham, Birmingham, United Kingdom

CONTENTS

16.1 INTRODUCTION

Due to the ever increasing population across the world, there has been growing concern to provide real time solutions to an aged population and patients through the provision of Internet of Things and computational intelligence in healthcare sector. The modern approach to medical care involves preventive collaborative and a personalized form of care, exploring the applications of IoT for surveillance, treatment and monitoring of different health conditions. For instance, Brain Computer Interface adopts different biosensors, such as resting state-functional magnetic resonance imaging (rs-fMRI) and electroencephalography (EEG) together with the Diffusion Tensor Imaging technique for medical data collection from patients with brain conditions like epilepsy, schizophrenia, Alzheimers, Parkinsons, and other neurodegenerative diseases. Several studies have revealed that low cost monitoring and treatment systems in the healthcare sector can be achieved with the utilization of different lightweight wearable sensors and Internet of things applications. These systems are used for real time rapid detection and diagnosis of diseases including analysis of the medical data of patients. The gadgets are developed and designed in such a way that they can collect, share, and store medical information, thus elimination in-efficiencies created by a manual approach. Edge computing has been revealed to provide qualitative real time solutions through noninvasive and invasive approaches for evaluating, monitoring, and controlling diverse brain conditions. Recent advances and the development of fast and rapid sensing technologies and connectivity adapted to medical and endpoint devices for cloud computing have brought improvement in the healthcare delivery system. Increased interest in Internet of things applications for healthcare like edge Computing and intelligence have increased the capacity in the provision of Real Time decision making, a solution to battery power consumption, real time responses, reduced bandwidth cost, data safety, and protection, including privacy (Kumar and Majumder, 2018). Edge Computing and intelligence adopts the utilization of connected devices and medical systems for data collection, processing, caching, and analysis in proximity to the place where the data is captured, through the use of artificial intelligence and internet of things. In the past few years, explosive growth has been witnessed in this area, particularly in line with medical research and the healthcare system. Studies have highlighted four different domains in edge computational intelligence for health care systems, like edge caching, edge inference, edge offloading, and edge training. This can provide real time information on the patient's facial recognition, natural language or voice processing, computer vision processing, movement or motor prediction, disease detection, and much more. Amin et al. (2020) examined the currently available edge computing techniques relating to smart healthcare and assayed a detailed breakdown of using edge intelligence-related procedures for prediction and classification. They noted that in spite of the benefits of edge intelligence, shortcomings exist. These include the availability of data, management of information processes, and local storage among others. They highlighted the importance of research in order to improve the quality of services to healthcare. Thus, this book chapter focused on the role of computational intelligence in the internet of things healthcare system, the existing shortcomings, and the future perspectives.

The application of computational Intelligence in IoT Healthcare is shown in Figure 16.1.

16.2   EDGE INTELLIGENCE IN HEALTHCARE SYSTEM

With Edge intelligence, data can be retrieved, analyzed, and insights delivered proximal to where it is captured in a network. Among the advantages of edge intelligence are that intelligence at low cost, data is stored at low bandwidth, there is a reduction in operational costs, and the model can be scaled linearly as Internet of Things (IOT). Su et al. (2020) unveiled the important components of edge intelligence as including edge training, edge caching, edge inference, and edge of floating. Amin et al. (2020) examined the currently available edge computing techniques relating to

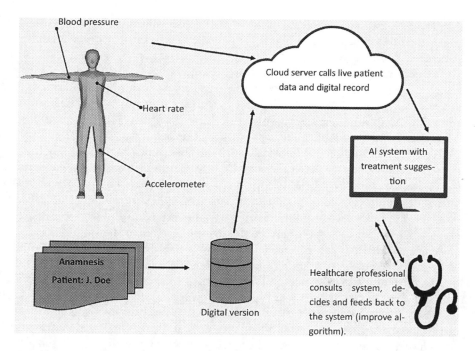

FIGURE 16.1 The application of computational Intelligence in IoT Healthcare.

smart healthcare and assayed a detailed breakdown of using edge intelligence-related procedures for prediction and classification. They noted that in spite of the benefits of edge intelligence, shortcomings exist. These include availability of data, management of the information process, and local storage among others. They highlighted the importance of research in order to improve the quality of services to healthcare.

Hayyolalam et al. (2021) shed light on how edge intelligence, together with artificial intelligence, can improve Medicare and quality care delivery. They noted that edge intelligence enhances smart Medicare courtesy of massive attenuation of the network burden, level of power consumed, and latency. They opined a better outcome with the integration of edge intelligence with artificial intelligence in improving Medicare. In the study, they introduced a new smart Medicare platform that pivots a variety of artificial intelligence techniques on edge intelligence in a parallel mode. Using the new model, shortcomings associated with previous smart Medicare platforms are addressed. However, issues like loss of data and management of an autonomous network continue.

Abdellatit et al. (2019) propose a smart Medicare using a multi-access computing monitoring platform that attempts to balance many access edge computing models, which is capable of improving the system effectiveness. With this large quantum of data produced by sensors, medical devices, and by individuals on the network, edge can be handled much more effectively. It also exhibits limitation of energy capabilities associated with devices at the network edge as a shortcoming. One merit of this invention is ease of delivering Medicare to healthcare seekers with minimal cost. The authors also suggest integration of components of the wireless network, acquired data features, and standard requirements of the exploited application, so as to bring about creation of quality and highly sustainable services for a smart Medicare system.

Tuli et al. (2020) designed a new platform named HealthFog which was meant to bring together an ensemble of Deep Learning in devices with edge computing and utilize it for automatic heart disease analysis in a real-life mode. With HealthFog, delivery of Medicare in the form of fog service courtesy of Internet of things devices is possible. It also demonstrates huge prowess in the management of cardiac patients' data at the beckon of the users. Fog-enabled cloud framework, shortened as 'FogBus,' is useful in evaluating the performance of HealthFog, especially in the aspect of the power consumption level, bandwidth of the network, latency, jitter, accuracy, and time of execution. HealthFog has another attribute of being configurable. Thus, it can be adapted to various modes of operation with assurance of quality service and high prediction precision, which are requisite for diverse fog computation situations.

Azar et al. (2019) proposed a model which was somewhat energy efficient for the analysis of Internet of Things using a fast error-bounded compressor on the collected data before transmission. Thereafter, the transmitted data on an edge node was developed and then processed using supervised Deep Learning techniques. The results of the design show that the quantity of transmitted data has been decreased by about one-hundred and three fold, with the quality of health data unaffected, and driver stress level prediction precision. Also information received from the compressed data was shown to be valid. Classification precision obtained from training the technique on characteristics that were derived from compressed data did not reduce.

Vatti et al. (2020) offer an account of how edge intelligence-based smart health (EISHP) can be used to determine the stress profile of a user. The novel EISHP measures and analyzes heart-rate variability with the help of Machine Learning algorithms and warns the users through Internet of Things devices concerning their stress level. This work indicates that with Edge Intelligence, predictive analysis and evaluation and the awareness of health status is possible.

16.3   SMART HEALTHCARE DELIVERY SYSTEM

The world is currently witnessing a rapid development and growth in the area of artificial intelligence, thus, computation intelligence in healthcare through the deployment and implementation of this technology has brought significant progress. Opportunities like cost effectiveness, an affordable healthcare delivery system, and improved health services are just a few of the benefits involved in the utilization of computational intelligence in the healthcare system. Bolhasani et al. (2021) reported that Deep and Machine Learning have received tremendous attention lately, due to their wide range of important applications in health care systems, such as natural language processing, computer vision, speech recognition, disease prediction, visual object detection, drug discovery, biomedicine, and bioinformatics. The growth of big data, Internet of Things, high-performance computers, connected devices, digital images, medical or health IoT, genomic data, medical databases, and electronic health record data are sources of data for Machine Learning applications. In medical imaging, like radiography, histology, and mammography, gene data and images taken on the mobile devices are sources of data for computational intelligence. Computational intelligence can be adopted in the monitoring, analyze, and prediction of pharmacogenomics, molecular diagnostics, identification of pathogenic variants, gene splicing, DNA sequencing, drug discovery, healthcare informatics, and personalized cancer care. The application of Machine and Deep Learning is useful in recognizing human health behavior through smartphone sensor data, learning patient physiological signals, patients with chronic multi-morbidity, smart dental Health-IoT systems, sports injury, fall detection models, and nutrition monitoring systems. Many studies have highlighted the role of computational intelligence in the diagnosis of cardiovascular diseases through HealthFog, Internet of Things, and Machine Learning approaches. Innovative techniques in edge and fog computing medical applications are effective and efficient data management services. Also, in brain disease, using electroencephalograms, neurodegenerative diseases diagnosis, and detection can be done through Machine Learning applications. Convolutional Neural Networks as a novel computational intelligence application are developed to diagnose and categorize stroke and other cerebral vascular accidents. Kumar and Majumder, 2018 reported that Internet of Things and cloud computing, like computing services involving the utilization of servers, databases, storage, networking, analytics and software through the internet, have emerged to be the solution to the dynamic problems in healthcare, environmental surveillance, and agricultural health. The authors noted that due to the high network latency, novel technology is emerging called edge computing, utilized to facilitate computational issues in the edge of the network and reducing the problem of network latencies. This technology also addresses the issue of real time responses, bandwidth cost, battery power consumption, privacy, and medical data safety.

16.4 AI ON EDGE ARCHITECTURE IN COMPUTATIONAL INTELLIGENCE FOR HEALTHCARE SYSTEM

Shah and Chircu (2018) reported that connectivity and technological advancement have resulted in an enormous increase and improvement in the healthcare system through the emergence of artificial intelligence and Internet of Things applications. The authors stated that over the years, there has been a steady increase in the number of research activities in this area, and numerous discoveries have been noted, such as internet connectivity, wearable devices, a sensor network for disease diagnosis, detection, treatment, and prevention. The authors suggested that this technology will be beneficial to physicians, patients, drug developers, and payers. Computational intelligence with the support of Machine Learning can be utilized in the development of Internet of Things devices, such as smart pills, sensors, and wearable monitors, can support healthcare professionals for collection of data and artificial intelligence systems, thus helping to detect changes in a patient's condition, identify trends, suggest alternative treatment options, increase patient adherence, accelerate drug discovery and help improve patient outcomes. Sathyasri et al. (2018) showed that e-health management systems based on computational intelligence can incorporate different sensor technologies for emerging multidimensional applications in the pathophysiology of diseases. The authors particularly noted that wearable sensors have shown to be more important in remote health maintenance systems, data collection, transfer, and treatment. Effective communication and monitoring of a patient's medical information from a database can be implemented. Lin (2020) highlighted the various important benefits of artificial intelligence technology, such as optimization, performance improvement, and intelligence. The author noted that integration and interaction between artificial intelligence technology with Internet of Things (IoT) and 5G in the health care system offer many more important services/applications. Zeadally et al. (2020) reported that many challenges witnessed in the healthcare system can be solved by utilizing smart healthcare through big data analytics and Internet of Things. These approaches involve the deployment of computational intelligence, such as electronic patient data, wireless technologies, mobile application, monitoring applications, smart phones, and wearable medical devices, to solve healthcare solutions by IoT biosensors monitoring equipment.

16.5 ROLE OF ARTIFICIAL INTELLIGENCE IN DIABETES MELLITUS MANAGEMENT

Diabetes mellitus is a debilitating condition resulting from absolute or relative impairment in insulin function. By the end of 2040, a hike in the population of people with diabetes mellitus reaching 600 million has been projected. Artificial intelligence has over time evolved as a technique with multifaceted applications, including determination and monitoring of diabetic patients' symptoms, owing to the fact that artificial intelligence supports creation of models which can be used in predicting hyperglycemia and other symptoms of diabetes mellitus (Ellahham, 2020).

Mishra and Shukla (2020) looked at programmed, diabetic location and analysis procedures. They unveiled the potential roles of artificial intelligence in the aspect of assisting health professionals in reaching accurate therapeutic dosing, courtesy of the Machine Learning model. Artificial intelligence can help remind patients about the schedule of their medicine. It can also allow them make right dietary choices on a daily basis. Contreras and Vehi (2018) reviewed the role of artificial intelligence in diabetes management, using 141 selected articles. They proposed a functional taxonomy which can be utilized for diabetes management.

Ellahham (2020) documented the promising roles of artificial intelligence in diabetes management. They noted that artificial intelligence can help diabetes patients be monitored remotely and virtually by healthcare professionals. This reduces the burden and cost of traveling to healthcare centers to receive care. *Diabetes mellitus* is a condition that does not only affect a blood glucose profile but also causes perturbation in vital signs. Today, many social media applications are available which can monitor vital signs, most especially heart rate and blood pressure. These devices can help in disease prognosis. With artificial intelligence, prevention and management of diabetes mellitus is possible because it supports models that can predict the likelihood of developing the disease.

Artificial intelligence supports platforms that can facilitate enlightenment, education, and orientation about diabetes mellitus causes, courses, and prevention. These platforms will create precautionary awareness, not only in diagnosed patients, but also in the general public. However, in order for this to be achieved, large quantities of data must be gathered. Updating existing data and knowledge must be a priority on a continuous basis (Li et al., 2020).

16.6 ROLE OF AI IN CARDIOVASCULAR DISEASE MANAGEMENT

Cardiovascular diseases (CVD) are health-threatening diseases which reportedly caused the death of 17.6 million people in 2016. Artificial intelligence intervention is welcome in the aspect of early detection of disease. Nowadays, there are many downloadable applications and smart devices designed using algorithms that can measure systolic blood pressure and diastolic pressure. They are also programmed to detect normal and abnormal blood pressure recordings.

Echocardiography is a diagnostic technique for heart disease which uses reflected sounds to produce real-time recordings of heart structure and function. Madani et al. (2018) developed and trained a neural network to classify fifteen standard views at the same time using labeled still images and videos from over two hundred transthoracic echocardiograms that picked a range of real-world clinical discrepancy. The authors showed that the model was able to classify among twelve video views, with 97.8% overall test accuracy, without any overfitting. This effort provides insight into the promising role of artificial intelligence.

Samad et al. (2019) investigated the use of Machine Learning to precisely predict survival following echocardiography in 171,510 patients by comparing Machine Learning models with linear logistic regression models. The results indicated a higher prediction precision and accuracy with Machine Learning models as far as common clinical risk scores are concerned, with logistic regression falling short of nonlinear random forest models. The study indicated supremacy of Machine Learning in fully utilizing large combinations of different input variables in predicting post echocardiography survival.

Suzuki (2017) utilized the artificial neural network (ANN) technique as the characteristic learning algorithm for detecting lumen and media adventitia in intravascular ultrasound images. They made use of spatial neighboring characteristics as input data. Two sparse autoencoders and a Softmax classifier were used to differentiate different vascular layers. Optimization of results was achieved through another ANN. The results of the study revealed a good relationship between the model and manual result. The study indicated the high efficiency with which the model can accurately and efficiently manage lumen and media advential detection in an intravascular ultrasound image.

Nakajima et al. (2017) trained an artificial neural network to classify potentially defective areas as true or false. They made use of nuclear cardiology expert interpretation of 1,001 gated stress/rest ^{99mTc}-MIBI images at twelve hospitals. The result indicates that artificial neural networks created a better area under receiver operating curve (AUC) than the summed difference score. The authors concluded that ANN was accurate in patients with previous myocardial infarction and coronary revascularization, diagnostically implying it's being used to help to diagnose coronary artery disease.

Betancur et al. (2018) worked on an automation-dependent prediction of obstructive disease from myocardial perfusion imaging (MPI) courtesy of Deep Learning in comparison with a total perfusion deficit (TPD). They used 1,638 patients who had not been diagnosed of coronary disease. They reported that the area under the receiver-operating characteristic curve for disease prediction by Deep Learning was higher than for TPD.

Tran (2016) in a bid to address challenges of automation with respect to left and right ventricle segmentation used deep, fully convoluted neural network. The authors pointed out that their novel model was trained efficiently end to end in a single stage of learning. The model is also shown to be robust enough to outshine previous fully automated techniques across many evaluation measures of cardiac datasets. Apart from this, the model is claimed to be fast, with a proven tendency to leverage commodity compute resources that will enable state-of-the-art cardiac segmentation at massive scales.

16.7   ROLE OF AI IN NEURODEGENERATIVE DISEASES

Cognitive decline, such as loneliness, agitation, and social isolation, which proved difficult to understand have been much more understood by artificial intelligence. In addition, with novel attributes such as handwriting (text) (Dash et al., 2018), sensor (digital), speech (audio) (Dash et al., 2019a) and a wearable activity monitor, it is possible to identify new indicators for identifying cognitive decline. Wearable cognitive devices exhibit the potentials for continuous monitoring and tracking of changes in cognitive function.

Machine Learning algorithms are useful in the detection, classification, and prediction of early pathological cognitive decline in older adults. Algorithms such as Logistic Regression, designed by Ewers et al. (2012), claims to be useful in discriminating Alzheimer's disease and neurocognitive disease, with Cox Regression modeling the conversion time from mild cognitive disease to Alzheimer's disease. Sparse autoencoders and 3D-CNN developed by Payan and Montana (2015) is also applicable in predicting Alzheimer's disease, with 95.39% accuracy for Alzheimer's disease versus a mild cognitive illness classification. An ensemble-based method using the Sparse representation-based classifier for predicting neurocognitive disease was developed by Liu et al. (2012). Gomar et al. (2011) examined the tendency of cognitive and biomarkers to predicting change from mild cognitive illness to Alzheimer's disease. Zhang et al. (2012) proposed a multimodal multitask learning that is capable of performing regression and classification simultaneously. In a bid to classify patients into mild cognitive impairment, Alzheimer's disease, and healthy, Moradi et al. (2015) were able to use an Alzheimer's disease neuroimaging initiative to train the classifier with accuracy of 77%.

The study by Tierney et al. (1996) investigated the likelihood of developing a neuropsychological-based test for the prediction of Alzheimer's disease onset. Using Analysis of Covariance, they were able analyze group differences of neurocognitive disease and Alzheimer's disease. Using a Dynamic Bayesian Network, causal relationship from functional magnetic resonance imaging was extracted and utilized in developing neural anatomical network. With this, Burge et al. (2004) were able to classify patients into healthy and dementia with an accuracy of 73%. With the help of Support Vector Machine model, Challis et al. (2015) were able to compare Bayesian Gaussian process logistic models on the basis of performance. They observed that using functional magnetic resonance imaging as input data; models were able to comparably predict mild cognitive illness and neurocognitive illness with an accuracy of 80%.

Fusion of data from magnetic resonance imaging and positron emission tomography using Deep machine method. This method involved integration of selected image derived-voxel patches into the Deep Boltzmann Machine. With the data fusion, there was not only an improvement in classification, but brain regions associated with Alzheimer's disease could be confirmed. Fusion of biomarkers from cerebrospinal fluid, positron emission tomography, and magnetic resonance imaging using intermediate data integration (early and intermediate) and novel multimodal and multitask learning resulted in better Alzheimer's disease versus neurocognitive disease classification.

Alzheimer's disease is the most prevalent neurodegenerative disease and with a high rate of wrong diagnosis. This necessitates highly intensified research efforts in a bid to surmount the challenge. For instance, in order to ease classification of Alzheimer's disease, Aljović et al. (2016) designed and presented a concept called artificial neural network system. This model makes use of biomarkers from cerebrospinal fluid, such as tau-phospho, Aβ40, albumin ratio, Aβ42 and tau-total.

The study by Tierney et al. (1996) investigated the likelihood of developing a neuropsychological-based test for the prediction of Alzheimer's disease onset. Using Analysis of Covariance, they were able analyze group differences of neurocognitive disease and Alzheimer's disease. Using Dynamic Bayesian Network, causal relationship from functional magnetic resonance imaging was extracted and utilized in developing a neural anatomical network. With this, Burge et al. (2004) were able to classify patients into healthy and dementia. With the help of Support Vector Machine model, Challis et al. (2015) were able to compare Bayesian Gaussian process logistic models on the basis of performance. They observed that using functional magnetic resonance imaging as input data, models were able to comparably predict mild cognitive illness and neurocognitive illness.

Machine Learning algorithms are useful in the detection, classification, and prediction of early pathological cognitive decline in older adults. Algorithms such as Logistic Regression designed by Ewers et al. (2012) claims to be useful in discriminating Alzheimer's disease and neurocognitive disease, with Cox Regression modeling the conversion time from mild cognitive disease to Alzheimer's disease. Sparse autoencoders and 3D-CNN developed by Payan and Montana (2015) are also applicable in predicting Alzheimer's disease, with 95.39% accuracy for Alzheimer's disease versus mild cognitive illness classification. An ensemble-based method using a Sparse representation-based classifier for predicting neurocognitive disease was developed by Liu et al. (2012), Gomar et al. (2011) examined the tendency of cognitive and biomarkers to predicting change from mild cognitive illness to Alzheimer's disease. Zhang et al. (2012) proposed a multimodal multitask learning that is capable of performing regression and classification simultaneously. In a bid to classify patients into mild cognitive impairment, Alzheimer's disease, and healthy, Moradi et al. (2015) were able to use an Alzheimer's disease neuroimaging initiative to train the classifier.

16.8 CHALLENGES OF COMPUTATIONAL INTELLIGENCE IN IoT HEALTHCARE

Bolhasani et al. (2021) reported that Quality of Service (QoS) optimization, privacy, and deployment represent some of the potential issues in the adoption and utilization of computational intelligence in the health care system. Numerous studies have highlighted so many challenges working against the wide spread of digital healthcare system. Some of them are issues of security and privacy. Generally, computational intelligence could pose a serious risk to the patient's information, particularly where medical information from gadgets is transferred to the cloud (Rahman et al., 2021), (Singh et al., 2021), which is known to be vulnerable due to spoofing, cloning, cloud polling, and RF jamming. Universal plug or Bluetooth low energy properties may be utilized to locate the gadgets and target the IoT devices (Rahman et al., 2019a), (Rahman et al., 2018). The risk posed by a denial of service attack is another serious problem of computational intelligence of IoT. A patient's safety index is low due to the complex nature of hardware and software vulnerabilities. Security protection from Sybil attacks, eavesdropping, sleep deprivation attacks, and sinkhole attacks is on the increase due to a large number of gadgets being connected to the internet. Medical information like centralized data sets of patient's information, electronic medical records, genomic data, and family history can be used by malicious software developers and hackers, thus bridging the confidentiality and privacy agreement. Other challenges include interoperability issues and Inter-realm authentication, medical information exchange barriers, and health care-related cybercrime, device communication, collation and management of medical data, complexity of data, and data integrity while updating information (Zeadally et al. 2020). Kumar and Majumder (2018) also report that the amount of data generation through sensor and computational intelligence is huge, and extraction from storage may pose a serious challenge (Dash et al., 2019b; Rahman et al., 2019b). Also, some other challenges observed are latency, response time, connectivity, and false predictions, which are very difficult for optimization of Quality of Service indices in real-time foggy environments.

16.9 ROLE OF AI IN *HELICOBACTER PYLORI DETECTION*

Huang et al. (2004) claimed that refined feature selection with neural network (RFSNN) is an effective computerized method for detecting and assessing the presence of *Helicobacter. pylori* infection and associated gastric irritation and inflammation and precancerous damage. This technique allows analysis and assessment of the stomach without involvement of invasive procedures. A study by Huang et al. (2008) recommended a computer-aided diagnosis system that utilizes sequential forward floating selection (SFFS) with support vector machine (SVM) for diagnosis of gastric *Helicobacter pylori* (H. pylori) histology from endoscopic images. *Helicobacter pylori* is known to play a role in peptic ulcer disease and chronic gastritis. Hence,

timely and accurate diagnosis of the microorganism is essential for disease management. Itoh et al. (2018) utilized a convoluted neural network (CNN) to develop a
learning tool that will recognize Helicobacter infection, as well as give an assessment of decision accuracy. They showed the sensitivity and specificity of the CNN
for the detection of *Helicobacter pylori* infection as 86.7 % and 86.7 %, respectively, with the AUC estimated as 0.956. The work revealed the usefulness of CNN
in *Helicobacter pylori* diagnosis.

16.10 CONCLUSION AND FUTURE PERSPECTIVES

In this chapter, we have provided a comprehensive review on the application and
challenges involved in the use of computational intelligence in the healthcare system
through the Internet of Things. We have identified various authors that have over the
years utilized computational intelligence and edge architecture for healthcare systems, particularly in the diagnosis, detection, prediction, monitoring, and treatment
of different disease conditions in patients. We have discussed various advantages and
drawbacks in the adoption of these technologies in the healthcare system, along with
the emerging developmental trend in artificial intelligence and Internet of Things.
Thus, in the future we expect more research efforts to be carried out in this direction
to mitigate some of the identified challenges and place computational intelligence on
the global pedestal for Internet of Things healthcare solutions.

REFERENCES

Abdellatif, A.A., Mohamed, A., Chiasserini, C.F., Tlili, M. and Erbad, A., 2019. Edge comput-
    ing for smart health: Context aware approaches, opportunities, and challenges. *IEEE
    Network*, 33(3), pp. 196–203.
Aljović, A., Badnjević, A., and Gurbeta, L. "Artificial neural networks in the discrimination
    of Alzheimer's disease using biomarkers data," *2016 5th Mediterranean Conference on
    Embedded Computing (MECO)*, Bar, Montenegro, 2016, pp. 286–289, doi: 10.1109/
    MECO.2016.7525762.
Amin, Syed Umar, M. Shamim Hossain et al. (2020). Edge intelligence and Internet of Things
    in healthcare: A survey. *IEEE Access*. doi:10.1109/ACCESS.2020.3045115.
Azar, J., Makhoul, A., Barhamgi, M. and Couturier, R., 2019. An energy efficient IoT data
    compression approach for edge machine learning. *Future Generation Computer
    Systems*, 96, pp. 168–175.
Betancur, J., Commandeur, F., Motlagh, M., Sharir, T., Einstein, A.J., Bokhari, S., Fish, M.B.,
    Ruddy, T.D., Kaufmann, P., Sinusas, A.J. and Miller, E.J., 2018. Deep learning for pre-
    diction of obstructive disease from fast myocardial perfusion SPECT: A multicenter
    study. *JACC: Cardiovascular Imaging*, 11(11), pp. 1654–1663.
Bolhasani, H., Mohseni, M. and Rahmani, A.M., 2021. Deep learning applications for IoT
    in health care: A systematic review. *Informatics in Medicine Unlocked*, 23, p. 100550.
Burge, J., Clark, V.P., Lane, T., Link, H. and Qiu, S., 2004. Bayesian classification of fMRI
    data: Evidence for altered neural networks in dementia. *University of New Mexico, Tech.
    Rep. TR-CS-2004-28.*
Challis, E., Hurley, P., Serra, L., Bozzali, M., Oliver, S. and Cercignani, M., 2015. Gaussian
    process classification of Alzheimer's disease and mild cognitive impairment from
    resting-state fMRI. *NeuroImage*, 112, pp. 232–243.

Contreras, I. and Vehi, J., 2018. Artificial intelligence for diabetes management and decision support: literature review. *Journal of Medical Internet Research*, *20*(5), p. e10775.

Dash, S., Abraham, A., Luhach, A.K., Mizera-Pietraszko, J., and Rodrigues, J.P.C., 2019a. Hybrid chaotic Firefly decision-making model for Parkinson's disease diagnosis, *International Journal of Distributed Sensor Networks*, 15(12).

Dash, S., Abraham, A., and Rehman, A., 2018. Kernel based Chaotic Firefly Algorithm for Diagnosing Parkinson's Disease. In: A. Madureira et al. (eds) *Advances in Intelligent Systems and Computing* 923, pp. 178–188, Springer Nature.

Dash, S., Biswas, S., Banerjee, D. and Rahman, A.U., 2019b. Edge and fog computing in healthcare–A review. *Scalable Computing: Practice and Experience*, *20*(2), pp. 191–206. http://www.scpe.org, doi: 10.12694/scpe.v20i2.1504, ISSN 1895-1767.

Ellahham, S., 2020. Artificial intelligence: the future for diabetes care. *The American journal of medicine*, *133*(8), pp. 895–900.

Ewers, M., Walsh, C., Trojanowski, J.Q., Shaw, L.M., Petersen, R.C., Jack Jr, C.R., Feldman, H.H., Bokde, A.L., Alexander, G.E., Scheltens, P. and Vellas, B., 2012. Prediction of conversion from mild cognitive impairment to Alzheimer's disease dementia based upon biomarkers and neuropsychological test performance. *Neurobiology of Aging*, *33*(7), pp. 1203–1214.

Gomar, J.J., Bobes-Bascaran, M.T., Conejero-Goldberg, C., Davies, P., Goldberg, T.E. and Alzheimer's Disease Neuroimaging Initiative, 2011. Utility of combinations of bio-markers, cognitive markers, and risk factors to predict conversion from mild cognitive impairment to Alzheimer disease in patients in the Alzheimer's disease neuroimaging initiative. *Archives of General Psychiatry*, *68*(9), pp. 961–969.

Hayyolalam, V., Aloqaily, M., Ozkasap, O. and Guizani, M., 2021. Edge Intelligence for Empowering IoT-based Healthcare Systems. *arXiv preprint arXiv:2103.12144*.

Huang, C.R., Chung, P.C., Sheu, B.S., Kuo, H.J. and Popper, M., 2008. Helicobacter pylori-related gastric histology classification using support-vector-machine-based feature selection. *IEEE Transactions on Information Technology in Biomedicine*, *12*(4), pp. 523–531. doi: 10.1109/TITB.2007.913128.

Huang, C.R., Sheu, B.S., Chung, P.C. and Yang, H.B., 2004. Computerized diagnosis of Helicobacter pylori infection and associated gastric inflammation from endoscopic images by refined feature selection using a neural network. *Endoscopy*, *36*(7), pp. 601–608. doi: 10.1055/s-2004-814519.

Itoh, T., Kawahira, H., Nakashima, H. and Yata, N., 2018. Deep learning analyzes Helicobacter pylori infection by upper gastrointestinal endoscopy images. *Endoscopy International Open*, *6*(2), pp. E139–E144. doi: 10.1055/s-0043-120830.

Kumar, S.M. and Majumder, D., 2018. Healthcare solution based on machine learning applications in IOT and edge computing. *International Journal of Pure and Applied Mathematics*, *119*(16), pp. 1473–1484.

Li, J., Huang, J., Zheng, L. and Li, X., 2020. Application of artificial intelligence in diabetes education and management: Present status and promising prospect. *Frontiers in Public Health*, *8*, p. 173. doi: 10.3389/fpubh.2020.00173.

Lin, B.S.P., 2020. Toward an AI-enabled SDN-based 5G & IoT network. *Network and Communication Technologies*, *5*(2), pp. 1–7.

Liu, M., Zhang, D., Shen, D. and Alzheimer's Disease Neuroimaging Initiative, 2012. Ensemble sparse classification of Alzheimer's disease. *NeuroImage*, *60*(2), pp. 1106–1116.

Madani, A., Arnaout, R., Mofrad, M. and Arnaout, R., 2018. Fast and accurate view classification of echocardiograms using deep learning. *NPJ Digital Medicine*, *1*(1), pp. 1–8.

Mishra, D.K. and Shukla, S., 2020. Role of artificial intelligence in diabetes management. *International Journal of Engineering Technologies and Management Research*, *7*(7), pp. 80–88.

Moradi, E., Pepe, A., Gaser, C., Huttunen, H., Tohka, J. and Alzheimer's Disease Neuroimaging Initiative, 2015. Machine learning framework for early MRI-based Alzheimer's conversion prediction in MCI subjects. *Neuroimage*, *104*, pp. 398–412. doi: 10.1016/j.neuroimage.2014.10.002.

Nakajima, K., Kudo, T., Nakata, T., Kiso, K., Kasai, T., Taniguchi, Y., Matsuo, S., Momose, M., Nakagawa, M., Sarai, M. and Hida, S., 2017. Diagnostic accuracy of an artificial neural network compared with statistical quantitation of myocardial perfusion images: a Japanese multicenter study. *European Journal of Nuclear Medicine and Molecular Imaging*, *44*(13), pp. 2280–2289.

Payan, A. and Montana, G., 2015. Predicting Alzheimer's disease: A neuroimaging study with 3D convolutional neural networks. *arXiv preprint arXiv:1502.02506*.

Rahman, A., Dash, S., Ahmad M., Iqbal T., 2021. Mobile Cloud Computing: A Green perspective, In: Udgata S. K., Sethi S., Sriram S. N. (eds), *Intelligent Notes in Networks and Systems*, 185, 523–533, Springer, Singapore. doi: 10.1007/978-981-33-6081-5-46.

Rahman, A., Dash, S., Kamaleldin, M., Abed, A., Alshaikhhussain, A., Motawei, H., Amoudi, N., Abahussain, W., 2019a. A Comprehensive study of mobile computing in Telemedicine, A. K. Luhach et al. (Eds):ICAICR 2018,CCIS 956, pp. 413–425, Springer.

Rahman, A., Dash, S., Luhach, A.K., Chilamkurti, N., Baek, S. and Nam, Y., 2019b. A Neuro-fuzzy approach for user behaviour classification and prediction. *Journal of Cloud Computing*, 8(1), pp. 1–15. doi: 10.1186/s13677-019-0144-9, Springer.

Rahman, A., Dash, S., Sultan, K. and Khan, M.A., 2018. Management of resource usage in mobile cloud computing. *International Journal of Pure and Applied Mathematics*, *119*(16), pp. 255–261, ISSN: 1314-3395 (on-line version).

Samad, M.D., Ulloa, A., Wehner, G.J., Jing, L., Hartzel, D., Good, C.W., Williams, B.A., Haggerty, C.M. and Fornwalt, B.K., 2019. Predicting survival from large echocardiography and electronic health record datasets: Optimization with machine learning. *JACC: Cardiovascular Imaging*, *12*(4), pp. 681–689.

Sathyasri, B., AntoBennet, M., Bhuvaneshwari, S. and Deepika, M., 2018. Artificial Intelligence Based e-Health Management System. *International Journal of Pure and Applied Mathematics*, *119*, pp. 31–44.

Shah, R. and Chircu, A., 2018. IoT and AI in healthcare: A systematic literature review. *Issues in Information Systems*, *19*(3). 33–41. doi: 10.48009/3_iis_2018_33-41

Singh, K. R., Dash, S., Deka, B., Biswas, S., 2021. Mobile Technology Solution for COVID-19. In Fadi Al-Turjamanet al., (Eds), *Emerging Technologies for battling COVID-19 applications and innovations*, Springer, pp. 271–294 ISBN: 978-030-60038-9

Su, X., Tarkoma, S., Jiang, T., Crowcroft, J., Hui, P. arXiv:2003.12172v2 [cs.NI] 12 Jun 2020.53 pages. Edge Intelligence: Architectures, Challenges, and Applications Dianlei Xu, Tong Li, Yong Li, Senior Member, IEEE.

Suzuki, K., 2017. Overview of deep learning in medical imaging. *Radiological Physics and Technology*, *10*(3), pp. 257–273.

Tierney, M.C., Szalai, J.P., Snow, W.G., Fisher, R.H., Nores, A., Nadon, G., Dunn, E. and George-Hyslop, P.S., 1996. Prediction of probable Alzheimer's disease in memory-impaired patients. A prospective longitudinal study. *Neurology*, *46*(3), pp. 661–665.

Tran, P.V. A fully convolutional neural network for cardiac segmentation in short-axis MRI. 2016. [(accessed April 27, 2017)]. arXiv.org.

Tuli, S., Basumatary, N., Gill, S.S., Kahani, M., Arya, R.C., Wander, G.S. and Buyya, R., 2020. Health fog: An ensemble deep learning based Smart healthcare system for automatic diagnosis of heart diseases in integrated IoT and fog computing environments. *Future Generation Computer Systems*, *104*, pp. 187–200.

Vatti, R.A., Vinoth, K. and Sneha, Y., 2020. Edge intelligence for predicting and detecting cardiac pathologies by analyzing stress and anxiety. *Journal of Critical Reviews*, *7*(18), pp. 2816–2822.

Zeadally, Sherali, Siddiqui, Farhan, Baig, Zubair and Ibrahim, Ahmed 2020, Smart healthcare: Challenges and potential solutions using internet of things (IoT) and big data analytics, *PSU research review*, 4(2), pp. 93–109, doi: 10.1108/prr-08-2019-0027.

Zhang, D., Shen, D. and Initiative, A.D.N., et al., 2012. Multi-modal multi-task learning for joint prediction of multiple regression and classification variables in Alzheimer's disease. *NeuroImage*, 59(2), pp. 895–907.

17 Machine Learning Techniques for High-Performance Computing for IoT Applications in Healthcare

Olugbemi T. Olaniyan, Charles O. Adetunji, and Mayowa J. Adeniyi
Edo State University Uzairue Iyamho, Nigeria

Daniel Ingo Hefft
University of Birmingham Birmingham, United Kingdom

CONTENTS

17.1 INTRODUCTION

With the significant growth and development seen in the field of artificial intelligence, Machine Learning (ML) and Internet of Things (IoT), a lot of technologies and applications are being deployed to healthcare systems to help resolve many challenges in the industry. Improved methods of data gathering, as well as procedures and analysis in medicine to resolve specific challenges, have significantly grown. The flexibility and versatility in telemedicine and artificial intelligence have provided endless possibilities in healthcare information technology, patient monitoring, information analysis collaboration, and intelligent assistance diagnosis (Bohr and Memarzadeh 2020). Machine Learning and artificial intelligence in the healthcare system provide a platform for medical data analysis, detection of sources of error, and development of computerized intelligence with medical tools and devices. These platforms provide accuracy and speed for the decision making process, prediction, diagnosis, and optimization through different algorithms and applications, like neural networks, fuzzy logic, Machine Learning, decision tree and Bayesian network. Telemedicine and patient monitoring provide fast and economic ways of conducting clinical investigation, clinical data generation through Wi-Fi connected digital medical devices, smart phone, 3D printed frame, wireless gyroscope platform, and robots. Several studies have shown that through artificial intelligence, medical record retrieving, and self-diagnosing can be done easily via neural networks, cloud computing, and big data analytics. For personalized medicine, the patient's information, like patient self-management, sickness progression, clinical involvement, and prediction of illness outcome, are collected through mobile and wearable gadgets connected to the internet through Internet of Things. The appropriate establishment of analytical models is developed for treatment options in the healthcare system. Artificial intelligence, such as hardware implementation, applications, and software algorithms, are useful for assisted living facilities, disease diagnostics, biomedical research, and biomedical information processing. These applications cover a wide range of healthcare solutions, like image processing, signal predictions, data extraction, pressure and volume prediction, and disease diagnostics. The challenge in the current electronic health records area is the problem of growing disease data in clinical settings with poor labels. It is known that health data plays a crucial role in the application of Machine Learning domains so as to provide accurate predictive analysis and performance on several tasks, including the interpretation of digitized images, diagnosis, monitoring streaming physiological data, gas detection, vital sign monitoring, motion tracking, and the tracking of diseases (Davenport and Kalakota 2019).

17.2   THE APPLICATION OF IoT IN THE HEALTHCARE SYSTEM

The work conducted by Kelly *et al.* (2020) analyzed the promising roles of IoT in Medicare, including the privilege of forecasting health challenges, timely and early diagnosis of diseases (Singh et al., 2021) and instituting treatment and prognosis. IoT is capable of being used in improving the delivery efficiency of Medicare and community health and safety. Owing to its growing popularity, IoT is foreseen as a potential successor for medical services (Dash et al., 2019a, 2019b). They noted that other areas of concern that future investigations will study are the relative efficiency of block-chain storage, the level of acceptance and consumer literacy, construction of standard protocol, and an operation link with state and international systems, among others. Harnessing of IoT relies on data management, cyber-security (Rahman et al., 2019a, 2019b) and safety, and confidentiality.

Pradhan *et al.* (2021) reviewed health IoT and noted that IoT has contributed immensely to assisting healthcare service providers to evaluate vital signs and biological parameters, such as pulse rate, blood pressure, peripheral oxygen saturation, electrocardiogram, lung volumes and capacities, and diagnosing diseases even at remote sites. They also noted the challenges associated with IoT (Rahman et al., 2021).

Merchant *et al.* (2016) and Merchant *et al.* (2018) noted that with IoT, there has been a reduction in rescue inhaler administration. In addition, adherence to medication has increased. They specifically evaluated the advantages of the technology-based intervention on asthma-associated hospitalizations, emergency units, and outpatient department utilization, using two hundred and twenty-four asthma patients. They reported a decline in the number of asthma-related emergency unit utilizations. The study indicated that integration of IoT into conventional healthcare protocols demonstrates a tendency of lowering hospitalizations relating to asthma.

Investigation by Naik *et al.* (2017) indicated that in patients with uncontrolled blood pressure, inclusion of an ingestible sensor in daily healthcare practice can help healthcare providers recognize latent factors that can contribute to sustained hypertension. It can also assist in determining specific interventions. In the study, 167 patients with poorly controlled hypertension were used and were requested to swallow pharmacological inactive drugs that contained an ingestible sensor anytime antihypertensive drugs are taken. An implantable sensor was placed on the patient's body, thus recording the times and date of drug ingestion.

Doryab *et al.* (2019) reported that passive sensing has the tendency to detect not only loneliness in college students but also identification of behavioral patterns that are associated with loneliness. They collected data from one-hundred and sixty college students by using smartphones and *Fitbits*. With a Machine Learning pipeline, a reported accuracy of 80.2% was attained in the detection of the binary level of loneliness, as well as 88.4% accuracy in the detection of change in the loneliness level. The study provided an insight into the potential of IoT and mobile technology in decreasing the impact of loneliness.

Sposaro and Tyson (2020) utilized an android-based smart phone with an incorporated three axial accelerometer. To detect a fall, data emanating from the accelerometer is assessed using many Machine Learning algorithms and position data. With the invention, a platform for detecting a fall was created. The device was reported to be cost effective, offering a realistic solution. This alert system is noted to reduce fall frequency and injuries, especially among the elderly, resulting in improvement in quality of life and comfort.

Electrocardiography is the recording of electrical activities taking place on the bare surface of the body. Usually, a typical electrocardiographic arrangement consists of a system of cables and electrodes connected to the electrocardiograph. Majumder *et al.* (2018) built an electrocardiography that uses no cables but has flexible and dry capacitive electrodes, for continuous measurement and evaluation of cardiovascular health. The model, which receives information from contact electrodes, transmits data to a computer via Bluetooth. The device was also reported to determine electrocardiographic signals even on textile-covered skin. Kakria *et al.* (2015) built a heart function monitoring platform for cardiac patients which can be operated remotely. They then assessed the model using forty people between eighteen and sixty-six years of age with the use of wearable sensors, with the android devices supervised by experts (Rahman et al., 2019a, 2019b).The important gains associated with this new device include high speed, convenience, and reliability.

Dojchinovski *et al.* (2019) extensively reviewed the benefits of patient-based automated voice and web services created by Google Assistant and Amazon. With voice assistants, patients' healthcare delivery was reported to improve. Patients were able to more easily access healthcare. The diagnosis and evaluation of medical conditions and improvement in communication are also feasible. Ilievski *et al.* (2019) also looked at how patient-based voice created by Google Assistant and Amazon can help in improving healthcare in the comfort of patient residences.

Jadczyk *et al.* (2019) created CardioCube platform, which is capable of collecting, indexing, and documenting health-related data, courtesy of an automated voice platform. Twenty-two people were recruited for the study, and both a voice-enabled, patient registration software package and a web-based, electronic health record platform were utilized. The study also highlights the ability of voice-enabled automated mechanisms to carry out complex, time-consuming activities.

A study by Chen *et al.* (2018) highlighted the roles of social robot utilization in the reduction in depression cases in older adults. They utilized a method known as systematic reviews and meta-analysis (PRISMA) for the identification of existing studies. According to the review, the reported social robot intervention includes communication, health monitoring, and companionship. The study provides insight into the clinical effectiveness of social robotics. Putte *et al.* (2019) utilized the social robot Pepper to carry out investigations concerning pain, memory, medical history, defecation, and sleep. Thirty-five patients averaging sixty-four years of age were recruited for the study. The robot was reported to be acceptable to both the healthcare providers and patients. Bauer *et al.* (2018) reviewed that robotic assistance and virtual reality can improve individualization through adaption of haptic conditions.

Ritschel *et al.* (2018) created a package known as Drink-O-Mender, a robot developed to offer drinks in social events. The focus of the invention is to earn consumers' confidence in terms of the healthiness of drinks being consumed. Other features of the robots include mechanisms that can sense the quantity and type of drinks that are being consumed. A review conducted by Wan *et al.* (2020) centered on human–robot interface, robust communication, and mobile healthcare robot, among others. They argued that robots can execute user interaction and time-requiring tasks, which include autonomous mobility and health receiver status recognition, without regular interactions with data centers (Rahman et al., 2018).

Polap *et al.* (2018) proposed a smart home platform which utilizes in-built sensors and artificial intelligence techniques to detect dermal disease onset and evaluate skin health conditions of the house residents. This device will assist in prevention and timely detection and management of dermal disease, especially melanoma, a disease that develops under excessive exposure to sunlight. Tear film breakup time is a biological parameter utilized in the diagnosis of dry eye disease. However, a robust method for examining tear film breakup time in murine models has yet to be established. In view of the enormous health burden constituted by diabetic ulcers, Bhelonde *et al.* (2015) suggest a new wound image platform, which functions only on a smart phone, with the assistance of an android application. The phone then conducts wound segmentation through an accelerated mean-shift algorithm. The device is able to evaluate the status of healing quantitatively, using trend analysis of time records. The paradigm seems cost-effective and quantitative. The technique improves wound care and accelerates wound healing tremendously.

We Shimizu *et al.* (2019) pioneered an invention named, the "Smart Eye Camera", which is capable of measuring tear film breakup time in a rat dry eye disease model. The images produced by the new technique were compared with that produced by conventional devices and they were found to be sufficient in quality for practical use.

17.3   DATA IN MACHINE LEARNING FOR HEALTHCARE

Machine Learning (ML) consists of methods that enable algorithms to update themselves on the basis of new information (Chen *et al.*, 2017; Nevin, 2018) and learn from the dataset. As far as Machine Learning algorithms are concerned, there is no need for priori hypothesis between independent and dependent variables, but there is a strong emphasis on accuracy prediction, with the tendency to identify non-linear interactions existing within the data (Graham *et al.*, 2019).

Machine Learning has been used in the diagnosis of several diseases. For instance, Machine Learning algorithms for classification of diseases include Linear Discriminant Analysis, Quadratic Discriminant Analysis, Multilayer Perceptions, Support Vector Machine, and Pixel/Voxel values in images. Specifically, Linear Discriminant Analysis and Quadratic Discriminant Analysis have been used for the classification of lung cancers (Rumelhart *et al.*, 1986; Vapnik, 1995). Naïve Bayesian Classifier is used in distinguishing cardiovascular US images with accuracy of up to 96.59% (Ding *et al.*, 2012). Support Vector Machine is used in ultrasound imaging (Prabusankarlal *et al.*, 2015). Multiclass AdaBoost has been utilized in distinguishing carcinomas, fibro adenomas, and cysts (Takemura *et al.*, 2010).

Machine Learning shares the same fate with data analysis in using conventional statistical techniques in the sense that they are both associated with bias. However, while small and known datasets seem easy to collect, transferring algorithms to another dataset is difficult. For large and unknown datasets, there may be a need to shift in the direction of Unsupervised Learning Methods, which are known to be difficult to train (Esteva *et al.*, 2019).

Fusion of data from magnetic resonance imaging and positron emission tomography using a Deep Machine method was developed by Suk *et al.* (2014). Their method involved integration of selected image derived-voxel patches into Deep Boltzmann Machine. With the data fusion, there was not only an improvement in classification, but brain regions associated with Alzheimer's disease could be confirmed. The fusion of biomarkers from cerebrospinal fluid, positron emission tomography, and magnetic resonance imaging using intermediate data integration (early and intermediate) and novel multimodal and multitask learning resulted in better Alzheimer's disease versus neurocognitive disease classification.

The study by Tierney *et al.* (1996) investigated the likelihood of developing a neuropsychological-based test for the prediction of Alzheimer's disease onset. Using Analysis of Covariance, they were able analyze group differences of neurocognitive disease and Alzheimer's disease. Using the Dynamic Bayesian Network, a causal relationship from functional magnetic resonance imaging was extracted and utilized in developing a neural anatomical network. With this, Burge *et al.* (2004) were able to classify patients into healthy and dementia. With the help of Support Vector Machine model, Challis *et al.* (2015) were able to compare The Bayesian Gaussian process logistic model on the basis of performance. They observed that by using functional magnetic resonance imaging as input data; models were able to comparably predict mild cognitive illness and neurocognitive illness.

Machine Learning algorithms are useful in the detection, classification, and prediction of early pathological cognitive decline in older adults. Algorithms such as Logistic Regression designed by Ewers *et al.* (2012) may be useful in discriminating between Alzheimer's disease and neurocognitive disease, with Cox Regression modeling the conversion time from mild cognitive disease to Alzheimer's disease. Sparse auto-encoders and 3D-CNN developed by Payan and Montana (2015) are also applicable in predicting Alzheimer's disease with 95.39% accuracy for Alzheimer's disease versus a mild cognitive illness classification. An ensemble-based method using the Sparse representation-based classifier for predicting neurocognitive disease was developed by Liu *et al.* Gomar *et al.* examined the tendency of cognitive and biomarkers to predict change from mild cognitive illness to Alzheimer's disease. Zhang *et al.* (2012) proposed a multimodal multitask learning that is capable of performing regression and classification simultaneously. In a bid to classify patients into mild cognitive impairment, Alzheimer's disease and healthy, Moradi *et al.*, (2015) were able to use Alzheimer's disease neuroimaging initiative to train the classifier.

A number of Machine Learning techniques have been designed to analyze data in the field of medicine (Bicciato, 2004). As far as cancer research is concerned, high throughput-based approaches have been designed. Hong-Qiang Wang *et al.* (2009) proposed a model known as Biomarker Association Networks for the purpose of classifying cancer. Their model was comparable to the existing ones, such as K-nearest neighborhood, Fisher discriminant analysis, radial basis function kernel, Bayesian network, and support vector machines with a linear kernel.

Alzheimer's disease is the most prevalent neurodegenerative disease and has a high rate of wrong diagnosis. This problem necessitates highly intensified research efforts in a bid to surmount the challenged. For instance, to ease classification of Alzheimer's disease, Aljović *et al.* (2016) designed and presented a concept called artificial neural network system. This model makes use of biomarkers from cerebrospinal fluid, such as tau-phospho, Aβ40, albumin ratio, Aβ42 and tau-total.

In a bid to classify contact dermatitis signature genes, eighty-nine patch-test positive reaction tissue samples against two irritants and four contact allergens were investigated and analyzed courtesy of random forest classification and co-expression network analysis, with the sole aim of discovering potential biomarkers. The results showed that, with random forest classification, '*PKMYT, CD47, HISTH1A, BATF, FASLG, RGS16, SYNPO, SELE, PTPN7, WARS, PRC1, EXO1, RRM2, PBK, RAD54L, KIFC1, SPC25, TPX2, TPX2, DLGAP5, IL37* and *CH25H*' were identified as potential biomarkers that can be used to distinguish allergic and irritant contact dermatitis in human skin.

With nearest shrunken centroid developed by Tibshirani *et al.* (2002), subsets of genes that typify leukemias and small, round blue-cell tumors was possible. Khan *et al.*, (2001) designed an approach for the classification of cancer on a particular strata, on the basis of their gene expression. Training of artificial neural networks with the aid of small round blue-cell cell tumors indicates a promising role of the method in diagnosis of tumors and therapeutic.

17.4 TRADITIONAL CENTRALIZED LEARNING: MACHINE LEARNING RUNS IN THE CLOUD, GATHERING DATA FROM DIFFERENT HOSPITALS

It has been stated that artificial intelligence in conjunction with edge intelligence can improve Medicare and quality care delivery. Also, edge intelligence enhances smart Medicare, courtesy of massive attenuation of network burden, level of power consumed, and latency. They opined there will be a better outcome with the integration of edge intelligence with artificial intelligence in improving Medicare. In the study, they introduced a new smart Medicare platform that pivots the variety of artificial intelligence techniques on edge intelligence in a parallel mode. Using the new model, shortcomings associated with previous smart Medicare platforms are addressed. However, issues like loss of data and management of autonomous network persist.

Abdellatif *et al.* (2019) propose a smart Medicare using a multi-access computer monitoring platform that attempts to balance many access edge computing models, which can improve the system effectiveness and efficacy. With this set-up, the large quantum of data produced by sensors, medical devices, and by individuals at the network edge can be handled much more effectively. It also shows that the imitation of energy capabilities associated with devices at the network edge is a shortcoming. One merit of this invention is ease of delivering Medicare to healthcare seekers with minimal cost. The authors also suggest integration of, components of wireless network, acquired data features, and standard requirements of the exploited application, so as to bring about creation of quality and highly sustainable services for a smart Medicare system.

Tuli *et al.* (2020) designed a new platform named as HealthFog, which initially was meant to bring together an ensemble of Deep Learning in devices with edge computing and utilize it for automatic heart disease analysis in a real-life mode. With HealthFog, delivery of Medicare in the form of fog service, courtesy of IoT devices is possible. It also demonstrates huge prowess in the management of cardiac patients' data at the beckon of the users. The Fog-enabled cloud framework shortened as 'FogBus' is useful in evaluating the performance of HealthFog, especially with power consumption levels, bandwidth of network, latency, jitter, accuracy, and time of execution. HealthFog has another attribute of being configurable. With this ability, it can be adapted to various modes of operation with the assurance of quality service and high prediction precision, which are requisite for diverse fog computation situations.

17.5 MACHINE LEARNING APPLICATIONS IN DISEASE PREDICTION

17.5.1 CANCER

Studies have revealed that artificial intelligence and Machine Learning are utilized to predict and diagnose disease at the early stages so as to commence appropriate treatment strategies. Machine algorithms and datasets are important factors to accomplish this task. Some of the highlighted predictive ability is the diagnosis and detection of cancer using nonionizing and noninvasive thermographic images, magnetic resonance imaging, ultrasound, mammogram, biopsy, computerized tomography, and X-ray reports, combined with Machine Learning algorithms like convolution neural network and Gaussian blur smoothing techniques to improve the picture quality (Wu and Zhao, 2017).

17.5.2 DIABETES

Diabetes has been described as a chronic disease that needs continual medical attention due to the effects on the body of elevated levels of blood sugar. There are different types: diabetes 1, diabetes 2, and gestational diabetes. Discriminant analysis, Machine Learning, deep neural networks, and algorithms like Gaussian Naive Bayes as predictive equations can be useful in the analysis and classification of this disease, thus helping to control the illness through appropriate treatment options (Al-Zebari and Sengur, 2019).

17.5.3 CARDIOVASCULAR DISEASES

The devastating effect of several heart disease is on the increase. Slow progress in preventive treatments has necessitated the development of prompt interventional strategies. Personalized treatment combined with Machine Learning algorithms like recurrent neural network, long short-term memory, and natural language processing can facilitate quick intervention and management (Ahmed et al., 2018).

17.5.4 CHRONIC KIDNEY DISEASE

Generally, chronic kidney diseases are diagnosed utilizing lab tests, clinical data, biopsy, and imaging studies. Some of these methods and techniques are invasive, expensive, and time consuming, thus Machine Learning algorithms could play a significant role in overcoming several of the aforementioned disadvantages (Amirgaliyev et al., 2018).

17.5.5 PARKINSON DISEASE

The progressive increase in the number of neurodegenerative diseases like Parkinson disease is becoming unimaginable, with limited treatment strategies. This disease develops due to reduced levels of dopamine, resulting in poor coordination, movement, and postural instability. Artificial intelligence and Machine Learning could be adopted in the differentiation, classification, and diagnosis of the disease severity and spread (Aich et al., 2019).

17.5.6 DERMATOLOGICAL DISEASES

Early diagnosis of several skin problems like herpes, eczema, psoriasis, and melanoma has been suggested to reduce the disease burden and outcomes. Data collection and augmentation using Machine Learning are very important features in analyzing any skin datasets. Machine Learning facilitates drug design and repurposing, poly pharmacology, event extraction, medical imaging and processing (Peyvandipour et al., 2018).

17.6 ISSUES AND CHALLENGES

Studies have shown that several challenges are found in the application and implementation of artificial intelligence in the healthcare system. Some of them are the costs of installation, personnel training, bandwidth, and internet connectivity status. The issues of privacy and a security bridge also has slowed the implementation of IoT and artificial intelligence in the healthcare system. Solutions to these problems are still in their infancy.

17.7 CONCLUSIONS

Recent advances in the development of computer intelligence technology have increased the capacities that exist in the healthcare system. This review analyzed the importance of big data (such as clinical data and healthcare utilization data generated from medical devices and wearable gadgets) in the healthcare system for the control of disease outbreak, treatment, diagnosis, monitoring, and management of the patient's health conditions. Our future goal is to improve understanding of the application of artificial intelligence so as to increase the implementation and utilization of Machine Learning in solving health problems and research using computational technologies and algorithms.

REFERENCES

Abdellatif, A. A., Mohamed, A., Chiasserini, C. F., Tlili, M., Erbad, A. (2019). Edge computing for smart health: Context-aware ap-proaches, opportunities, and challenges, *IEEE Network*, 33 (3):196–203.

Ahmed, M. R., Hasan Mahmud, S. M., Hossin, M. A., Jahan, H., Haider Noori, S. R. (2018). "A cloud based four-tier architecture for early detection of heart disease with machine learning algorithms," in *Proceedings of the IEEE 4th International Conference on Computer and Communications*, Chengdu, China.

Aich S, H. Kim, K. Younga, K. L. Hui, A. A. Al-Absi, and M. Sain, (2019) "A supervised machine learning approach using different feature selection techniques on voice data-sets for prediction of Parkinson's disease," in *Proceedings of the 21st International Conference on Advanced Communication Technolog*, PyeongChang Kwangwoon_Do, Korea (South).

Aljović, A., Badnjević, A. and Gurbeta, L. (2016). "Artificial neural networks in the discrimination of Alzheimer's disease using biomarkers data," in *2016 5th Mediterranean Conference on Embedded Computing (MECO)*, Bar, Montenegro, pp. 286–289, doi: 10.1109/MECO.2016.7525762.

Al-Zebari, A., Sengur, A. (2019). "Performance comparison of machine learning techniques on diabetes disease detection," in *Proceedings of the 1st International Informatics and Software Engineering Conference*, Ankara, Turkey.

Amirgaliyev, Y., Shamiluulu, S., Serek, A. (2018). "Analysis of chronic kidney disease dataset by applying machine learning methods," in *Proceedings of the IEEE 12th International Conference on Application of Information and Communication Technologies*, Almaty, Kazakhstan.

Baur, K., Schättin, A., de Bruin, E.D., Riener, R., Duarte, J.E., Wolf, P. (2018). Trends in robot-assisted and virtual reality-assisted neuromuscular therapy: a systematic review of health-related multiplayer games. *Journal of NeuroEngineering and Rehabilitation*, 15(1):107.

Bhelonde, A., Didolkar, N., Jangale, S., Kulkarni, N. (2015). Flexible Wound Assessment System for Diabetic Patient Using Android Smartphone. *International Conference on Green Computing and Internet of Things; ICGCIoT'15*; October 8–10, 2015; Noida, India.

Bicciato, S. (2004). Artificial neural network technologies to identify biomarkers for therapeutic intervention. *Current Opinion in Molecular Therapeutics*, 6(6):616–23. PMID: 15663326

Burge, J., Clark, V.P., Lane, T., Link, H., Qiu, S. (2004). Bayesian classification of FMRI data: Evidence for altered neural networks in dementia. University of New Mexico, Tech. Rep.TR-CS-2004-28.

Bohr, A., Memarzadeh, K. (2020). The rise of artificial intelligence in healthcare applications. *Artificial Intelligence in Healthcare*, 25–60. doi:10.1016/B978-0-12-818438-7.00002-2

Challis, E., Hurley, P., Serra, L., Bozzali, M., Oliver, S., Cercignani, M. (2015). Gaussian process classification of Alzheimer's disease and mild cognitive impairment from resting-state fmri. *Neuro Image* 112, 232–243.

Chen, M., Hao, Y., Hwang, K., Wang, L., Wang, L. (2017). Disease prediction by machine learning over bigdata from healthcare communities. *IEEE Access* 5, 8869–8879.

Chen, S., Jones, C., Moyle, W. (2018). Social robots for depression in older adults: a systematic review. *Journal of Nursing Scholarship* 50(6):612–22. doi: 10.1111/jnu.12423.

Dash, S., Abraham, A., Luhach, A.K., Mizera-Pietraszko, J., Rodrigues, J.P.C. (2019a). Hybrid chaotic Firefly decision-making model for Parkinson's disease diagnosis.

Dash, S., Biswas, S., Banerjee, D., Rahman, A. (2019b). Edge and Fog Computing in Healthcare – A Review, *Scalable Computing: Practice and Experience*, 20(2):191–205. http://www.scpe.org, doi: 10.12694/scpe.v20i2.1504, ISSN 1895-1767

Ding, J., Bashashati, A., Roth, A., Oloumi, A., Tse, K., Zeng, T., Haffari, G., Hirst, M., Marra, M.A., Condon, A., Aparicio, S. (2012). Feature-based classifiers for somatic mutation detection in tumour–normal paired sequencing data, *Bioinformatics*, 28(2):167–175.

Dojchinovski, D., Ilievski, A., Gusev, M. (2019). "Interactive home healthcare system with integrated voice assistant," in *2019 42nd International Convention on Information and Communication Technology, Electronics and Microelectronics (MIPRO)*, 2019, pp. 284–288, doi: 10.23919/MIPRO.2019.8756983.

Doryab, A., Villalba, D.K., Chikersal, P., Dutcher, J.M., Tumminia, M., Liu, X., Cohen, S., Creswell, K., Mankoff, J., Creswell, J.D., Dey, A.K. (2019). Identifying behavioral phenotypes of loneliness and social isolation with passive sensing: statistical analysis, data mining and machine learning of smartphone and FitBit data. *JMIR mHealth and uHealth*, 7(7):e13209.

Davenport, T., Kalakota, R. (2019). The potential for artificial intelligence in healthcare. *Future Healthcare Journal*, 6(2), 94–98. https://doi.org/10.7861/futurehosp.6-2-94

Esteva A., Robicquet, A., Ramsundar, B., Kuleshov, V., Depristo, M., Chou, K., Cui, C., Corrado, G., Thrun, S., Dean, J., (2019). A guide to deep learning in health care. *Nature Medicine* 25. doi:10.1038/s41591-018-031

Ewers, M., Walsh, C., Trojanowski, J.Q., Shaw, L.M., Petersen, R.C., Jack, C.R., Feldman, H.H., Bokde, A.L., Alexander, G.E., Scheltens, P., et al. (2012). Prediction of conversion from mild cognitive impairment to alzheimer's disease dementia based upon biomarkers and neuropsychological test performance. *Neurobiology of Aging* 33(7):1203–1214.

Graham, S., Depp, C., Lee, E., Nebeker, C., Tu, X., Kim, H., Jeste, D., (2019). Artificial intelligence for mental health and mental illnesses: an overview. *Current Psychiatry Reports* 21 (11), 116.

Hong-Qiang Wang, Hau-San Wong, Hailong Zhu, Timothy T.C. Yip. (2009). A neural network-based biomarker association information extraction approach for cancer classification. *Journal of Biomedical Informatics* 42(4), 654–666.

Ilievski, A., Dojchinovski, D., Gusev, M. (2019). Interactive Voice Assisted Home Healthcare Systems, in *Proceedings of the 9th Balkan Conference on Informatics; BCI'19*; September 26–28, 2019; University of Sofia, Bulgaria.

Jadczyk, T., Kiwic, O., Khandwalla, R.M., Grabowski, K., Rudawski, S., Magaczewski, P., Benyahia, H., Wojakowski, W., Henry, T.D. (2019). Feasibility of a voice-enabled automated platform for medical data collection: Cardio Cube. *International Journal of Medical Informatics* 129:388–93. doi: 10.1016/j.ijmedinf.2019.07.001.

Kakria, P., Tripathi, N.K., Kitipawang, P. (2015).A real-time health monitoring system for remote cardiac patients using smartphone and wearable sensors. *International Journal of Telemedicine and Applications*, 2015:373474.

Kelly, J.T., Campbell, K.L., Gong, E., Scuffham, P. (2020). The Internet of Things: Impact and Implications for Health Care Delivery. *Journal of Medical Internet Research*, 22 (11) e20135.

Khan, J., Wei, J.S., Ringner, M., Saal L.H., Ladanyi M., Westermann F., Berthold F., Schwab M., Antonescu C.R., Peterson C., Meltzer P.S. (2001). Classification and diagnostic prediction of cancers using gene expression profiling and artificial neural networks. *Nature Medicine*, 7: 673–679. 10.1038/89044.

Majumder, S., Chen, L., Marinov, O., Chen, C., Mondal, T., Deen, M.J. (2018). Noncontact wearable wireless ECG systems for long-term monitoring. *IEEE Reviews in Biomedical Engineering*, 11:306–21.

Merchant, R.K., Inamdar, R., Quade, R.C. (2016). Effectiveness of population health management using the propeller health asthma platform: a randomized clinical trial. *The Journal of Allergy and Clinical Immunology. In Practice*, 4(3):455–63.

Merchant, R., Szefler, S.J., Bender, B.G., Tuffli, M., Barrett, M.A., Gondalia, R., Kaye, L., Van Sickle, D., Stempel, D.A. (2018). Impact of a digital health intervention on asthma resource utilization. *World Allergy Organization Journal*, 11 (1):28.

Moradi, E., Pepe, A., Gaser, C., Huttunen, H., Tohka, J., (2015) Alzheimer's Disease Neuroimaging Initiative. Machine learning framework for early MRI-based Alzheimer's conversion prediction in MCI subjects. *Neuro Image*, 104, 398–412. doi:10.1016/j.neuroimage.2014.10.002.

Naik, R., Macey, N., West, R.J., Godbehere, P., Thurston, S.C., Fox, R., Xiang, W., Kim, Y., Singh, I., Leadley, S., DiCarlo, L. (2017) First use of an ingestible sensor to manage uncontrolled blood pressure in primary practice: the UK hypertension registry. *Journal of Community Medicine and Health Education*, 07(1).

Nevin L. (2018). Advancing the beneficial use of machine learning in healthcare and medicine toward a community understanding. *PLoSMed* 15, 4–7.

Payan, A., Montana, G. (2015). Predicting alzheimer's disease: a neuroimaging study with 3dconvolutional neural networks. arXiv preprint arXiv:1502.02506

Peyvandipour, A., Saberian, N., Shafi, A., Donato, M., Draghici, S. (2018) "A novel computational approach for drug repurposing using systems biology," *Bioinformatics*, 34, 16:2817–2825.

Połap, D., Winnicka, A., Serwata, K., Kęsik, K., Woźniak, M. (2018). An intelligent system for monitoring skin diseases. *Sensors* (Basel), 18(8). doi: 10.3390/s18082552.

Prabusankarlal, K., Thirumoorthy, P., Manavalan, R. (2015). Assessment of combined textural and morphological features for diagnosis of breast masses in ultrasound. *Human Centric Computing Information Science* 5, 12. doi:10.1186/s13673-015-0029-y.

Pradhan, B., Bhattacharyya, S., Pal, K. (2021). IoT-Based Applications in Healthcare Devices. *Journal of Healthcare Engineering* 2021.

van der Putte, D., Boumans, R., Neerincx, M., Rikkert, M., de Mul, M.M. (2019). A Social Robot for Autonomous Health Data Acquisition Among Hospitalized Patients: An Exploratory Field Study. *14th ACM/IEEE International Conference on Human-Robot Interaction (HRI)*; HRI'19; March 11–14, 2019; Daegu, Korea (South), Korea (South). 2019.

Rahman, A., Dash, S., Ahmad M., Iqbal T., (2021). Mobile Cloud Computing: A Green perspective, In: Udgata S. K., Sethi S., Sriram S. N. (eds), *Intelligent Notes in Networks and Systems*, 185, 523–533, Springer, Singapore. doi:10.1007/978-981-33-6081-5-46.

Rahman, A., Dash, S., Kamaleldin, M., Abed, A., Alshaikhhussain, A., Motawei, H., Amoudi, N., Abahussain, W., (2019a). A Comprehensive study of mobile computing in Telemedicine, A. K. Luhach et al. (Eds):ICAICR 2018,CCIS 956, pp. 413–425, Springer.

Rahman, A., Dash, S., Luhach, A.K., Chilamkurti, N., Baek, S., Nam, Y., (2019b)A Neuro-Fuzzy Approach for User Behavior Classification and Prediction, *Journal of Cloud Computing: Advances, Systems and Applications*, 8(17), doi:10.1186/s13677-019-0144-9, Springer.

Rahman, A., Dash, S., Sultan, K., Khan, M.A., (2018). Management of Resource Usage in Mobile Cloud Computing, *International Journal of Pure and Applied Mathematics*, 119 (16), 255–261, ISSN: 1314-3395 (on-line version)

Ritschel, H., Seiderer, A., Janowski, K., Aslan, I., André, E. (2018). Drink-O-Mender: An Adaptive Robotic Drink Adviser. *Proceedings of the 3rd International Workshop on Multisensory Approaches to Human-Food Interaction; MHFI'18*; October 16, 2018; Boulder, Colorado.

Rumelhart, D.E., Hinton, G.E., Williams, R.J. (1986). Learning representations by back-propagating errors, *Nature*, 323:533–536.

Shimizu, E., Ogawa, Y., Yazu, H., Aketa, N., Yang, F., Yamane, M., Sato, Y., Kawakami, Y., Tsubota, K. (2019). Smart Eye Camera: An innovative technique to evaluate tear film breakup time in a murine dry eye disease model. *PLoS One*, 14(5):e0215130. doi: 10.1371/journal.pone.0215130.

Singh, K. R., Dash, S., Deka, B., Biswas, S. (2021). Mobile Technology Solution for COVID-19. In Fadi Al-Turjaman et al., (Eds), *Emerging Technologies for battling COVID-19 applications and innovations*, Springer, pp. 271–294 ISBN: 978-030-60038-9.

Sposaro, F., Tyson, G. (2020). iFall: an Android application for fall monitoring and response. *Annual International Conference of the IEEE Engineering in Medicine and Biology Society* (1557–170X (Print)) :a. doi: 10.1109/iembs.2009.5334912.

Suk, H.I., Lee, S.W., Shen, D., Initiative, A.D.N, et al. (2014). Hierarchical feature representation and multimodal fusion with deep learning for AD/MCI diagnosis. *Neuro Image* 101, 569–582.

Takemura, A., Shimizu, A., Hamamoto, K. (2010). Discrimination of Breast Tumors in Ultrasonic Images Using an Ensemble Classifier Based on the AdaBoost Algorithm With Feature Selection, *IEEE Transactions on Medical Imaging*, 29 (3):598–609, doi: 10.1109/TMI.2009.2022630.

Tibshirani, R., Hastie, T., Narasimhan, B., Chu, G. (2002). Diagnosis of multiple cancer types by shrunken centroids of gene expression. *PNAS*, 99: 6567–6572. doi: 10.1073/pnas.082099299

Tierney, M., Szalai, J., Snow, W., Fisher, R., Nores, A., Nadon, G., Dunn, E., George-Hyslop, P.S. (1996). Prediction of probable alzheimer's disease in memory-impaired patients aprospective longitudinal study. *Neurology* 46(3), 661–665.

S. Tuli, N. Basumatary, S. S. Gill, M. Kahani, R. C. Arya, G. S. Wander, and R. Buyya (2020). Healthfog: An ensemble deep learningbased smart healthcare system for automatic diagnosis of heartdiseases in integrated iot and fog computing environments," *Future Generation Computer Systems*, 104: 187–200.

Vapnik V.N., (1995). *The Nature of Statistical Learning Theory*, Springer-Verlag, Berlin.

Wan, S., Gu, Z., Ni, Q. (2020). Cognitive computing and wireless communications on the edge for healthcare service robots. *Computer Communications*, 149:99–106. doi: 10.1016/j.comcom.2019.10.012.

Wu, Q., Zhao, W. (2017). "Small-cell lung cancer detection using a supervised machine learning algorithm," in *Proceedings of the International Symposium on Computer Science and Intelligent Controls*, Budapest, Hungary.

Zhang, D., Shen, D., Initiative, A.D.N., et al. (2012). Multi-modal multi-task learning forjoint prediction of multiple regression and classification variables in alzheimer's disease. *Neuro Image* 59(2), 895–907.

18 Early Hypertensive Retinopathy Detection Using Improved Clustering Algorithm and Raspberry PI

Bhimavarapu Usharani
Koneru Lakshmaiah Education Foundation,
Vaddeswaram, India

CONTENTS

18.1 INTRODUCTION

Because of hypertension, hypertensive retinopathy (HR) harms the retina, then, harms the veins. It is vital for an early diagnosis of HR because of the possibility of damage to the cardiovascular or retinal microcirculation. These two diseases caused by HR usually are found in profoundly hypertensive patients. At the point when HR signs show up, most individuals have lost theirs sight. Thorough investigations suggest that retinal miniature vascular alterations might be discovered through

DOI: 10.1201/9780367548445-21

an advanced fundus camera. Hypertension (HPT) damages the blood vessels of the retina, back of the eye, where the image focus. if the pressure increases in the blood vessel then there is the chance of permanent sight loss. It's a non-contagious illness that is in the family of illness that includes the heart, kidneys. Also hypertension can lead to stroke [1]. While some hypertension victims show signs and indications like migraines, high energy, or temper problems, some show no outward signs [2]. Concurring with the World Health Organization, 1.13 billion people worldwide have hypertension, or under one in five are hypertension victims [3, 4]. Hypertension is capable of causing pivotal issues such as heart issues, etc. Medical expenditures are classified by different health service types, and medical condition provide information on the burden of direct medical expenditure imposed by a disease, the manner in which that burden is distributed, and also point to potential savings through changes in risk factors and reductions in incidence and prevalence of disease [5]. Hence, it is fundamental to make an early diagnosis of hypertension to limit the impact of hypertension issues for these patients. Notwithstanding, victims of hypertension don't show clear signs at the beginning of the disease. Then, at that point, secondary diagnoses of the hypertension may increase expenditures for other health services besides hospital care. The person is assumed fit in regard to arterioles throughout the body and consistently overextends them, so the disease advances to gradually hurt the heart, brain, or kidney organs. Consequently, it is hard to forecast any health issues in regard to hypertension. One pressing factor is an anomaly on the retina perceived as hypertensive retinopathy (HR). Early awareness in regard to the permanency of HR is fundamental when using alternative ways of prevention at the exclusion of medical treatment. The main objective is to reduce the cost of illness, but studies show there has been reliance on the first listed diagnosis to specify health services attributed to a disease, with the consequences of evaluating and dealing with the appearance of retinopathy, or subsequently dismissing a fundus picture assessment by consulting ophthalmologists.

In this way, the essential reason for that request is to reinforce a blatant exercise design for a completely robotized assessment of retinal photos, particularly with the law about HR eye-related infection. To see selective anomalies alongside fundus pictures, a preparation approach achieves awareness of potential savings attainable through effective targeted interventions than have more than four elite HR-related injuries. The classification of hypertension is discussed in Table 18.1.

There are quite a few grading systems because hypertensive retinopathy has been proposed in accordance with classification of its severity. One common classification

TABLE 18.1
Hypertensive Retinopathy (DR) Severity Levels

Severity Levels	Signs
Acute hypertension	Comprehensive arteriolar narrowing
Chronic hypertension	Focal Narrowing, arteriovenous nicking
Severe hypertension	Hemorrhages, exudates, cotton wool spots
Accelerated hypertension	Optic disc swelling, choriocapillaris, choroidal arterioles

TABLE 18.2

Keith Wagener Barker Classification of Hypertensive Retinopathy

Grade	Severity Specifications
Grade 1	Mild generalized retinal arteriolar narrowing
Grade 2	Definite focal narrowing and arteriovenous nipping
Grade 3	Signs of grade 2 retinopathy plus retinal hemorrhages, exudates, and cotton wool spots
Grade 4	Severe grade 3 retinopathy plus papilledema

TABLE 18.3

Wong and Mitchell Classification of Hypertensive Retinopathy

Grade	Severity Specifications
Mild	Generalized arteriolar narrowing, focal arteriolar narrowing, arteriovenous nicking, arteriolar wall opacity
Moderate	Retinal hemorrhage (blot, dot, flame shaped), microaneurysms, cotton wool spot, hard exudates
Severe	Retinal hemorrhage (blot, dot, flame shaped), microaneurysms, cotton wool spot, hard exudates, optic disc swelling

is the Keith Wagener Barker array proposed in 1939 [6] and mentioned in Table 18.2. Wong-Mitchell recently posed a simplified grading provision in 2004 [7] mentioned in Table 18.3.

The vascular changes, along with the blood pressure, are observable between the retinas as hypertensive retinopathy [8]. Early ends contain broad narrowing on the retinal arteriolar vessels proper, in similarity with vasospasm than with improved vascular tone. Ongoing hypertension drives as per primary adjustments inside the boat divider certain so intimal thickening yet hyaline degeneration. These are clearly umbilical and disseminate areas of veins railing opacification, known as copper silver wiring. Serious hypertension creates focal regions on ischemia of the retinal nerve fiber layer, and show up specifically as love coat spots. Disappointment regarding the carnage remains unresolved, with no proper treatment for this hypertension that may not be detected because of seeing lipids as hard exudates, thus with scant or no recovery for the patient. Extreme hypertensions do organization as per a dim intracranial pressing factor, incurring optical nerve ischemia, then optic ring expanding [9–12]. Figure 18.1 shows the hypertension ex retinopathy.

The rest of this paper is organized as follows. Section 18.2 discusses the preliminary part of the research. Section 18.3 presents the necessary background for this study to employ hypertension detection and classification. Section 18.4 discusses the proposed work along with the improved clustering technique. Section 18.5 shows the experimental results and comparisons to state-of-the-art HR systems. Finally, the Section 18.6 concludes the paper.

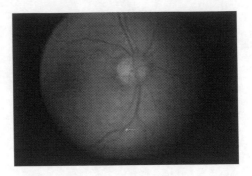

FIGURE 18.1　Retina image of hypertensive retinopathy.

18.2　PRELIMINARIES

18.2.1　Particle Swarm Optimization Clustering

Clustering is an innovation that bunches comparative information focuses into a similar group from the given informational collection [13]. A few uses of the bunching calculations are picture division [14], design investigation [15], picture pressure [16]. Bunching dependent on swarm insight, i.e., PSO can be utilized for grouping examination [17]. Particle Swarm Optimization-based calculations consider datapoints as particles. The underlying bunches of particles are gathered from other grouping calculations. The bunch of particles is refreshed ceaselessly, dependent on the focal point of the group's groups, speed, and position until the group place unites.

Particle Swarm Optimization Clustering Algorithm [17]

1. Evaluate the fitness measure for each particle in each cluster. $\forall x_i \epsilon C_j$ (C_j is a cluster j)
2. Evaluate the velocity and position updates for each particle in a cluster $\forall C_j$.
3. Replace Pbest with individual i^{th} cluster centre and Gbest with neighbor cluster centres.

Particle Swarm Optimization clustering algorithm iterates until it reaches maximum iteration, or it converges. The formulas for velocity updates, position updates, Pbest, Gbest are already discussed in the above section.

The cluster centroids used for the Particle Swarm Optimization image clustering algorithm [18] are:

(a) $vi(t+1) = wvi(t) + c1r1(t)\big(yi(t) - xi(t)\big) + c2r2(t)\big(\hat{y}(t) - xi(t)\big)$

(b) $xi(t+1) = xi(t) + vi(t+1)$

18.2.2　Raspberry PI

Raspberry pi is a small computer and uses many kinds of processors. By using the raspberry pi, one performs net browsing, sends emails, and many other actions.

FIGURE 18.2 Raspberry Kit.

Raspberry PI is a very low-cost embedded system and provides many features, like amazing image quality, quality audio, capable of 3D games and many more. As the raspberry pi is low cost and uses the open-source software components, it works efficiently for real time applications. Raspberry pi processes the images and optimizes the results. Raspberry pi improves the quality of the output and reduces the cost of the processing for the medical image's applications. Raspberry pi supports programming languages like python, C, C++, Java, and Ruby languages. The Raspberry Kit is shown in Figure 18.2.

18.3 RELATED WORK

The author has studied existing articles approving neural networks patterns after a conclusion of hypertension, and also half discreditable studies presented with classification accuracy but with logistic regression [19–21]. In each paper, the expansion procedure, characteristic collection, floor truth definition, training information sets, look at statistics sets, overfitting avoidance, confusion matrix, and exactness records were assessed. Also assessed agreement as to whether the types had been supported yet not, both by a lawful record accepted, yet with the aid of a panel of authorities within the area [22–25].

La Freniere et al. [26] introduced an artificial neural network (ANN) in conformity with foretell hypertensive sufferers employing the Canadian Primary Care Sentinel Surveillance Network (CPCSSN) data set. The impartial services constants were age, femininity, BMI, systolic and diastolic blood pressure, excessive yet low-density lipoproteins, triglycerides, cholesterol, microalbumin, then water albumin–creatinine ratio. Confusion casting or Receiver Operating Characteristic (ROC) nook have been employed accordingly to calculate the correctness on the paradigm.

Polak and Mendyk [27] enhanced and tried an artificial neural network because of the risk of excessive gore strain with the use of statistics out of the Center of Disease Control and the National Center for Health Statistics (CDC-NCHS). The impartial persistent purposes of its model have been age group, gender, nutrition, smoke, or consumption patterns, physical activity level, and BMI index. ROC turn was applied in conformity with the accurateness of the model, or it honored a contrast along a logistic regression alignment standard.

Tang et al. [28] used an artificial neural network because of cardiovascular ailment together with hypertension; that bill of exchange employed Chinese inhabitants. Statistical analysis revealed that fourteen danger issues confirmed statistical regard together with cardiovascular disease. The ROC curve is employed after tolerating the presentation regarding the standard.

Ture et al. [29] employed a multilayer perceptron for the array on hypertensive patients. The unbiased aspects employed had been stage, masculinity, household description over hypertension, smoking tendencies, lipoprotein, triglyceride, uric acid, quantity cholesterol, and physique matter index (BMI). ROC turn is applied in conformity with absolving the precision over the version.

Lynn et al. [30] built a neural network version in accordance with the assumption of the gene endophenotype-disease connection for Taiwanese hypertensive males, using the aspects age group, crossing gore, hypertension treatment, then lung. Categorization rigor is employed after assessing the overall execution on the standard.

Sakr et al. [31] formed a neural network to evaluate and implement the overall execution, including special computing device instruction strategies on forecasting the threat about flourishing hypertension. Age group, masculinity strength exams, and scientific history is old because of old categorization. ROC is employed according to metering the propriety on the standard.

The AVR algorithm has been applied on the retina color fundus images on datasets like DCCT, ETDRS; the AVR ratio has been considered by gauging the blood vessel diameter through the edge detection procedure by segmenting the optic disk area and blood vessel organization into arteries and veins [32–35].

18.4 METHODOLOGY

The workflow of the detection and the classification of the hypertension retinopathy is depicted in Figure 18.3.

18.4.1 PRE-PROCESSING

The pre-preparing stage helps improve the nature of the retina picture. The spatial space methods are the procedures that work on pixels. The upside of the spatial space strategies is proficient calculation and requiring fewer handling assets. In a pixel-based methodology, the information data is simply the pixel. After applying the pixel-based

FIGURE 18.3 Procedure flow for detecting the hypertension.

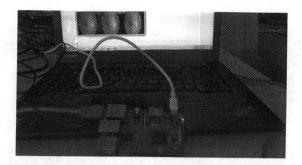

FIGURE 18.4 Preprocessed retina Image (a) original Image (b) Gray Image (c) After Preprocessing.

methodology, the picture created is of high contrast, and these improvement procedures rely upon the Gray levels [36]. In this investigation, the pre-preparing step utilizes histogram handling, as this method chips away at pixels. The stages in pre-handling utilized are resizing, transformation of Red Blue Green (R GB) picture to Gray picture, applying the Contrast Limited Adaptive Histogram Equalization (CLAHE) [37], strategies utilized [38–40]. The pre-handling period of the retina picture is shown in Figure 18.4.

18.4.2 SEGMENTATION

The primary target of the division is to revise or change the retina picture's appearance to make it more exact, easier to separate, and to concentrate the highlights of the retina picture. In this work, improved Particle Swarm Optimization (PSO) grouping is utilized to fragment the pre-prepared picture. The division stage helps for the identification of the microaneurysms from the retina picture. After applying the improved particle swarm optimization enhancement bunching, the outcomes are shown in Figure 18.5.

18.4.2.1 Elevated Continuous Particle Swarm Optimization Clustering

Elevated continuous particle swarm optimization clustering is used in this paper to effectively segment the microaneurysms.

ALGORITHM 1:   ELEVATED CONTINUOUS PARTICLE SWARM OPTIMIZATION (ECPSO)

Result: G_{best} of the whole swarm
Initialization: Generate 2*Swarm size particles

1. Generate particles based on swarm size.
2. Generate clusters using a centroid updating scheme. Evaluate fitness measures for the particles.
3. For each particle, apply the neighborhood search scheme. Iterate for all particles.
4. For each particle, perform update position, velocity vectors. Apply a centroid updating scheme and evaluate fitness measures. Apply a neighbor search scheme for every particle.
5. Return Gbest.

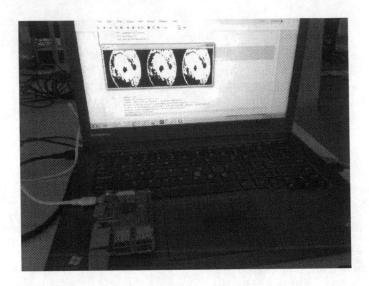

FIGURE 18.5   Segmented retina image.

18.4.2.2   Feature Extraction

In the element extraction phase of the proposed framework, separate the injuries from the portioned picture. A bunch of highlights, including region, border, circularity, and width, are utilized to remove sores. Immaterial and excess highlights ought to be disposed of from the portioned picture to improve precision. The highlights are recorded in Table 18.4. The fundus picture after applying the feature extraction is depicted in Figure 18.6.

TABLE 18.4
List of Features

S.No	Feature Description	Formula
f1	Entropy	$\sum_{i=0}^{\infty}\sum_{j=0}^{\infty}\dfrac{p}{\log p}$
f2	Max. Probability	$\mathrm{Max}\{p_{ij}(x,y)\}$
f3	Energy	$\sum\sum P_{ij}{}^2$
f4	Mean	$\sum_{i=0}^{\infty}\sum_{j=0}^{\infty}\dfrac{p}{NXM}$
f5	Variance	$\sum_{i=0}^{\infty}\sum_{j=0}^{\infty}(P(I\text{-}J)\text{-}\mu)^2$
f6	Dissimilarity	$\sum_{i=0}^{\infty}\sum_{j=0}^{\infty}P(I\text{-}J)$
f7	Contrast	$\sum_{i=0}^{\infty}\sum_{j=0}^{\infty}P(I\text{-}J)^2$

FIGURE 18.6 Feature extraction of the retina image.

18.5 EXPERIMENTAL RESULTS

Assessed the detection of the microaneurysms identification method on the freely accessible informational collections: Inspire, Drive, and STARE. For experimentation, hardware used is Intel core i5 3.4 GHz processor, and to implement the experiments IDE Anaconda was used for python. The metric levels used in this study are given in Table 18.5.

Execution estimates utilized in the proposed method are examined in Table 18.6. Looked at the precision, affectability, and explicitness of the current strategy with other research in Table 18.7. The measurement properties used to gauge the similitude and the conveyance of the information are kurtosis and skewness. Kurtosis affirms the evenness of the information conveyance. To recognize the evenness of the information dispersion in the bunch, we are utilizing the actions kurtosis and skewness to evaluate the expectation about the dissemination of the information in the groups.

The current work is contrasted and the current methodologies are addressed in Figure 18.8. The kurtosis and skewness measures are utilized as the precision measures. These factual measures for the current work are contrasted and the current PSO calculation is given these qualities in Table 18.7.

TABLE 18.5
Metric Levels

Metric	Description
True Positive (TP)	Count of correctly classified microaneurysms pixels
True Negative (TN)	Count of correctly classified non-microaneurysms pixels
False Positive (FP)	Misclassified count of non-microaneurysms treated as microaneurysms
False Negative (FN)	Misclassified count of microaneurysms treated as non

TABLE 18.6
Performance Metrics

Measure	Formulae
Sensitivity	$\dfrac{TP}{(TP+FN)}$
Specificity	$\dfrac{TN}{(TN+FP)}$
False positive rate	$\dfrac{FP}{(FP+TN)}$
False Negative Rate	$\dfrac{FN}{(TP+FN)}$
Accuracy (Acc)	$\dfrac{(TP+FN)}{(TP+FP+TN+FN)}$

TABLE 18.7
Comparison of Similar Measures of EPSO and PSO Clustering

Image Id	EPSO				PSO			
	Entropy	Kurtosis	Skewness	Runtime	Entropy	Kurtosis	Skewness	Runtime
Image001	4.28	5.56	2.44	15.64	5.34	6.45	3.54	17.49
Image002	3.25	2.41	2.54	14.45	4.67	3.45	3.65	18.93
Image003	3.59	3.15	2.13	16.49	4.76	4.64	3.56	20.48
Image004	3.21	0.12	0.99	16.58	4.85	2.45	2.65	21.89
Image005	1.13	0.65	1.09	15.84	3.67	2.76	2.67	16.85
Image006	1.07	0.99	0.13	16.19	3.39	1.94	3.57	19.56
Image007	2.54	0.99	0.98	14.01	3.67	3.56	2.67	18.76
Image008	3.31	1.15	1.37	13.11	4.87	3.75	2.49	18.45
Image009	3.86	−1.85	0.35	14.38	4.83	0.56	2.69	19.76
Image010	2.45	−1.71	0.44	16.31	3.92	0.27	2.57	19.56
Image011	3.56	−1.79	0.48	15.47	4.25	0.16	2.18	19.62
Image012	3.68	−1.87	0.45	18.94	4.67	0.75	2.45	23.56
Image013	2.78	−1.91	0.41	16.62	3.27	0.35	2.64	20.53
Image014	3.59	−2.11	0.65	18.16	4.73	0.77	2.69	22.64
Image015	3.45	0.99	0.98	17.85	4.83	2.59	2.92	21.68
Image016	3.47	−1.63	0.15	16.54	4.52	0.58	2.74	19.65
Image017	3.84	−1.97	0.07	15.69	4.79	0.63	2.35	18.48
Image018	3.45	−1.99	0.28	16.61	4.72	0.67	2.45	20.31
Image019	3.47	2.12	2.01	15.62	4.62	4.26	3.54	19.56
Image020	3.18	−1.69	0.53	15.69	4.16	0.52	2.55	19.45

18.5.1 PERFORMANCE ANALYSIS

For assessing the proposed model, a successive model was utilized for EPSO clustering. The presentation of the EPSO clustering model has been approved on the freely accessible datasets Inspire, Drive, and STARE. The exactness of various existing classifiers is contrasted and the current work is outlined in Figure 18.7. Every one

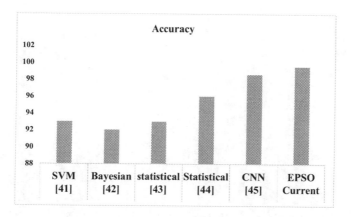

FIGURE 18.7   Comparison of various classifiers with the EPSO algorithm.

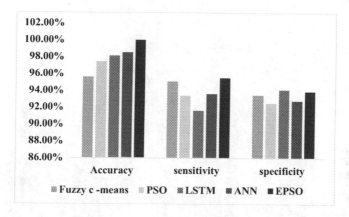

FIGURE 18.8   Comparison of different classifiers on the INSPIRE dataset.

of the strategies that are considered in the INSPIRE information bases are given in Figure 18.8. Every one of the techniques are reviewed in the DRIVE information base that is given in Figure 18.9. Every one of the strategies are examined in the STARE data sets that are given in Figure 18.10.

The proposed approach incredibly is more accurate than other methods; on the other hand in our future work, we would enhance the truth along, and encompass, a gore ark of retina, namely an input characteristic regarding EPSO clustering; However, the array now is not solely a few classification that seem normal but with signs and symptoms concerning HR, but also the array about HR values stretched over IV grades over HR.

The consequences of trials on datasets that are freely accessible show that the proposed procedure outperforms prestigious strategies as far as high request insights measures. To analyze bunches, utilize high request measurements, for example, kurtosis and skewness. These two high request factual measures give a more exact portrayal of the state of the bunch. The primary test in picture bunching is the way to control groups dependent on the past complex portrayals and to distinguish the new bunch. Existing bunch calculations utilized the mean, middle, change, and

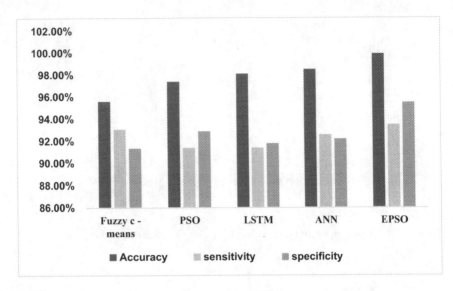

FIGURE 18.9 Comparison of different classifiers on the DRIVE dataset.

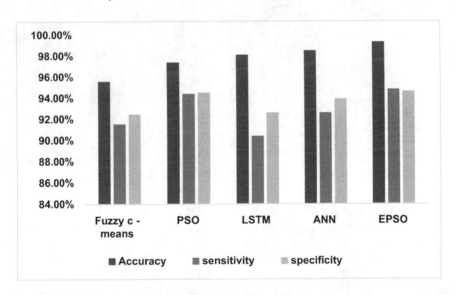

FIGURE 18.10 Comparison of different classifiers on the STARE dataset.

covariance to address the group. This work utilized kurtosis and skewness for the portrayal of the groups. The fundamental benefit of utilizing these two high request insights is to distinguish a low intricacy bunch. The skewness addresses the unevenness of the group, while the kurtosis addresses the centralization of the bunch in the event that, assuming the two unique groups have comparative measurable properties, the bunches are blended. The difficult assignment is the location of new bunches and converging among old and new groups. Conglomeration of bunches is by inspecting the measurements of the entropy, kurtosis, and skewness of the groups. The closeness

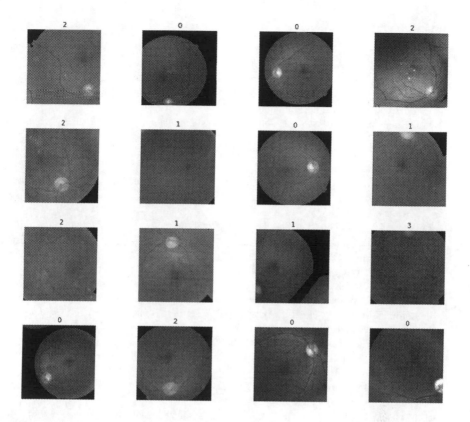

FIGURE 18.11 Classification of severity levels of the hypertensive retinopathy.

measures are examined in Table 18.7. The EPSO classifier is used to arrange the various grades of the hypertensive retinopathy. The various grades of the hypertensive retinopathy are displayed in Figure 18.11.

18.6 CONCLUSION

An unsupervised five class order of hypertensive retinopathy procedure is introduced in this chapter. This chapter has introduced the recognition and grouping model for the fundus pictures utilizing the EPSO clustering algorithm. For execution investigation the separated highlights have given to the classifiers Fuzzy C-Means, PSO to assess the presentation. As per the exhibition investigation, the EPSO classifier beats the wide range of various classifiers for the Inspire, Drive, and STARE data sets. Hence, the normal accuracy of the framework for multi characterization is 98.47%. The outcomes guarantee that the discovery and the order of the hypertensive retinopathy execution of the introduced EPSO clustering algorithm on the freely accessible dataset will be delivered with extreme accuracy, sensitivity, and the specificity of 97.99. Later on, the current framework can be utilized to naturally distinguish the other retinal problems at the beginning phase of the hypertensive retinopathy.

REFERENCES

1. Drozdz, D., Kawecka-Jaszcz, K., Cardiovascular changes during chronic hypertensive states, *Pediatric Nephrology* 29 (9) (2014) 1507–1516.
2. Goodhart, A.K., Hypertension from the patient's perspective, *The British Journal of General Practice* 66 (652) (2016), 570.
3. Hypertension, World Health Organisation
4. Williams, B., et al., ESC/ESH Guidelines for the management of arterial *Hypertension*, 39, 2018, 33.
5. Hodgson, T. A., & Cai, L. Medical care expenditures for hypertension, its complications, and its comorbidities *Medical Care*, 2001: 599–615.
6. Keith, N.M., Wagener, H.P., Barker, N.W. (1939) Some different types of essential hypertension: Their course and prognosis. *The American Journal of the Medical Sciences*, 197:332–343
7. Wong, T.Y., Mitchell, P. (2004). Hypertensive retinopathy. *The New England Journal of Medicine*, 351:2310–2317.
8. López-Martínez, F., Schwarcz, M. D. A., Núñez-Valdez, E. R. & García-Díaz, V. Machine learning classifcation analysis for a hypertensive population as a function of several risk factors. *Expert Systems with Applications* 110, 206–215.
9. Wong, T.Y., Mitchell, P. (2007). The eye in hypertension. *Lancet*, 369:425–435.
10. Keith, N.M., Wagener, H.P., Barker, N.W. (1939). Some different types of essential hypertension: their course and prognosis. *The American Journal of the Medical Sciences*, 197:332–343. doi:10.1097/00000441-193903000-00006
11. Wong, T.Y., Mitchell, P. (2004). Hypertensive retinopathy. *The New England Journal of Medicine*, 351:2310–2317
12. Fraser-Bell, S., Symes, R., Vaze, A. (2017). Hypertensive eye disease: A review. *Clinical & Experimental Ophthalmology*, 45(1):45–53.
13. Michael, R.A. (1973). The broad view of cluster analysis. *Cluster Analysis for Applications*, 1–9.
14. Ray, S. and Turi, R.H. "Determination of number of clusters in k-means clustering and application in colour image segmentation". In *Proceedings of the 4th international conference on advances in pattern recognition and digital techniques*, pages 137–143. Calcutta, India, 1999.
15. Li, X. and Fang, Z. (1989). Parallel clustering algorithms. *Parallel Computing*, 11(3):275–290.
16. Richard, C.T. (1981). Clustering analysis and its applications. In *Advances in Information Systems Science*, 169–292. Springer.
17. Kennedy, J. Stereotyping: Improving particle swarm performance with cluster analysis". In *Proceedings of the 2000 Congress on Evolutionary Computation*. CEC00 (Cat. No. 00TH8512), volume 2, pages 1507–1512. IEEE, 2000.
18. Mahamed, G., Engelbrecht, A.P., and Salman, A. (2004). Image classification using particle swarm optimization. In *Recent advances in simulated evolution and learning*, 347–365. World Scientific.
19. Seidler, T. et al. A machine learning approach for the prediction of pulmonary hypertension. *Journal of the American College of Cardiology* 73, 1589. (2019).
20. Ambale-Venkatesh, B. et al. Cardiovascular event prediction by machine learning: the multi-ethnic study of atherosclerosis. *Circulation Research* 121, 1092–1101. (2017).
21. Mortazavi, B. J. et al. Analysis of machine learning techniques for heart failure readmissions. *Circulation: Cardiovascular Quality and Outcomes* 9, 629–640. (2016).

22. Debray, T. P. A. et al. A new framework to enhance the interpretation of external validation studies of clinical prediction models. *Journal of Clinical Epidemiology* 68, 279–289 (2015).

23. Tengnah, M. A. J., Sooklall, R. & Nagowah, S. D. A predictive model for hypertension diagnosis using machine learning techniques. In *Telemedicine Technologies* (eds Jude, H. D. & Balas, V. E.) 139–152 (Academies Press, Elsevier, 2019).

24. Clim, A., Zota, R. D. & Tinica, G. Te Kullback–Leibler divergence used in machine learning algorithms for health care applications and hypertension prediction: a literature review. *Procedia Computer Science* 141, 448–453 (2018).

25. Singh, N., Singh, P. & Bhagat, D. A rule extraction approach from support vector machines for diagnosing hypertension among diabetics. *Expert Systems with Applications* 130, 188–205.

26. LaFreniere, D., Zulkernine, F., Barber, D., & Martin, K. Using machine learning to predict hypertension from a clinical dataset. In *2016 IEEE Symposium Series on Computational Intelligence (SSCI)*, 1–7 (2016).

27. Polak, S. & Mendyk, A. Artificial neural networks-based Internet hypertension prediction tool development and validation. *Applied Soft Computing* 8, 734–739. (2008).

28. Tang, Z.-H. et al. Comparison of prediction model for cardiovascular autonomic dysfunction using artificial neural network and logistic regression analysis. *PLoS One* 8, e70571. (2013).

29. Ture, M., Kurt, I., Turhan Kurum, A. & Ozdamar, K. Comparing classification techniques for predicting essential hypertension. *Expert Systems with Applications* 29, 583–588. (2005).

30. Lynn, K. S. et al. A neural network model for constructing endophenotypes of common complex diseases: an application to male young-onset hypertension microarray data. *Bioinformatics* 25, 981–988. (2009)

31. Sakr, S. et al. Using machine learning on cardiorespiratory fitness data for predicting hypertension: Te Henry Ford exercise testing (FIT) Project. *PLoS One* 13, e0195344.

32. Manikis, G.C., Sakkalis, V., Zabulis, X., Karamaounas, P., Triantafyllou, A., Douma, S., Zamboulis, C., Marias, K. (2011) An image analysis framework for the early assessment of hypertensive retinopathy signs. *E-Health and Bioengineering Conference (EHB)* 2011:1–6

33. Muramatsu, C., Hatanaka, Y., Iwase, T., Hara, T., Fujita, H. (2011) Automated selection of major arteries and veins for measurement of arteriolar-to-venular diameter ratio on retinal fundus images. *Computerized Medical Imaging and Graphics: The Official Journal of the Computerized Medical Imaging Society* 35(6): 472–480.

34. Narasimhan, K., Neha, V.C., Vijayarekha, K. (2012) Hypertensive retinopathy diagnosis from fundus images by estimation of AVR. *Procedia Engineering* 38:980–993.

35. Saez, M., González-Vázquez, S., Penedo, M.G., Barceló, M.A., Pena-Seijo, M., Tuero, G.C., Pose-Reino, A. (2012) Development of an automated system to classify retinal vessels into arteries and veins. *Computer Methods and Programs in Biomedicine* 108(1):367–376.

36. Burger, W., Burge, M. J., Burge, M. J., and Burge, M. J. *Principles of digital image processing*, volume 54. Springer-Verlag, London, 2009.

37. Duan, J. and Qiu, G. Novel histogram processing for color image enhancement. In *Third International Conference on Image and Graphics (ICIG'04)*, 55–58. IEEE, 2004.

38. Zimmerman, J. B., Pizer, S. M., Staab, E. V., Perry, J. R., McCartney, W., and Brenton, B. C. An evaluation of the effectiveness of adaptive histogram equalization for contrast enhancement. *IEEE Transactions on Medical Imaging*, 7 (4):304–312, 1988.

39. Lin, T. M. Du, and J. Xu. The preprocessing of subtraction and the enhancement for biomedical image of retinal blood vessels. Sheng wu yi xue gong cheng xue za zhi=. *Journal of Biomedical Engineering=Shengwu yixue gongchengxue zazhi*, 20(1):56–59, 2003.

40. Kerre, E. E. and Nachtegael, M. Fuzzy techniques in image processing. *Physica*, 52 2013.

41. Narasimhan, K., V.C. Neha, K. Vijayarekha (2012) "Hypertensive Retinopathy Diagnosis from fundus images by estimation of AVR", *International Conference on Modeling Optimization and Computing (IC- MOC), Procedia Engineering*, 38: 980–993

42. Noronha, K., Navya, K.T., Nayak, K.P. (2012) "Support System for the Automated Detection of Hypertensive Retinopathy using Fundus Images", *International Conference on Electronic Design and Signal Processing (ICEDSP)*, pp. 7–11.

43. Manikis, Georgios C., Sakkalis, Vangelis, Zabulis, Xenophon, Karamaounas, Polykarpos, Triantafyllou, Areti, Douma, Stella (2011) "An Image Analysis Framework for the Early Assessment of Hypertensive Retinopathy Signs", *Proceedings of the 3rd International Conference on E-Health and Bioengineering – EHB*, 24–26 November, 2011, Iaşi, Romania.

44. Mirsharif, Q., Tajeripour, F., Pourreza, H. (2013) Automated characterization of blood vessels as arteries and veins in retinal images, *Computerized Medical Imaging and Graphics*, Vol. 37, pp. 607–617.

45. Triwijoyo, B.K., Budiharto, W., Abdurachman, E., (2017) The classification of hypertensive retinopathy using convolutional neural network, *Procedia Computer Science*, 116 166–173.

19 IoT Based Elderly Patient Care System Architecture

K. Rupabanta Singh and Sujata Dash

Maharaja Srirama Chandra Bhanja Deo University,
Odisha, India

CONTENTS

19.1 INTRODUCTION

The modern healthcare system has been changed by the IoT and wireless technologies. As the demands of creating a smooth progression of treatment for patients are increased day-by-day, the IoT-based healthcare system has been developed. The IoT-based healthcare system provides different types of services to fulfil the elderly and lonely patients' needs. Many IoT sensors are used to collect data from the patients, and the sensor's data is sent to the server for performing data analysis and ensuring the remote monitoring of the patients. Many authors have proposed variants of healthcare architecture, depending on the data transmission and data analysis is performed in the system. The customized 6LoWPAN network is used to share data from sensors to the cloud through a gateway [1]. The system uses an Analog-to-Digital converter; it convert the analog data to digital data, Micro-controller unit (MCU), which performs the preprocessing task and making 6LoWPAN packets, and RF transceivers; thus, it transmits the 6LoWPAN packets of data to the gateway. The major role of the gateway is to deliver the data to the cloud server. The cloud server stores the health data and performs the data analysis. It provides an interface to display the health information to the client.

The healthcare system architecture integrates the architectures of the IoT and the cloud. The IoT is the interconnection of the objects or the devices in the network. The objects in the network are communicated to each other, for short-rage communication devices (like Bluetooth, RFID, ZigBee, etc.) that have been used in the network. The IoT framework enables humans to interact with the devices around us. Some of these devices use sensors, and they can generate data and share it in the network. The architecture of the IoT framework has four layers, and it is shown in the Figure 19.1. The perception layer includes all the sensors, RFID, Bluetooth, etc.

The function of this layer is to generate the sensors' data and transmit it to the network layer, which includes different networking protocols that enable the transmission of the sensors' data to the middleware layer. WIFI, GPS, ISDN, 5G, etc., are available in the network layer. The middleware layer performs the data pre-processing and transmits the data to the application layer. The application layer provides the interface to the users and enables the user to interact with the system. The application may be a Smart Home application, Smart City application, e-health application, and so on. The popularity of the IoT is increased by the help of cloud computing. Cloud computing provides different types of services, and some of the services are worthy of the healthcare system. A cloud server stores data, performs data processing and analysis, and reduces the management tasks and cost of the maintenance of the cloud user. Cloud computing plays a vital role in the healthcare system, increasing the QoS of the healthcare system. Remote monitoring of patients is possible in a healthcare system because of the integration of cloud computing technology into the healthcare system. The health data availability is also increased, and the physician can access the patient information any time from any place. The cloud computing model is shown in the Figure 19.2. The cloud server provides three services: Software as a Service (SaaS), Platform as a Service, and Infrastructure as a Service (IaaS). Software as a Service provides the cloud software that can be accessed by the client, including database software that can store a huge amount of healthcare data for processing. Platform as a Service ensures the client can manage the application, as well as develop the application. Collaborative work can be performed with the help of PaaS.

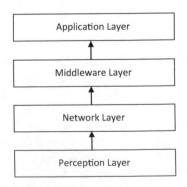

FIGURE 19.1   Four layer IoT framework.

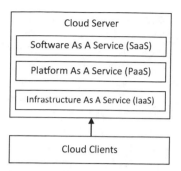

FIGURE 19.2   Cloud computing model.

Infrastructure as a Service provides the virtual machine, thus, the client can own the high configuration machine at low cost. The infrastructure provided by the cloud server is scalable, and the storage capacity can be increased or decreased anytime.

Arduino is a less-expensive IoT platform and many authors have been using Arduino to develop an IoT-based healthcare system. The author [2] has used an Intel Curie for a healthcare system, and it is attached to the Arduino. The curie module is able to interact directly with the server or via PC. The curie module can perform basic data processing and forward the health data to the server for further processing. The server will store the health information, and the end user can retrieve the health information either by using the web application on the PC or through a mobile application. The mobile health system [3] can collect health data, using mobile devices from patients in real-time, and send the health data to the server through the internet. The mobile health system architecture has three layers, as shown in the Figure 19.3. All the health sensing devices are included in the first layer, and these sensors have collected the health data from the patients. The responsibility of the second layer is to store the health data for further processing, and it ensures the end user to get the health information through a mobile application.

The third layer involves all the data processing and analyzing algorithms. Nowadays, machine learning based algorithms are widely used for data analysis activities. In these three layers of the healthcare system architecture, data storage and data processing are the parts of the cloud server. The sensors collect a huge amount of data, which needs to be transmitted to the cloud server for data storing and

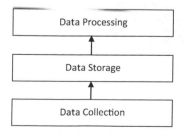

FIGURE 19.3   Three layers mobile health system architecture.

processing. Consequently, high bandwidth on the Internet is needed to transmit health data to the cloud server. As the cloud server is located far from the sensor devices, the data transmission latency is high, which can create failure in real-time processing. In order to reduce the latency, the author [4] used Fog computing in the healthcare system. A fog server is placed near the end users, so that it can collect and store health data from the sensors. The end user is equipped with different medical sensors, which collect health data. The sensors send the health data to the Fog server or primary health center through the LoRaWAN gateway. The LoRaWAN gateway enables the sensors to transmit data within tens of kilometers. Although the villages reached don't have a proper internet facility, they can still implement this fog-based healthcare system. The Fog node will store the data and can also perform the analysis task. When the end user requires emergency services or the illness becomes serious, the health center provides the services. The physicians are also directly connected with the Fog node, so they can easily track their patient health information. If the patient needs emergency treatment, then the health center will send the ambulance immediately. The Fog-based system architecture is shown in Figure 19.4. The system architecture has three layers: the IoT layer, the Fog computing layer, and the cloud layer. The bottom layer, i.e., the IoT layer, involves all the medical sensors that collect health data and share it with the above layers. The middle layer is the Fog computing layer, involving the Fog nodes and the Fog servers, which are placed near the end users. The Fog node can give responses to the end user request. Physicians can also directly access the health information which is stored on the Fog node. The top layer is the cloud computing layer; its server ensures the remote monitoring of the patients through the Internet. It stores the health data and can also perform the data analysis operation.

Today, the smart phone plays a vital role in the communication and computing architecture. Many authors are beginning to be interested in applying mobile computing in the healthcare sector [20, 21]. Varieties of wearable IoT health sensors have been developed, and the Body Sensors Network (BSN) can be constructed. This BSN can connect with a personal smart phone and share health data. Since the smart phone has Internet facility, it acts as the Edge node and transmits the health data to the cloud for remote monitoring and data processing. Nowadays, Deep Learning is commonly used for medical image processing to diagnose a disease [22]. Very recently, Edge computing has been applied in the IoT-based healthcare system, so that the system reduced the latency and ensured the ability to perform real-time operations. Edge extends the healthcare services, as the users may be working in an industry, playing sports, or are patients are staying in the hospital. The main advantages of using Edge computing in the healthcare system are reducing latency, enabling real-time processing, and increasing the patients' mobility. The three-tier architecture of the Edge computing based health care system is shown in the Figure 19.5 [5].

The rest of the chapter is organized as follows: Section 19.2 defines the healthcare system without patient mobility support, Section 19.3 defines the healthcare system with patient mobility support, Section 19.4 explores the existing architectures of the healthcare system, Section 19.5 describes the proposed architecture, Section 19.6 is the discussion about the healthcare system, and Section 19.7 concludes the chapter.

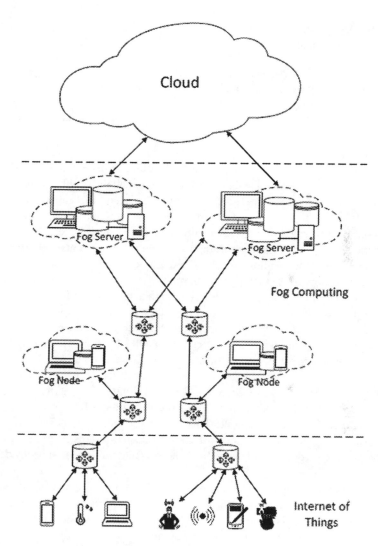

FIGURE 19.4 Fog based healthcare system architecture.

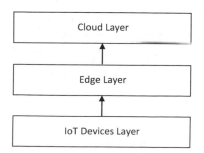

FIGURE 19.5 Edge based healthcare system architecture.

19.2   HEALTHCARE SYSTEM WITHOUT PATIENT MOBILITY SUPPORT

The lifestyle of our generation has changed from the last few decades. Most people are living independently and separately from their parents. When the parents become old and sick, they need family to take care of them. But some people have to live alone, although they are getting infirm. The IoT based health monitoring system will help these lonely people when they are sick. Many authors have proposed variants of the health monitoring systems.

The author [9] developed a system that can monitor a patient continuously. The system uses Raspberry pi as a local processing unit and all the health sensors are connected with Raspberry pi. These sensors automatically collect health data from the patient and send it to the server for further processing. The system monitors the patient continuously and transmits the health data in real-time. Patient data, which is store in the server, can be access through a web application. This web application performs the user authentication process in order to keep the patient data secure. The three tier architecture of the system without mobility support is shown in Figure 19.6. The bottom tier is the wired sensor network, which is responsible for collecting samples from the patient. The patient is directly interactive with this wired sensor network, and the collected sample is transferred above, to the middle tier. The middle tier is the processing unit. It includes different model like processing model, decision model, etc. Raspberry pie is used in this middle tier to perform a variety of data processing tasks. The topmost tier is the hardware control tier, which includes the web application that ensures remote access of the patient data, like heartbeat rate, body temperature, etc. This tier enables remote monitoring of the patient so that the patient will get emergency service if required. This system is suitable for those heart patients who are staying alone at home. Some of the healthcare systems don't allow patient mobility, so this model is applicable to those patients who must stay inside their homes.

The health monitoring system can use Fog computing for data pre-processing and short-term storage [8]. Cloud computing is used for data processing, analyzing, and long-term storage [23]. This system requires little data transmission to the cloud, as the preprocessing task is performed at the Fog node. Latency is also reduced, as the fog node is placed near to the end user. As the latency is reduced, real-time operation

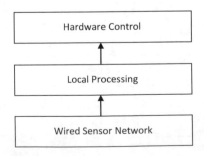

FIGURE 19.6   Three tier architecture of the healthcare system without mobility support.

can be performed effectively and efficiently. The system will increase the availability and quality of e-health service.

19.3   HEALTHCARE SYSTEM WITH PATIENT MOBILITY SUPPORT

The wireless Sensor Networks (WSN) play a vital role in almost all technical areas. Now has now become one of the crucial parts of an IoT based health care system. The health sensors are available in the form of wireless wearable devices, and they can be easily fixed with the patient's body. As the smartphone is commonly used, it and the WSN network have allowed the healthcare system to become more flexible. Today, people want to monitor their health status continuously, although they travel away from home or a healthcare center. Thus, the traditional health monitoring system couldn't fulfill their mobility requirements. The users want to go to their workplaces, play sports, and visit different places, etc. The healthcare system has been changed in order to fulfill the users' requirements and maintain their satisfaction.

The author [10] proposed a fog based healthcare system architecture that allows the user mobility. To ensure the patient's mobility, different wearable sensors (like a smart watch, fitness band, smart phone, etc.) have been used in this health monitoring system. There are three tiers in the proposed architecture: cloud computing tier, Fog computing tier, and sensors. The architecture of the system is shown in the Figure 19.7.

The Sensor tier: This tier includes all the devices that collect health information from the user. These devices can gather internal and external information of the patient. Internal information includes blood pressure, blood sugar level, oxygen level, heartbeat rate, etc. External information includes temperature, location, etc. The external environmental information is also gathered because it may affect the patient's health information. With the help of the smartphone, the location of the user can be easily tracked and sent to the fog node for further processing. Here wearable sensors (like a fitness band, smart watch, etc.) are used to gather internal information of the user and send this health information to the fog node.

The Fog computing tier: The sensors' data received from the sensor tier are huge. This data may have irrelevant data that needs to be filtered for data analysis. So, in this tier, the fog node performs data preprocessing, then stores this preprocessed data for further processing. The fog node reduces the latency as it is placed near the end user, and the efficiency of the real-time processing is increased.

The Cloud computing tier: The cloud computing tier performs the major controlling task of the health monitoring system. The fog node performs different tasks under the supervision of the cloud server. The remote monitoring of the patient's health status is ensured by the cloud tier. The cloud server implements various computations and stores the health information for long-term availability. It provides a patient's information to the physician or healthcare center, depending on their request. The health monitoring system is managed by the cloud server. In this healthcare system, the patient

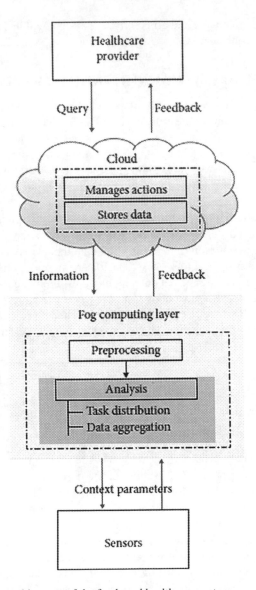

FIGURE 19.7 The architecture of the fog based healthcare system.

can move to different places along with the health sensors. A patient can go to their workplace or can play sports, and the patient is free from the necessity to stay in one place to be monitored.

19.4 EXISTING ARCHITECTURES OF IoT BASED HEALTH CARE SYSTEM

These days, many authors have expressed interest in studying the health monitoring system. Various health monitoring systems based on IoT have been proposed.

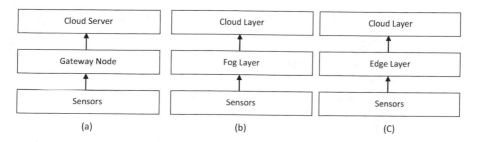

FIGURE 19.8   (a) IoT based system architecture (b) Fog based system architecture (c) Edge based system architecture.

Almost all systems use three-layer architecture, but different authors use different components in the middle layer. The variant three-layer architectures are shown in the figure 19.8.

The following is the list of health monitoring systems based on the above architecture:

i.   An IoT based health monitoring system has been developed which supports real-time analysis, and can manage the network of sensor devices and the gateways [6]. The system is using cloud deep neural network architecture for the classification of heartbeat sensor's data in multiple cardiac conditions.

ii.   The author [7] proposed a system for the cardiovascular disease patient by providing patient path estimation, patient table, and emergency alert service. The efficiency of the patient health monitoring, disease diagnosis, and the treatment system for heart patients is increased in the proposed system.

iii. The health monitoring system uses a Fog node for data pre-processing and short-term storage [8]. The cloud server is used for data processing, analyzing the sensor's data, and storing the health information. The system will increase the availability and quality of e-health service.

iv.   A healthcare system has been proposed that can monitor and diagnose a serious disease like diabetes, based on a cloud and IoT application [13]. The author proposed a new classification algorithm called Fuzzy rule based on a neural classifier, for diagnosing the disease and its severity.

v.   A health monitoring system has been developed by integrating the concepts of IoT and Machine Learning [14]. The proposed system can provide efficient health monitoring for a patient, determine whether the patient has a particular disease or not, as well as provide immediate relief in the case where the patient is facing a critical situation requiring immediate attention by the doctor. The KNN algorithm is used to classify the current status of the patient.

vi.   A smart healthcare system has been proposed for real-time analyzing and predicting human activities by using the latest technologies: IoT, Edge of Things, and the cloud of things [15]. The Recurrent Neural Network uses an edge node like a laptop or PC for activity prediction. For fast computation of experimental data, the Graphics processing unit of the edge node is used.

vii. The author [11, 16] proposed a system called Health Fog, which implements the Deep Learning algorithm in the Edge node. This system can perform automatic heart disease analysis. The proposed architecture will overcome the limitations of the Fog based health monitoring system. As the system performs the data processing and analysis in the edge node, it gives an efficient output of real-time processing, with very low latencies.

viii. The author [17] proposed an IoT based healthcare system that will improve the accuracy level of the prediction of cardiovascular disease. This novel novel method aims to find significant features by applying Machine Learning techniques that improve accuracy in the prediction of cardiovascular disease. The proposed method uses the combination of Random Forest and Linear Method. The system is quite accurate in the prediction of heart disease.

ix. A secure IoT based healthcare system has been proposed that can guarantee the fulfillment of privacy requirements during the transmission of sensitive information [18]. The proposed system ensures context privacy by encrypting the data exchanged using symmetric keys that are in turn encrypted according to the L-IBE scheme. This system is compatible with the heterogeneity of the IoT environment.

19.4.1 COMPARISON OF THE EXISTING HEALTH CARE SYSTEMS

Nowadays, many varieties of healthcare systems have been developed. A few of the healthcare systems have been compared in Table 19.1.

19.5 PROPOSED ARCHITECTURE OF IOT BASED ELDERLY PATIENT CARE SYSTEM

Edge Computing for IoT is an emerging technology that helps to develop smart systems. As in the advancement of wireless technologies, the application of IoT is growing very fast in the Healthcare sector. Many IoT sensors are used in the e-health system, which generates a huge amount of data. In the traditional e-health system, this data needs to be transmitted to the cloud server for processing and remote accessing. There are chances of exposing this information to an unauthorized third party. Again, a huge amount of data needs to be uploaded to the cloud server. A remote e-health telemedicine system has been developed consisting of a portable sensing unit comprising of pulse Oximeter, ECG, EMG, GSR, Body temperature, and Blood pressure that is assigned to the work of acquiring various medical data, along with images of the eye, tongue, and area of impact of the patient's body [6]. The system works on a slow Internet connectivity, and data acquisition can be done without the Internet. It has the features of storing and processing the sensed data locally for storage and diagnosis of the patient. The system is capable of working in both online and offline modes, so that it can be used efficiently in a limited Internet area. But the system needs to transfer a large amount of data to the cloud server for remote accessing. The latency time will depend on the data transfer rate from the local computer to the cloud server.

TABLE 19.1

Comparison of different healthcare system

Ref.No	Authors	Devices/Technologies/ protocols	Computation	Applications Domain	Data Analysis / Machine Learning	Patient Mobility	Implementation
[6]	J. Granados, et. al.	GPU, GATT protocol, CoAP, Bluetooth, IoT sensors	Cloud computing	Realtime analysis of ECG data for heart patient	Convolutional neural network	No	Yes
[7]	M. Neyja, et. al.	Wi-Fi, Cellular network, IoT censors	Local server	Smart healthcare system for heart patients	Hidden Markov Model Predictor	Yes	Yes
[8]	K. Monteiro, et.al.	Wi-Fi, Arduino, Raspberry Pi	Cloud computing, Fog computirg	Fall detection and heart beat monitoring for heart patients	Not mentioned	Yes	No
[9]	P. M. Kumar, et. al.	5G, Hadoop, IoT sensors	Cloud computing	It is applied on diabetes patient.	Fuzzy Rule based Neural Classifier	No	Yes
[14]	S. Reddy K, et. al.	PIC microcontroller, GSM, IoT sensors	Cloud computing	It is applied on diabetes patient.	KNN algorithm	No	
[15]	Md. Zia Uddin	Graphics Processing Unit (GPU), IoT sensors	Cloud computing, Edge computing	Realtime analysis of ECG data for heart patient in a smart home.	Recurrent Neural Networks	No	Yes
[16]	S. Tuli, et. al.	FogBus, Samsung Galaxy S7, Dell XPS 13, Raspberry Pi, GPU, IoT sensors	Cloud computing, Fog computing	Automatic heart disease detection.	Deep Learning	No	Yes
[17]	S. Mohan, et. al	Not mentioned	Local server	Heart disease prediction for a patient.	Random Forest	No	Yes
[18]	R. Boussada, et. al.	L-IBE, IoT sensors	Local server	Application of data encryption on healthcare system.	Elliptic Curve Discrete Logarithm	No	Yes

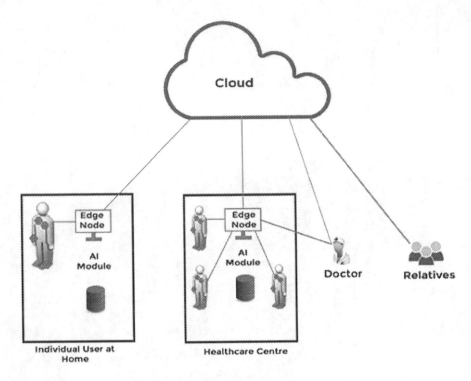

FIGURE 19.9   Proposed system architecture.

The application of IoT based Deep Learning and Edge computing in the Healthcare system will overcome the limitations of the present Healthcare system [10, 12, 19]. This system implements an edge node and can gather and store data from the IoT sensors. The multiple health sensors and a smart phone will be used in the proposed system, so that the system will support patient mobility. Mobile technology will help to gather health information from the patient [24]. The edge node will perform the processing and analyzing of the data, which are collected from the sensors. To create remote availability, fewer amounts of data will be transmitted to the cloud server. The edge node will perform the pre-processing and analyzing of the sensor's data with the help of a Machine Learning algorithm. The edge node will store the sensor's data for the short-term, and the duration will be controlled by the user. The cloud server will store selected data for long-term, and it will be made available to the physicians. The edge node will send an emergency signal message to the doctor and relatives through the cloud server. The architecture of the system is shown in Figure 19.9.

19.5.1   FEATURES OF THE PROPOSED ARCHITECTURE

The proposed architecture the edge node can be placed in both the healthcare center and the patient's residence. For implementing the proposed system in the healthcare center, a computer with high configuration may be required, as the data storage and data processing have to be performed at the edge node. The system will support the patient mobility. An AI module will be used to monitor the patient's health status continuously. It can set an alarm signal when the patient needs emergency service. The features of the system are given below:

- The end users are equipped with different medical sensors, as well as wearable sensors, like Smart watch, Smart band, etc.
- The information generated by the sensors is directly stored on the edge node.
- The edge node will store data for a short-term period.
- The data will be uploaded to the cloud only when the edge node completes the preprocessing and analysis tasks using the AI module.
- The cloud will be used for long-term storage.
- At the Edge node, AI will assist continuously; if the patient needs emergency service, the system will send an alarm signal to the consult doctor and the relatives through android apps.
- The senior patient who lives alone can be monitored using the AI module, and emergency services will be provided whenever required.
- The system will also support patient mobility.

19.5.2 ADVANTAGES OF THE PROPOSED SYSTEM

The proposed system will overcome some of the limitations of the present healthcare system. The major advantage of the proposed system is to perform the data processing and analysis tasks at the edge node, and the edge node will be placed very near to the end users. The following list includes the advantages of the system:

- The requirement of data transmission will be reduced as the data preprocessing and analyzing tasks will be done on the edge node.
- The latency of the analysis will be reduced as the analyzing task performs on the edge node.
- Data is stored at the edge of the network, near IoT devices, thus, fewer amounts of data will need to be transferred. This arrangement reduces the risk of data violation.
- A smaller amount of data needs to be transferred and stored at the cloud server. So It reduces both the cloud storage cost and the internet bandwidth cost.
- The senior patient living alone can be monitored at their residence.
- The patients will receive emergency services without delay.
- Doctors can monitor their patients from remote places.
- The patient's caretaker and the doctor will get an alarm message through Android apps.
- The patient can move to different places without restriction.

19.6   DISCUSSION

The architecture of the healthcare system has been changing along with the development of IoT health sensors and the other technologies related to the healthcare system. In the beginning, people were satisfied when the healthcare system provided AI to monitor and diagnose of various diseases. Then, cloud computing became popular, and many researchers were interested in working on cloud computing. Therefore, the capability of the cloud computing increased, and the healthcare center also started using cloud computing [25]. Many patients don't want to stay at the hospital for several days of treatment. They want to stay at home during treatment, so the healthcare system started a remote health monitoring system. As the development of IoT health sensors increased, the health monitoring system improved, and many authors [26–29] have developed IoT based healthcare system. The health monitoring part of the healthcare system was successful, but the control part failed as real-time processing couldn't perform efficiently. Therefore, the challenges encountered are follows:

- A huge amount of data needs to be transferred from IoT sensors to the cloud server.
- As the cloud server is located far from the IoT devices, it takes more latency.
- In remote places, the e-Health monitoring system can't be deployed, due to poor Internet speed. To work, the system needs high speed Internet bandwidth.
- Due to high latency, the real-time controlling system of e-Health also fails.

Many authors [30, 31] have started using Fog computing and have developed healthcare systems by using Fog computing and cloud computing. In the Fog based healthcare system, the Fog node is placed near the health sensors, and the sensors need to send health data to the fog node. So, the latency is reduced since the fog node is placed near the patient. Then, the fog based healthcare system has overcome some of the challenges. The real-time processing can be performed efficiently since the latency is reduced. The analytical power of the Machine Learning algorithm becomes greater, and it is applied in the healthcare system [33, 35–38, 42]. The authors [32, 41] proposed a healthcare system that can deploy the Machine-Learning algorithm on the fog node. So the system enhanced the capabilities of the fog node and reduced the cloud dependency. As the availability of the wearable IoT health devices increases, researchers have used these wearable IoT devices and Edge computing in the health monitoring system [34, 39, 40].

The computing power of the smart phone also increases continuously, and the number of the elderly patients who live alone also increases. So, the IoT Based Elderly Patient Care has been proposed in this study. The proposed system uses Edge computing for the data processing and the analysis. The edge node gathers health data from the sensors and stores it at the local storage. The sensors don't require transmitting the health data to a long distance server. Thus, it will reduce the power consumption of the sensors. The efficiency of the real-time processing will also increase as the latency is decreased. The edge node may be a PC, a laptop, or a smart phone. Using the wearable health sensors and a smart phone, the proposed system supports the patient's mobility. The patient can move to different places and allows exercising freely. The Internet requirement of high bandwidth is also reduced, and the AI module is deployed on the edge node. The system monitors the patient continuously with the help of the AI module. For the remote monitoring facility, the edge node will transmit the processed data to the cloud server. The IoT sensors generate a huge amount of data, but the edge node will transmit fewer amounts of data. So the proposed system will overcome the mentioned challenges.

19.7 CONCLUSION

The conceptual Healthcare system architecture has been proposed, and this system will overcome some challenges of the present e-Health monitoring system. As the data processing and analyzing tasks of the proposed model will be performed on the edge node, it can perform the real-time processing very effectively. It will reduce data consumption since the information generated by the sensors is directly stored on the edge node. The data will be uploaded to the cloud for long-term storage. The patient will get proper monitoring as the AI assists continuously on the edge node. The proposed system will not require high speed internet bandwidth. The system will be free from resource limitation since edge computing will be performing the data analysis task.

REFERENCES

1. T.N. Gia, et al., *Customizing 6LoWPAN Networks towards Internet of Things Based Ubiquitous Healthcare Systems, NORCHIP*, 2014.
2. N. Kumar, IoT Architecture and System Design for Healthcare Systems, *International Conference On Smart Technologies For Smart Nation*, 2017.
3. S. H. AlMotiri, et al., Mobile Health (m-health) System in the context of IoT, *IEEE 4th International Conference on Future Internet of Things and Cloud Workshops (FiCloudW)*, 2016.
4. J. Kharel, et al., An architecture for smart health monitoring system based on fog computing, *Journal of Communications*, 12(4): 228–233, 2017.
5. P. Pasquale, et al., An edge-based architecture to support efficient applications for healthcare industry 4.0, *IEEE Transactions on Industrial Informatics*, 15(1): 481–489, 2018.
6. J. Granados, et al., *IoT platform for real-time multichannel ECG monitoring and classification with neural networks*, Springer, 2017.
7. M. Neyja, et al., *An IoT-based E-health monitoring system using ECG signal*, IEEE, 2017.
8. K. Monteiro, et al., *Developing an e-health system based on IoT, fog and cloud computing*, IEEE, 2018.
9. N. Kamal, et al., Three Tier Architecture for IoT Driven Health Monitoring System Using Raspberry Pi, *IEEE International Symposium on Smart Electronic Systems (iSES)*, 2018.
10. A. Paul, et al., Fog computing-based IoT for health monitoring system, *Journal of Sensors, Hindawi*, 2018, 1386470, 2018.

11. S. Dash, et al., (eds), Deep Learning Techniques for Biomedical and Health Informatics in *Studies in Big Data Series*, Springer, ISSN:2197-6511, ISBN:978-3-030-33966-1, 2020.

12. A. Subasi, et al., *IoT based mobile healthcare system for human activity recognition*, IEEE, 2018.

13. P. M. Kumar, et al., *Cloud and IoT based disease prediction and diagnosis system for healthcare using Fuzzy neural classifier*, Elsevier, 2018.

14. S. Reddy, et al., IoT based Health Monitoring System using Machine Learning, *IJARIIE*, 5, (3), 2019, 381–386, ISSN(O):2395-4396.

15. M. D. ZiaUddin, *A wearable sensor-based activity prediction system to facilitate edge computing in smart healthcare system*, Elsevier, 2019.

16. S. Tuli, et al., *HealthFog – An ensemble deep learning based smart healthcare system for automatic diagnosis of heart diseases in integrated IoT and fog computing environments*, Elsevier, 2020.

17. S. Mohan, et al., *Effective heart disease prediction using hybrid machine learning techniques*, IEEE, 2019.

18. R. Boussada, et al., *Privacy-preserving aware data transmission for IoT-based e-health*, Elsevier, 2019.

19. S. Dash, et al., Edge and fog computing in healthcare – A review, *Scalable Computing: Practice and Experience*, 20 (2), 191–205. http://www.scpe.org, doi: 10.12694/scpe. v20i2.1504, ISSN:1895-1767, 2019.

20. A. Rahman, et al., A comprehensive study of mobile computing in telemedicine, A. K. Luhach et al. (Eds): *ICAICR 2018, Advanced Informatics for Computing Research, CCIS 956*, pp. 413–425, 2019. Springer, Singapore.

21. A. Rehman, et al., Management of resource usage in mobile cloud computing, *International Journal of Pure and Applied Mathematics*, 119(16), 2018, 255–261, ISSN: 1314-3395 (on-line version) 2018.

22. P. Mishra, et al., Deep learning based biomedical named entity recognition systems, In: S. Dash et al., (eds), *Deep learning techniques for biomedical and health informatics* published by Studies in Big Data Series, Springer, pp. 23–40, 68, 2020. ISBN: 978-3-030-33965-4.

23. A. Rahman, et al., (2021), Mobile cloud computing: A green perspective, In: Udgata S. K., Sethi S., Sriram S. N. (eds), *Intelligent notes in networks and systems*, vol. 185, pp. 523–533, Springer, Singapore. doi:10.1007/978-981-33-6081-5-46.

24. K. Rupbanta Singh, et al., (2020), Mobile technology solution for COVID-19. In Fadi Al-Turjaman et al., (Eds), *Emerging technologies for battling COVID-19 applications and innovations*, Springer, pp: 271–294 ISBN: 978-030-60038-9, 2021.

25. D. S. R. Krishnan, et al., *An IoT based patient health monitoring system*, IEEE, 2018.

26. A. Abdelgawad, et al., *IoT-Based health monitoring system for active and assisted living*, Springer, 2017.

27. J. Cabra, et al., *An IoT approach for wireless sensor networks applied to e-health environmental monitoring*, IEEE, 2017.

28. C. Raj, et al., *HEMAN: Health monitoring and Nous An IoT based e-Health Care System for Remote Telemedicine*, IEEE 2017.

29. I. Alihamedi, et al., *Proposed architecture of e-health IOT*, IEEE, 2019.

30. K. Monteiro, E. Rocha, E. Silva, G. L. Santos, W. Santos, P. T. Endo, *Developing an e-health system based on IoT, fog and cloud computing*, IEEE (2018).

31. O. Debauche, *Fog IoT for Health- A new architecture for patients and elderly monitoring*, Elsevier, 2019.

32. Tuli, et al., *HealthFog- An ensemble deep learning based smart healthcare system for automatic diagnosis of heart diseases in integrated IoT and fog computing environments*, Elsevier, 2020.

33. K.G. Sheela, et al., *IoT based health monitoring system using machine learning*, Elsevier, 2018.

34. M. D. ZiaUddin, *A wearable sensor-based activity prediction system to facilitate edge computing in smart healthcare system*, Elsevier, 2019.

35. S. Mohan, et al., *Effective heart disease prediction using hybrid machine learning techniques*, IEEE, 2019.

36. A. Souri, et al., *A new machine learning-based healthcare monitoring model for student's condition diagnosis in Internet of Things environment*, Springer, 2020.

37. H. Fouad, et al., *Analyzing patient health information based on IoT sensor with AI for improving patient assistance in the future direction*, Elsevier, 2020.

38. K.C.H. Suneetha, et al., *Disease prediction and diagnosis system in cloud based IoT – A review on deep learning techniques*, Elsevier, 2020.

39. L. Greco, et al., *Trends in IoT based solutions for health care – Moving AI to the edge*, Elsevier, 2020.

40. D. Mrozek, et al., *Fall detection in older adults with mobile IoT devices and machine learning in the cloud and on the edge*, Elsevier, 2020.

41. W. Li, et al., *A comprehensive survey on machine learning-based big data analytics for IoT-Enabled smart healthcare system*, Springer, 2021.

42. H. Bolhasani, et al., *"Deep learning applications for IoT in health care – A systematic review"*, Elsevier, 2021.

Index

Printed in the United States
by Baker & Taylor Publisher Services